SOFT COMPUTING

by

Dr. NEETA AWASTHY
Director
Raj Kumar Goel Institute of Technology for Women
Ghaziabad

S.K. KATARIA & SONS®
Publisher of Engineering Books
4885/109, Prakash Mahal, Dr. Subhash Bhargav Lane,
Opposite Delhi Medical Association, Daryaganj, New Delhi–110002 (INDIA)
Phone: +91-11-43551243, Mobile: +91-9871775858, 7678425209
e-mail: katariabook@gmail.com, katariabooks@yahoo.com
Website: www.skkatariaandsons.com
(e-books, subscriptions and bulk buying also available on our website)

AF662821

Published by:

S.K. Kataria & Sons®

4885/109, Prakash Mahal, Dr. Subhash Bhargav Lane,
Opposite Delhi Medical Association, Daryaganj, New Delhi-110002 (INDIA)
Phone: +91-11-43551243, Mobile: +91-9871775858, 7678425209
e-mail: katariabook@gmail.com, katariabooks@yahoo.com
Website: www.skkatariaandsons.com

© ALL RIGHTS STRICTLY RESERVED

STATUTORY WARNING

Information contained in this book has been obtained by author from sources believed to be reliable and are correct to the best of his knowledge. Every effort has been made to avoid errors and omissions and ensure accuracy. Any error or omission noted may be brought to the notice of the publisher which shall be taken care of in forthcoming edition of this book. However, neither the publisher nor the author guarantee the accuracy or completeness of any information published herein, and neither the publisher nor author take any responsibility of liability for any inconvenience, expenses, losses or damage to anyone resulting from contents of this book.

The book is meant for educational and learning purposes and there is no attempt on the part of publisher/author to render engineering and other professional services. If such services are required, the assistance of an appropriate professional should be sought.

The author of the book has taken all possible care to ensure that the contents of the book do not violate any existing copyright or other intellectual property rights of any person in any manner whatsoever. In the event, the author has been unable to track any source and if any copyright has been inadvertently infringed, the facts may be brought to the notice of the publisher in writing for corrective action.

Send all correspondence to : **M/s S.K. Kataria & Sons,** New Delhi
Email: katariabook@gmail.com

First Edition: July 2011-12, Reprint: 2026

Price : ₹ 395/-

Laser Typesetting by : Plus Computers, Delhi   *Printed at :* New A.S. Printing Press, Delhi

Preface

Earlier the topic of Neural Network, Fuzzy Logic and Genetic Algorithm used to be a cake for Research scholars only. Then, last decade , it became part of M.Tech. program in different universities. The Research and Post Graduate scholars are differently trained to learn. Please mind the word used is 'learn' and not 'study'. You teach them or not... they learn by themselves. Under Graduate student is a different lot all together. They are fresh from schools. They are accustomed for annual system and semester system comes to them as a surprise. Secondly, they are accustomed to find a syllabus in one single text book and fail to map the syllabus in different books as prescribed by the university syllabus.

Last semester, I realized, that, my son ran from pillar to post to find a reasonably good book on 'Soft Computing', but find only a couple of options. Even in those books, there were problems in notations and mixed up fundamentals; not to mention typos and small mistakes. This semester my students are facing the same problem. So, quickly, I put all my resources together and geared up for the challenge.

Being a natural fighter, I like the word 'challenge'. When I take it, I think, "If I can't. Who can?". On being successful I deceive the whole spirit by saying 'If I can, Who can't?", with all the conducive environment around myself, I think, the book was already written and I had to give only the finishing touch.

The book has been designed unit wise and synchronized as per the syllabus of UPTU. Each chapter is each unit of UPTU syllabus.

Unit-I and Chapter I, therein is the Introduction and Architecture of Artificial Neural Networks (ANN). The fundamentals are sailed through from What a Neuron is, the motivation of making a ANN as per the working of brain, History of ANN, Nerve structure and synapse, Artificial Neuron and its model, activation functions. The architecture of ANN is explored by discussing single layer, multilayer feed forward networks, recurrent networks. Also discussed are various heuristics (experimental results) to design convergent ANN with no overtraining. Once we are through with hardware/ software implementation, Various learning techniques are discussed. The perception rule, convergence rule, Auto and Hetro Associative memory is discussed.

Unit-II and Chapter II, has details of different Back propagation networks. The perceptron model and its single layer network is discussed. Then Multilayer perceptron model with multi layer neural network is discussed. The algorithm of back propagation learning methods, affect of various factors like learning rule co-efficient; adding a momentum factor and others to improve learning process though back propagation training is discussed with applications.

Unit-III and Chapter III, Covers the introduction of Fuzzy Logic with Basic concepts of fuzzy logic, Fuzzy sets and Crisp sets, Fuzzy set theory, operations, Properties of fuzzy sets, Fuzzy and Crisp relations, Fuzzy to Crisp conversion.

Unit-IV and Chapter IV goes into the details of Fuzzy Membership, Rules, Membership functions, interference in fuzzy logic, fuzzy if-then rules, Fuzzy implications and Fuzzy algorithms Also covered are Fuzzyfications & Defuzzificataions and the industrial applications of Fuzzy Logic in form of Controller and other Industrial applications.

Unit-V and Chapter V, covers Basics of Genetic Algorithm (GA), working principle, procedures of GA, flow chart of GA, Genetic representations, (encoding) Initialization and selection, Genetic operators, Mutation, Generational Cycle and applications.

However, I understand, that a graduate must have a good seminar report, a good Industrial Training Report, an ingenious project and a couple of papers to his/her credit to define a cutting edge over their other counterparts to be that 25% employable graduate', that we are able to produce each year. The annexure are just for creating that cutting edge. You need peep through research monograms, books, periodicals and papers given in the bibliography to master the subject.

Dr. Neeta Awasthy

Acknowledgement

As a researcher on pattern recognition and classification, it is only natural that I classify everything… including my acknowledgments.

I express my sincere gratitude towards Dr. J.P.Saini and Dr. D.S.Chauhan, my Ph.D. Supervisors, for their valuable guidance and constant encouragement throughout, to endeavor deeper to find out how much more is to be explored. I thank GOD for giving me mental, physical, emotional and spiritual strength required to complete this work.

I am thankful to the management of Raj Kumar Goel Group of Institutions for providing me the much required environment and peace of mind, without which it was never possible for me to write a book.

I am thankful to my colleague Ms. Kadambari Agarwal (HoD, CS) for making this work possible in form of a book, from the material lying randomly in my closet for years together gathering dust. I am thankful to my research scholars Mr. Bhaskar, Ms. Gauri Katiya and Ms. Jaspreet for their inputs in Annexure 3 (Research Areas). I am thankful to scores of books and technical papers, from where various ideas were drawn but origin of which is now obscure.

I wish to mention my son, Tanmay Awasthy, who ran from pillar to post to find a reasonably good book to write his elective paper of 'Soft Computing', but, could find none. His effort thereby, gave me a feel of the demand of such a book and my worthiness to write one, but, lack of desire to do so.. I take pleasure to acknowledge the continuous motivation and encouragement of my husband Mr. Rajesh Awasthy.

I am thankful to my friends and relatives for their appreciation and criticism, respectively, a perfect blend of which keeps me performing and sustaining all odds.

"Lastly, the painstaking efforts made by Mr. Manoj Tyagi and his team in typing the manuscript is duly acknowledged. I am also thankful to **Mr. Sanjeev Kataria** and his team at **S.K. Kataria and Sons**, New Delhi, for their meticulous efforts in converting the manuscript to a book."

Dr. Neeta Awasthy

Foreward

It gives immense pleasure that Dr. Neeta Awasthi, authored a new book that focuses on UPTU syllabus and provides an extra mental challenges for students with more drive towards the subject. Proper and apt usage of dimensions, consistency of variables, scrutiny for errors removal and a model to create quick grasping text book on the subject.

A special effort has been made to provide proper illustrations in the text book as the demand of the syllabus. A set of review questions reinforcing of the learning after reading the chapter. Information on crucial values tabulated in addition,

The Neural Network concepts have been illustrated with the learning algorithm details, fuzzy variables, memberships function and genetic algorithm cycle leaving the reader to amalgamate all its for a particular application.

Dr. Neeta has painstakingly collected the material from the lecture notes from her under graduate and post graduate teaching experience in various universities over the several years. She deserves appreciation on extension of her Ph.D. research in Neural Networks. The text book can be considered clarity on fundamentals, step by step building of concepts with illustrated examples which are the of the hour in the of soft computing. This book can provide a new sight on contents.

Prof. D.S. Chauhan

Dedicated

to

My Parents

Sri J.S. Shukla

&

Late Smt. Girish Shukla

CONTENTS

1. NEURAL NETWORKS-I (INTRODUCTION AND ARCHITECTURE) 1–47

1.1. Neural Networks Introduction 1
1.2. Biological Neural Network 4
1.3. Artificial Neural Networks 5
1.4. An Artificial Neuron 9
1.5. Artificial Neuron Model 11
1.5.1. McCulloch and Pitts Model 11
1.5.2. Rosen Blatt's Perceptron Model 14
1.5.3. Adaline 17
1.6. Activation Functions 19
1.6.1. Linear Functions 19
1.6.2. Binary Step Function 20
1.6.3. Hardlim Function 20
1.6.4. Saturated Linear Activation Function 21
1.6.5. Sigmoidal Functions 21
1.7. Neural Network Architecture 22
1.7.1. Single Layer Network 22
1.7.2. Multi Layers Feed Forward Neural Network 23
1.7.3. Radial Basis Network 23
1.7.4. Time Delay Neural Networks (TDNN) 23
1.7.5. Recurrent Neural Network (RNN) 25
1.8. Learning Methods 26
1.9. Various Learning Techniques 30
1.9.1. Supervised Learning Algorithms 30
1.10. Perceptron and Convergence Rule 33
1.11. Auto Associative and Hetro Associative Memory 37
➠ *Short Questions with Answers* 40
➠ *Long Questions with Answers* 41
➠ *Summary* 45
➠ *Review Questions* 45
➠ *Numerical Problems* 46

2. NEURAL NETWORKS-II (BACK PROPAGATION NETWORKS) 48–72

2.1. The Perceptron 48
2.2. Solution to the Neuron Model 49
2.2.1. Single Layer Artificial Neural Network 50
2.2.2. Model for Multilayer Perceptron 51
2.2.3. Backpropagation Learning 51
2.3. Multi Layer Perceptron (MLP) 55
2.3.1. Problems with MLP 56
2.3.2. Important Heuristics for Training MLP 56
2.4. Back Propagation Learning Methods 59
2.5. Back Propagation Algorithm 60
2.6. Factors Affecting Back Propagation Learning 62

2.7. Some Application Domains 64
➡ *Short Questions with Answers* 67
➡ *Long Questions with Answers* 68
➡ *Summary* 70
➡ *Review Questions* 71
➡ *Numerical Problems* 71

3. FUZZY LOGIC-I (INTRODUCTION) 73–111
3.1. Basic Concepts of Fuzzy Logic 73
3.2. Crisp Sets 75
3.2.1. Characteristics of Crisp Sets 75
3.2.2. Operations of Crisp Sets 77
3.2.3. Properties of Crisp Sets 79
3.2.4. Partition and Covering 81
3.3. Fuzzy Sets: Theory and Operations 83
3.3.1. Membership Function 84
3.3.2. Basic Fuzzy Set Operation 85
3.4. Properties of Fuzzy Sets 90
3.5. Crisp and Fuzzy Relation 92
3.5.1. Crisp Relation 92
3.5.2. Fuzzy Relations 95
3.5.3. Operations on Fuzzy Relations 96
3.6. Fuzzy to Crisp Conversion 98
➡ *Short Questions with Answers* 103
➡ *Long Questions with Answers* 104
➡ *Summary* 110
➡ *Review Questions* 110
➡ *Numerical Problems* 110

4. FUZZY LOGIC-II (FUZZY MEMBERSHIP, RULES) 112–158
4.1. Membership Function 112
4.2. Inference in Fuzzy Logic 116
4.3. Fuzzy IF- Then Rules 120
4.4. Fuzzy Implications and Fuzzy Algorithm 122
4.5. Fuzzification and Defuzzification 129
4.6. Fuzzy Controller 134
4.6.1. Conventional Control System 134
4.6.2. Fuzzy Control 136
4.6.3. Fuzzy Logic Controller: Basic Concept 137
4.6.4. Classification of Fuzzy Controllers 137
4.6.5. Fuzzy PI Controller 138
4.6.6. Fuzzy PD Controller 139
4.6.7. Design of Fuzzy Logic Controller 139
4.7. Analysis and Design for *T-S* Fuzzy Model 140
4.7.1. Introduction 140
4.7.2. Fuzzy Modeling 141
4.7.3. Fuzzy logic Control System 142
4.7.4. Design of Stable Fuzzy Logic Controllers 143
4.7.5. Algorithm for Design of Stable Fuzzy Logic Controllers 144

4.7.6. Illustrative Example 144
4.7.7. Results (and Discussion) 148
➠ *Short Questions with Answers* 150
➠ *Long Questions with Answers* 151
➠ *Summary* 157
➠ *Review Questions* 157
➠ *Numerical Problems* 158

5. GENETIC ALGORITHM (GA) 159–183
5.1. Basic Concepts 159
5.2. Working Principle 162
5.3. Procedure of Genetic Algorithm 170
5.4. Genetic Algorithm Flowchart 171
5.5. Genetic Representation 172
5.6. Encoding, Initialization and Selection 172
5.6.1. Binary Encoding 172
5.6.2. Value Encoding 173
5.6.3. Permutation Encoding 174
5.6.4. Tree Encoding 175
5.7. Genetic Operators 176
5.8. Mutation 177
5.8.1. Mutation Rate P_m 178
5.9. Common Applications of Genetic Algorithm 179
➠ *Short Questions with Answers* 182
➠ *Summary* 182
➠ *Review Questions* 183
➠ *Numerical Problems* 183

ANNEXURES 185–258
Annexure–A: History and Summary (For your Leisure Reading) 187–217
Annexure–B: More ANN Networks (For more Enthusiastic Students) 218–224
Annexure–C: Research Areas in Soft Computing (For more Ambitious Students) 225–241
Annexure–D: Ann Applications 242–250
Annexure-E : Multiple Choice Questions 251–258

MODEL QUESTION PAPERS 259–265

PREVIOUS YEARS QUESTION PAPERS 266–284

REFERENCES 285–292

INDEX 293–294

1 CHAPTER ONE

NEURAL NETWORKS-I (INTRODUCTION AND ARCHITECTURE)

CHAPTER OUTLINE

1.1. Neural Networks Introduction
1.2. Biological Neural Network
1.3. Artificial Neural Networks
1.4. An Artificial Neuron
1.5. Artificial Neuron Model
1.6. Activation Functions
1.7. Neural Network Architecture
1.8. Learning Methods
1.9. Various Learning techniques
1.10. Perceptron and Convergence Rule
1.11. Auto Associative and Hetro Associative Memory

1.1. NEURAL NETWORKS INTRODUCTION

I ask you a question: See the following black diagram and find out the function of the black box.

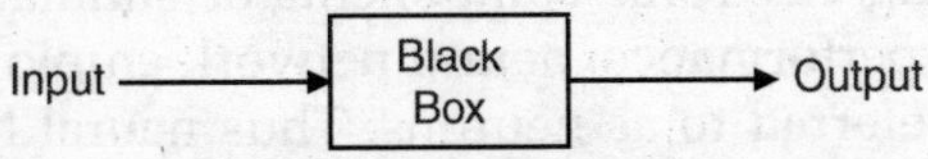

Let us say, the input-output table is as follows:

Input	Output
1	1
2	4
3	9
4	16
5	?

Correctly your answer will be 25 and the function of the black box is to square the input. This is our Biological Neural Network; where we guess the process/output just by observing previous results. Designing this network in hardware or creating a program (software) is the work of Artifical neural Network.

Neural Networks are, in essence, biologically inspired networks since they are based on the current understanding of the biological nervous system. In essence they are comprised of a network of densely interconnected simple processing elements which perform in a manner analogous to the most elementary functions of a biological neuron.

Artificial Neural Networks (ANN) represent an emerging technology rooted in many disciplines. These networks are endowed with some unique attributes:

1. Universal Approximation, *i.e.*, Input-Output mapping,
2. The ability to learn from and adopt to their environment.
3. The ability to invoke weak assumptions about the underlying physical phenomenon to responsible for the generation of the input data.
4. The work on ANN is motivated right from its inception by the recognition that the brain computes in an entirely different way from the conventional digital computers.
5. Being a highly complex, nonlinear and parallel computer, the brain has the capability of organizing neurons so as to perform certain computations many times faster than the fastest digital computer in existence.
6. To be specific, the brain routinely accomplished perceptual recognition tasks in something of the order of 100-200 ms. Whereas tasks of much lesser complexity will take days on a conventional computer.

A developing neuron is synonymous with a plastic brain and plasticity permits the developing nervous system to adapt to its surrounding environment. Neurons exist in wide variety of shapes and sizes in different parts of the brain. The interactions between neurons are mediated through synapse. A synapse is assumed to be a simple connection that can impose excitation or inhibition.

Just as plasticity appears to be essential to the functioning of neurons as information processing units in the human brain, so it is with neural networks made up of artificial neurons. In the most general form, a neural network is a machine that is designed to model the way in which the brain performs a particular task or function of interest. The network is usually implemented using electronic components or simulated in software on a digital computer. To achieve good performance, neural network employ a massive interconnection of simple computing cells referred to as neurons. Thus neural Network may be defined as follows:

Definition

A neural network is a massively parallel distributed processor that has a natural propensity for storing experimental knowledge and making it available for use. It resembles the brain in two respects:

1. Knowledge is acquired by the network through learning process.
2. Interneuron connection strengths known as synaptic weights are used to store the knowledge.

Therefore, neural network may be viewed as an adaptive machine [140]. The procedure used to perform the learning process is called learning algorithm, the function of which is to modify the synaptic weights of network in an orderly fashion so as to attain a desired objective. The modification of synaptic weights provides the traditional method for the design of neural networks. Such an approach is the closest to linear adaptive filter theory, which is already well established and successfully applied in such diverse field as communications, control and radar. However, it as also possible for a neural network to modify its own topology, which is motivated by the fact that neurons in the human brain can die and that new synaptic connections can grow.

A neural network derives its computing power through its massively parallel distributed structure and its ability to learn. These two information processing capabilities make it possible for ANN to solve complex problems that are currently intractable. Specifically, a complex problem of interest is decomposed into a number of relatively simple tasks, and neural networks are assigned a subset of the tasks that match their inherent capabilities.

ANN offers the following features:

1. Neuron is basically a non linear device. Consequently, a neural network, made up of an interconnection of neurons, itself nonlinear. Nonlinearity is an important property, particularly if the underlying physical mechanism responsible for the generation of an input signal is inherently nonlinear.
2. A popular paradigm of learning called supervised learning involves the modification of the synaptic weights of a neural network by applying a set of labelled training samples.
3. Neural Networks have a built-in capability to adapt their synaptic weights to changes in surrounding environments. The network trained to operate in a specific environment can be easily retrained to deal with minor changes in the operating environment conditions.
4. In context of pattern classification, a neural network can be designed to provide information not only about which particular pattern to select, but also about the confidence in the decision making.
5. Knowledge is represented by the very structure and activation state of a neural network. Every neuron in the network is potentially affected by the global activity of all other neurons in the network.
6. Network has the potential to be inherently fault tolerant in sense that performance is degraded gracefully under the adverse operating environment .
7. Massively parallel nature of a neural network makes its potentially fast for the computation of certain tasks. This feature makes it ideally suited for implementation using VLSI technology.
8. Neural Networks enjoy universality as information processors. This feature manifests itself in different ways:
    - Neurons in any form, represent an ingredient common to all networks.
    - The commonality makes it possible to share theories and learning algorithms in different applications of Neural Networks.
    - Modular networks can be built by a seamless integration of modules.
9. The design of a network is motivated by analogy with brain, which is a living proof that fault- tolerant parallel processing is not only physically possible but also powerful.

To summarise, the unique features of ANN are:

1. Parallel structure with distributed storages and processing of information.

2. Learning ability by adjusting their network interconnection weights and biases based on learning algorithm.

1.2. BIOLOGICAL NEURAL NETWORK

In everyday life, brain performs tasks (*e.g.*, recognizing hand-written character, classifying objects to category) through massive-parallelism of highly connected neurons. Brain also recognizes character, whose some part may be missing. All these tasks cannot be performed by computer (classic), but can be performed by neural network with higher accuracy.

An Artificial Neural Network (ANN) is an information processing paradigm that is inspired by the way biological nervous systems, such as brain process information. The key element of this paradigm is novel structure of the information processing system. It is composed of highly interconnected processing elements (neurons) working in union to solve specific problems. Artificial Neural Networks, like people learn by examples.

A neuron (or nerve cell) is a special biological cell that processes information Fig. 1.1 [7]. It is composed of a cell body, or soma, synapses and two types of out-reaching tree like branches: the axon and the dendrites. The building blocks are discussed as follows:

Soma: The cell body has a nucleus that contains information about hereditary traits and plasma that holds the molecular equipment for producing material needed by the neuron. Soma receives information through the dendrites, sums them and produces an output if the sum is greater than a threshold value. The input signals are passed on to the cell body through the cell body which may accelerate or retard an arriving signal. It is this acceleration or retardation of the input signals that is modelled by the weights.

Dendrites: A neuron receives signals (impulses) from other neurons through its dendrites (receivers) and transmits signal generated by its cell body along the axon (transmitter), which eventually branches into strands and sub strands. At the terminals of these strands are the synapses.

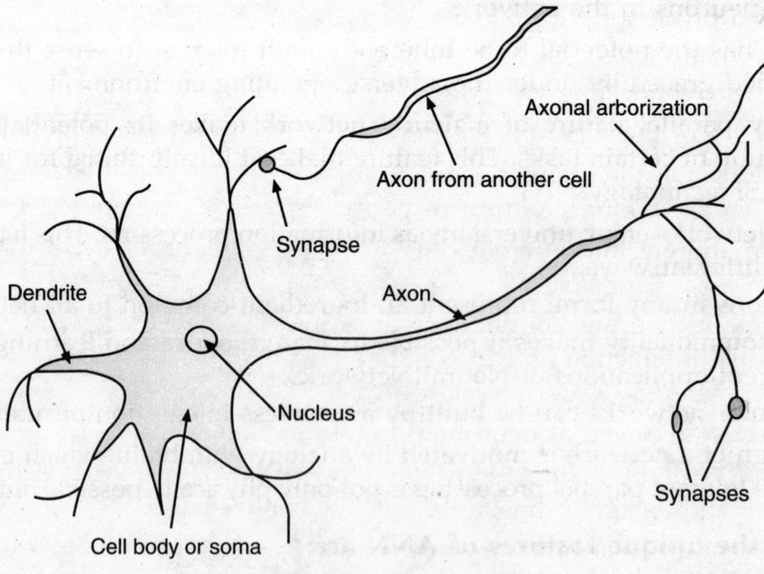

Fig. 1.1. A Biological Neuron

Axon: The Axon is a long slender projection emanating from the neurons soma. The cell body generates the output neuronal signal, a spike, which is transferred along the axon to the synaptic terminals of other neurons.

Synapse: A synapse is an elementary structure and functional unit between two neurons (axon strands of one neuron and a dendrite of another). When the impulse reaches the synapse's terminal, certain chemicals called neurotransmitters are released. The neurotransmitters diffuse across the synaptic gap, to enhance or inhibit, depending on the type of the synapse, the receptor neuron's own tendency to emit electrical impulses.

1.3. ARTIFICIAL NEURAL NETWORKS

The brain consists of 10^{10} neurons of 1.5×10^{-9} gms wt. and where each of the neurons is connected with 10^4 neurons. Such a massive network is there inside the 90 cubic inches of volume. These are much slower than the gates in computing power but due to its massively network, it becomes powerful processing unit.

A neuron is a special cell that conducts an electro chemical signal. It consists of cell body. There are filament like structures known as dendrites. The dendrites behave as input channels. There is only one axon attached to a cell which acts as an output channel. This axon is a non-linear threshold device which produce a voltage pulse called 'action potential'. Brain acts in a bottom-up approach. If the cumulative inputs received by some raise the internal elective potential of the cell then the neuron in said to be fired by propagating action potential; which further excites or inhibits other neuron.

Action Potential: The axons terminates to synapse. It is connected to other synapse by small gaps. This gap contain neuro-transmitted fluid. This fluid accelerates or retards the electric changes to soma.

Properties of a Brain

1. Each neuron has small computational ability but its massively parallel network makes the brain a processor with massive computational ability.
2. There are many ways in which the neurons are interconnected, thereby, giving rise to many interconnecting variables.
3. The strength of these parameter and their connection pattern are variable.
4. The potentials of the brain are continuous which change in time and space whereas the neurons have binary output response guided by threshold of synables.
5. There are afferent and efferent cells. The brain gets input by receptors or afferent neurons.

   These afferent neurons continuously combine signals from receptors with effectors for adaptive interaction and to generate nemal signals from train to other form of energy, enabling action or responses to receptors.
6. The impulses received at neurons is highly nonlinear.
7. The brain is divided into many subnetworks, where distant neurons are rarely connected and local neurons are massively connected.
8. The different areas of brain are responsible for various kinds of work.

Generation and Propagation of Action Potential

Across the membrane of neuron, there exists a negative interior and positive exterior is an exterior to a interior potential of – 70 mV, which when excited goes upto 30 mV.

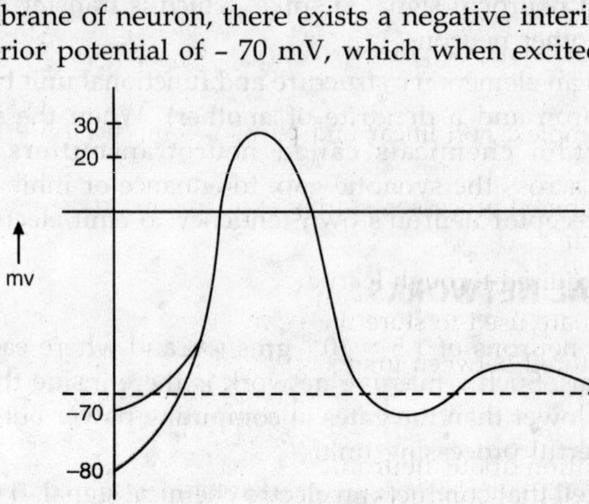

Fig. 1.2. Generation and Propagation of action potential

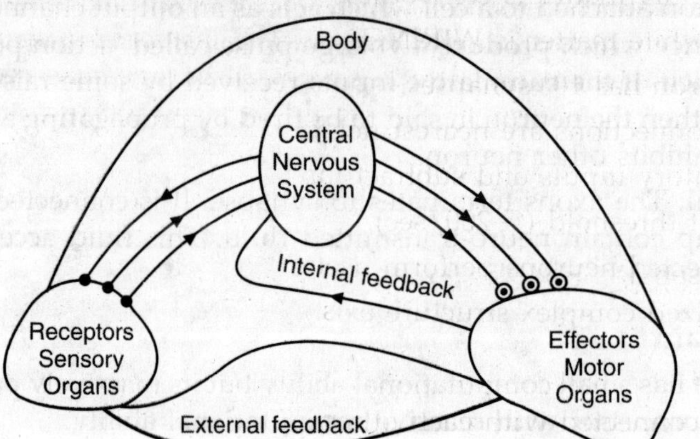

Fig. 1.3. Information flow in Nervous system

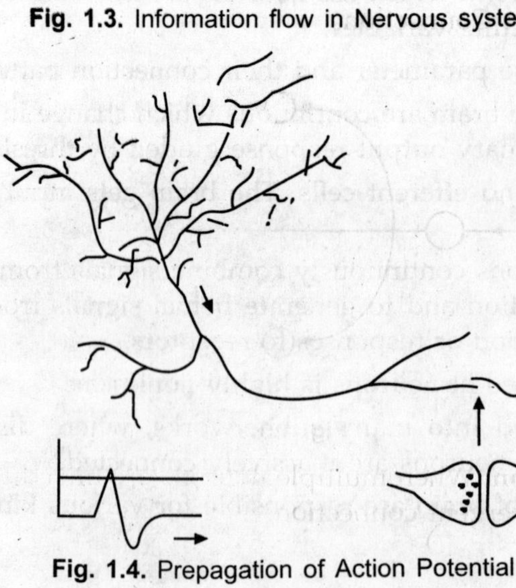

Fig. 1.4. Prepagation of Action Potential

NOTE

Incoming impulse: excitatory cause firing inhibitory- hindle-firing.

Threshold: 40 mV

EPSP: Excitatory post synaptic potential

IPSP: Inhibitory

Brain: A highly complex, non linear and parallel computer (information processing system).

ANN: Parallel distributed processor, storing experimental knowledge. It resembles the brain in two respect:

- Knowledge is acquired through learning,
- Interconnections are used to store the knowledge.
- Information is stored between interconnection of neurons known as SYNAPTIC.

Neuro Science

Some Biological Information about neuron:

- All neurons are not alike (10^{10} to 12^{12}).
- Much of the Gray matter is cell bodies.
- Much of the white matter is WIRING.
- A typical neuron has 10^4 synapses.
- Most of the connections are nearest neighbors type.
- Add all excitatory inputs and subtract inhibitory inputs.
- Neuron fire at threshold (@ 200/sec).
- Densely connected neurons perform useful computational functions.
- Many specialized complex structure exist, such as Vision, Hearing center.

Connection Patterns

The neurons are connected with each other as:

(*a*) **Divergent connection:** When a single neuron is connected with many neurons this is known as Divergent Connection

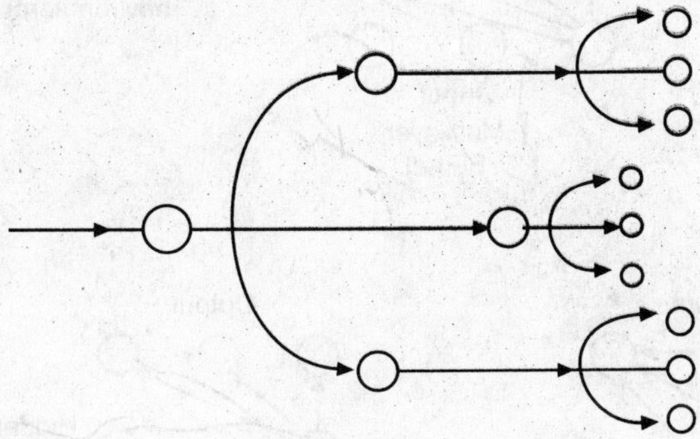

Fig. 1.5.

(*b*) **Convergent connection:** When multiple neurons are linked to one common neuron, this is known as convergent connection.

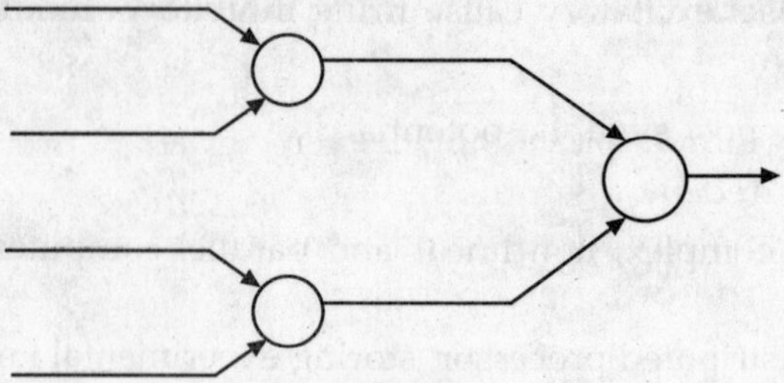

Fig. 1.6.

(*c*) **Chain connection:** Many loops and subloops with no feedback is said to be a chain connection.

(*d*) **Loop connection:** Many loops and subloops with feedback is the characteristic property of a loop connection.

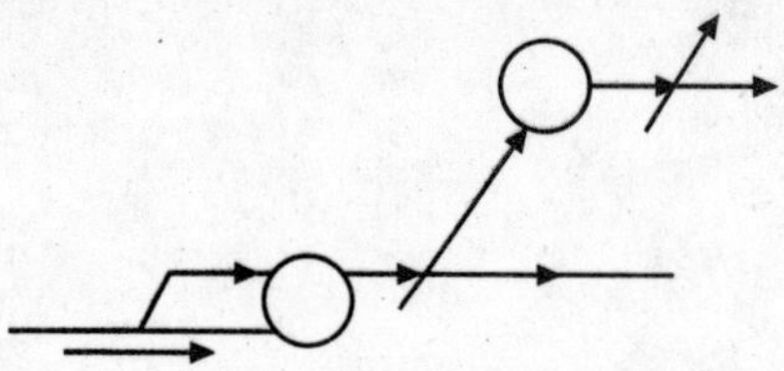

Fig. 1.7.

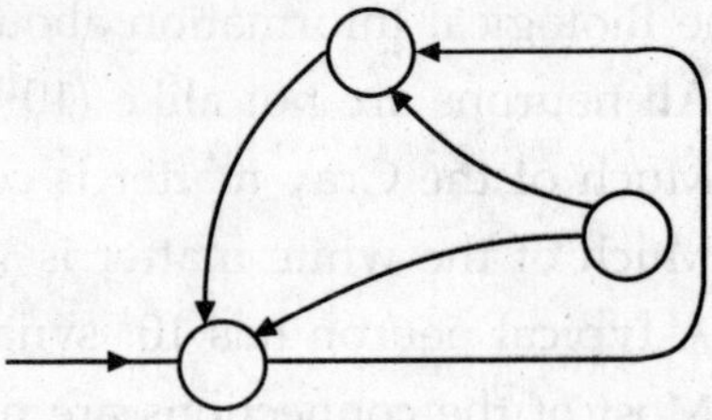

Fig. 1.8.

> **NOTE** Some common nets for Artifical Neural Network.

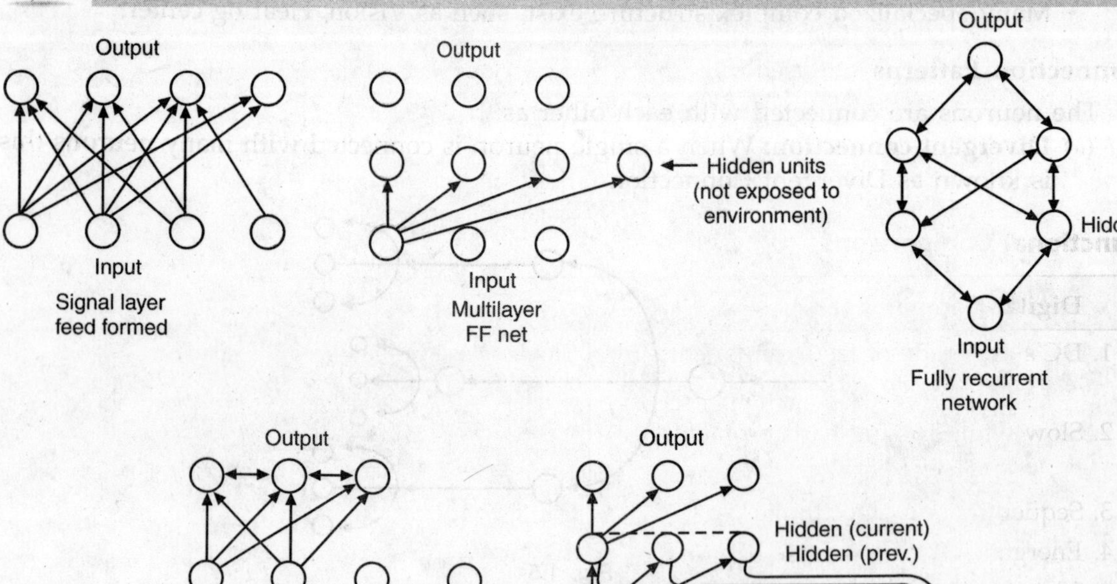

Fig. 1.9.

1.4. AN ARTIFICIAL NEURON

The human brain no doubt highly complex structure viewed as a massive, highly interconnected network of simply processing elements called neurons. However, the behaviour of a neuron can be captured by a simple model as shown in Fig. 1.9. Every component of the model bears a direct analogy to the actual constituents of a biological neuron and hence is termed as artificial neuron or a perception.

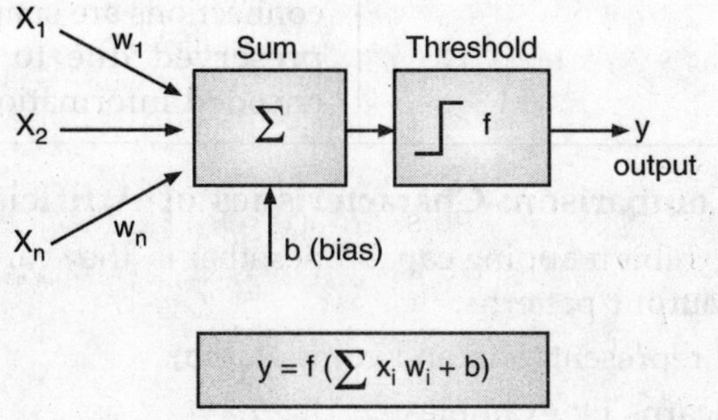

Fig. 1.10. Simple Model of an Artificial Neuron

Here, x_1, x_2, x_3 ... x_n are the n inputs to the artificial neuron. w_1, w_2.......w_n are the weights attached to the links.

Table 1.1. Associated Terminology of Biological and Artificial Neural Network

Biological Neural Network	Artificial Neural Network
Cell Body	Neurons
Dendrite	Weight or interconnection
Soma	Net input
Axon	Output

Functional Comparisons

Digital Computer (DC)	Biological Computer (BC)
1. DC's are made of fast components. (ns)	Slow noisy and unlike components (speed ratio: 10^6)
2. Slow while solving hard problems.	Get approximate solutions very fast. (recognize: 200 ms)
3. Sequential and algorithmic.	Solve problems collectively
4. Energetic efficiency: 10^{-16} joule/operation/sec.	10^{-16} joule/operation/sec.
5. Unable to realize the power, size and complexity.	Have large number of computing elements and computing is not restricted within neurons

6. Information stored in memory. New information in same loction destroys old information.	Interconnections. New information is added by adjusting interconnection strengths.
7. Non fault tolerant in the sense that information corrupted in memory can not be retrived.	Exhibit fault tolerance since information is distributed in connections throughout the N/Ws. Even if few neurons are not working or few connections are snapped, the information is still preserved due to distributed nature of the encoded information.

NOTE

Functional Comparison: Characteristics of Artificial Neural Network

- The ANN exhibit mapping capabilities, that is, they can map input patterns to their associated output patterns.
- Distributed representation and computation.
- The **ANN** learns by examples.
- The **ANN** possesses the capability to generalize. Thus they can predict new outcomes from past trends.
- The **ANN** is robust systems and is fault tolerant. They can, therefore, recall full patterns from incomplete, partial or noisy patterns.
- The **ANN** can process information in parallel, at high speed, and in a distributed manner.
- Low energy consumption.
- Graceful degradation of performance.

Devices based on biological neural networks will posses some of these desirable characteristics. Modern digital computers outperform humans in the domain of numeric computation and related symbol manipulation. The biological neural system architecture is completely different from modern digital computer (see Table 1.2).

Table 1.2. Modern Digital Computer V/s Biological Neural System

Paradigm	Digital computer	Biological Neural System
Processor	Complex High speed One or a few	Simple Low speed A large number
Memory	Separate from a processor Localized Noncontent addressable	Integrated into processor Distributed Content addressable
Computing	Centralized Sequential Stored programs	Distributed Parallel Self-learning
Reliability Expertise	Very vulnerable Numerical and symbolic manipulation	Robust Perceptual problems
Operating environment	Well-defined, well-constrained	Poorly defined, unconstrained

1.5. ARTIFICIAL NEURON MODEL

Three perceptron models of a neuron are discussed as follows:

1.5.1. McCulloch and Pitts Model

The idea of the simple neuron model first emerged in the 1940s with the work of McCulloch and Pitts. The cybernetics movement, attempted to combine biology, psychology, engineering and mathematics resulting in architectures for networks of neurons which would perform a number of tasks. In 1949, Hebb put forward the theory of neural networks developing *internal representation* related to experience [2].

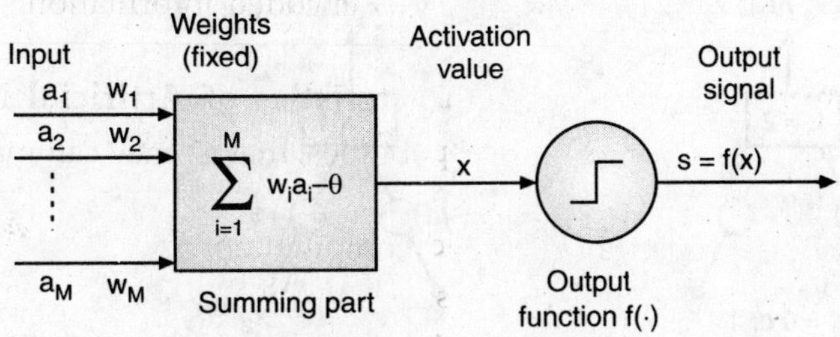

Fig. 1.11. McCulloch and Pitts Model of Neuron

In the 1950s, research continued initially into the development of networks to perform specific tasks but this changed and the goal became to develop machines that could learn. By the end of that decade there had been a lack of significant developments and work in this field diminished considerably.

The perceptron effectively splits the input patterns into two distinct regions with one region being represented by a 1 on the output and the other a 0. Rosenblatt's training algorithm for the perceptron would converge if the input patterns to the perceptron were linearly separable. The perceptron would therefore approximate the decision boundary between the two classes of outputs.

Perceptrons were successfully trained to perform certain tasks but there were failures that could not be overcome. Minsky and Papert pointed out the serious shortcomings of perceptrons [36] and interest in the study of neural networks again declined.

The Exclusive-OR (Ex-OR) function is a major illustration of the limitation of perceptrons. For the Ex-OR function an output of 1 is generated if the inputs are {0,1} or {1,0} and an output of 0 is generated if the inputs are {0,0} or {1,1}. This is not a linearly separable function so the perceptron cannot learn it. A more complicated decision surface is required here and it was found that a curved decision surface is required to separate the two classes of inputs.

Characteristics of this model are:

(*a*) Output 0 or 1

(*b*) Threshold = integer

(*c*) Weights + 1 or – 1

(*d*) Input 0 or 1

Output $y = 1, \text{ If } \Sigma W_i X_i \geq L$

$= 0, \text{ if } \Sigma W_i X_i < L$

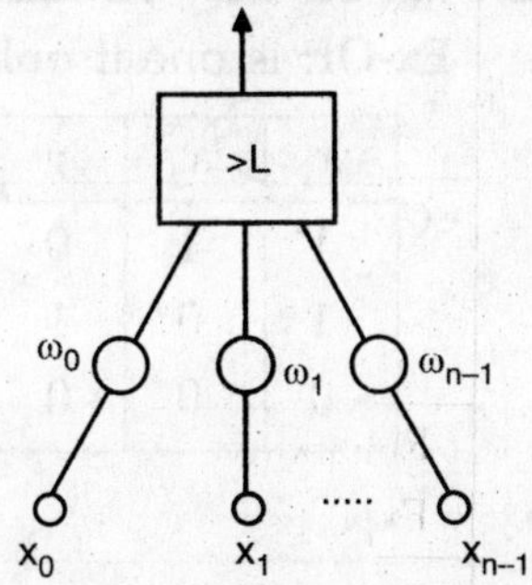

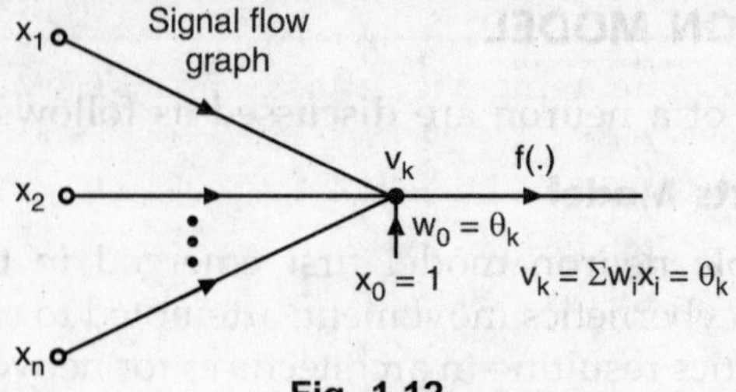

Fig. 1.12.

Logic Gates: Can be represented by McCulloch and Pitts Model as follows:

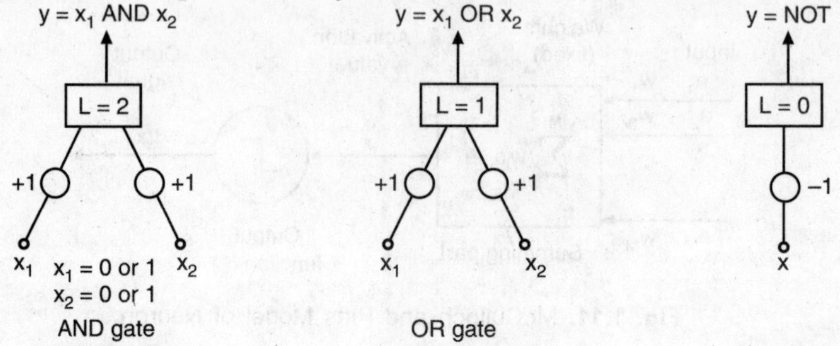

x_1	x_2	y
0	0	0
0	1	0
1	0	0
1	1	1

x_1	x_2	y
0	0	0
0	1	1
1	0	1
1	1	1

x	y
0	1
1	0

Fig. 1.13.

EXAMPLE 1.1.

Generate 'AND' not using Mc Culloh and Pitt Model.

Solution: The truth table for 'AND NOT' shall be as follows:

Return a 1 when x_1 is 1 and x_2 is 0

x_1	x_2	y
1	0	1
0	1	0
0	0	0

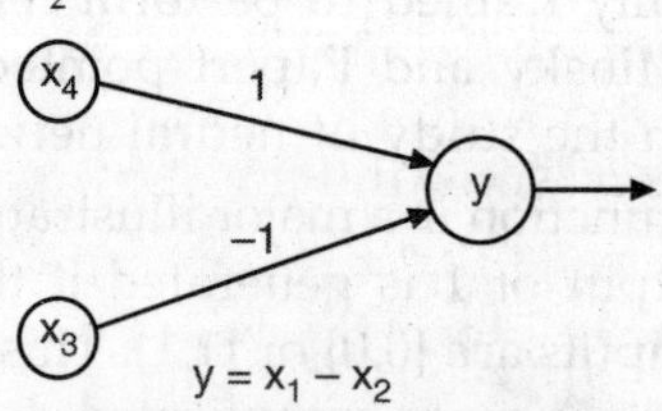

Fig. 1.14.

EXAMPLE 1.2.

Ex-Or: is one if only one input is true, so the trut table becomes.

x_1	x_2	y
1	1	0
1	0	1
0	0	0

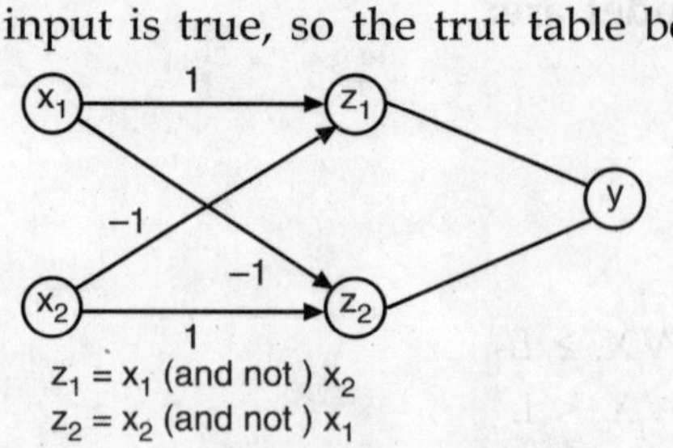

Fig. 1.15.

EXAMPLE 1.3.

Matlab program for NOR gate using McCullock and Pitt Model shall be as follows:

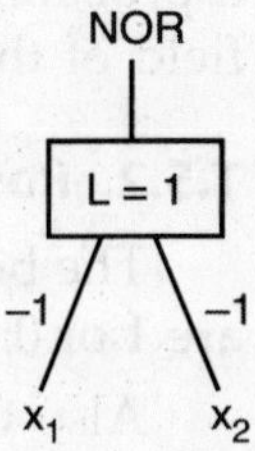

Fig. 1.16.

```
clear;
clc;
disp ('Enter weights');
w1 = input ('weight w1 = ')
w2= input ('weight w2 =')
theta = input ('theta =')
y = [0 0 0 0];
x1 = [0 0 1 1];
x2 = [0 1 0 1];
z = [0 0 1 0];
Con = 1;
While con
Zin = x1 * w1 + x2 * w2;
For I = 1: 4
    If Zin (i) = theta
        y(i) = 1;
    else
        y(i) = 0;
    end
end
disp ('Output of Net');
disp (y)
if y = = z
    Con = 0
Else
    disp ('Net is unable to learn enter different weights threshold');
            w1 = input ('weight w1 =') ;
            w2 = input ('weight  w2 =') ;
            theta = input ('theta =') ;
    end
end
disp ('Network learned') ;
disp ('weight of nemon') ;
disp (w1) ;
disp (w2) ;
disp ('threshold value = ') ;
disp (theta)
```

Stocastic Model of a Neuron

The models of neurons discussed so far are deterministic *i.e.*, its output is precisely predictable for all inputs. In this neuron; the activation function of McCulloh-Pitts model is a probability function $P(x)$ which decides whether the neuron will fire. Here m is the local field of the neuron.

1.5.2. Rosen Blatt's Perceptron Model

The binding with McCulloh and Pitts model is that the weight and fixed 1/0 and inputs are 1 or 0; in Rosenblatt's models weights are adjusted by error (between target and output).

Also the outputs from sensory units to a fixed is fed to a fixed set of association units, the output of which is fed to McCulloch and Pitts neuron. The association units perform predetermined manipulations on their inputs.

$$x = \begin{cases} 1 \text{ with probability } P(v) \\ -1 \text{ with probability } 1 - P(v) \end{cases}$$

So, for probability $P(v)$ to be sigmoid

$$P(v) = \frac{1}{1 + \exp(-v/T)}$$

Where v is local field induced in neuron and T is pseudo temperature, which controls noise level and thus, uncertainty in firing. T is not temperature. It is just a parameter controlling the thermal fluctuations. When $T \to 0$; $P(v) \to 1$ is the neuron converts to deterministic McCulloh-Pitts form.

The algorithm is as follows:

1. 1.All weights and biases are set to be zero. (It can be randomly selected also). Set learning rate a (between 0 and 1). The random selection can be made using Fuzzy rule or Genetic Algorithm also.
2. Continue the loop till stopping criterion is matched, stopping can be decided based on number of epachs (reperations) or permissible error or target matching. The repeation loop is given in 3 to 7.
3. For each training pair do 4 – 6.
4. Set activation of input units $x_i = s_j$ for $i = 1$ to n.
5. Calculate output response

$$y_{in} = b + \sum_i x_i w_i$$

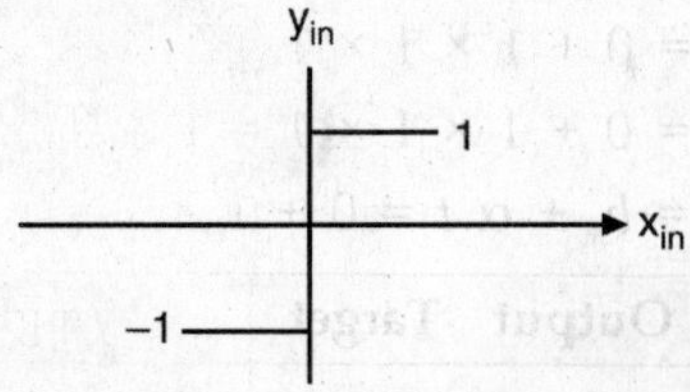

Fig. 1.17. Hardlimiter activation function

and $$y = f(y_{in}) = \begin{cases} 1 \rightarrow y > \theta \\ 0 \rightarrow -\theta \leq Y \leq \theta \\ -1 \rightarrow Y_{in} < -\theta \end{cases}$$

Perceptron

6. Weights and bias are updated if target is not equal to output response:

   If $t \neq y$ and x_i is non zero {active inputs}; where t is target and y is output.

   then $w_i(n) = w_i(o) + \alpha t\, x_i$ show updation in weights

   $b_{(n)} = b + \alpha t$ {here $x_i = 1$} show updation in bias

   else $w_i(n) = w_i(o)$

   $b_{(n)} = b_{(o)}$

7. Test for stopping condition.

Use a perceptron for AND gate operation with bipolar inputs and targets

Inputs		Bias	Targets
x_1	x_2	b	t
1	1	1	1
- 1	1	1	- 1
1	- 1	1	- 1
- 1	- 1	1	- 1

1. Say $w_1 = w_2 = 0$; $b = 1$; $\alpha = 1$, $\theta = 0$
2. Do while stopping condition is false
3. For (1, 1) ; do 4 – 6
4. Set activation of input $x_i = (1, 1)$
5. Calculate net input

$$Y_{in} = b + \Sigma x_i\, \alpha_i = 0 + 1 \times 0 + 1 \times 0 = 0$$

Activation $$Y = f(y_{in}) = \begin{cases} -1 & Y_{in} > 0 \\ 0 & - \leq Y_{in} \leq 0 \\ 1 & Y_{in} < -0 \end{cases} \Rightarrow Y = 0$$

6. Output $y = 0$ whereas target was 1

$$w_{i(\text{new})} = w_{io} + \alpha\, t\, x_i$$
$$w_1 = 0 + 1 \times 1 \times 1 = 1$$
$$w_2 = 0 + 1 \times 1 \times 1 = 1$$
$$b_n = b_o + \alpha\, t = 0 + 1 \times 1 = 1$$

Input		Bias	Net	Output	Target	Weight changes			Weights		
x_1	x_2	B	Y_{in}	Y	t	Δw_1	Δw_2	Δb	w_1	w_2	B
1	1	0	0	0	1	1	1	1	1	1	1

–1	1	1	1	1	–1	1	–1	–1	2	0	0
1	–1	1	2	1	–1	–1	1	–1	1	1	–1
–1	–1	1	–3	–1	–1	0	0	0	1	1	–1

So, finally after first epoch $\omega_1 = 1$, $\omega_2 = 1$, $b = -1$

MATLAB Program for AND gate using Perceptron Mode

Input		Output
x_1	x_2	y
+ 1	+ 1	+ 1
+ 1	– 1	– 1
– 1	+1	– 1
– 1	– 1	–1

```
Clear;
Clc ;
x = [1 1 – 1 – 1; 1 – 1 1 – 1];
t = [1 -1 – 1 – 1]
w = [0 0]
b = 0
Alpha = input ('Enter learning rate');
Theta = input ('Enter threshold value =');
Con = 1
While con
   For i = 1: 4
      Y_in = b + x(1, i) * w(1) + x(2, i) * w(2) ;
      If Y_in > theta
         Y = 1
      End
      If y_in ≤ theta and y_in ≥ – theta
         Y = 0
      End
      If y_in ≤  theta – theta
         Y = 1
      End
      If Y – t(i)    {If y – t is some none zero value}
      Con = 1
   For j = 1: 2
      w(j) = w(j) + alpha * t(i) * x(j, i);
      end
      b = b + alpha * t(i)
      end
   end
   epoch = epoch + 1;
```

```
end
disp (final weight matrix);
disp(w);
disp ('final bias');
disp (b);
```

1.5.3. Adaline

ADAptive LINear Element (ADALINE) is a computing model proposed by Windrow. The main difference between Rosenblatt's perceptron model and the Windrow Adaline model is that, in the Adaline model the analog activation value (x) is compared with the target output (b). The output is a linear function of the activation value (x).

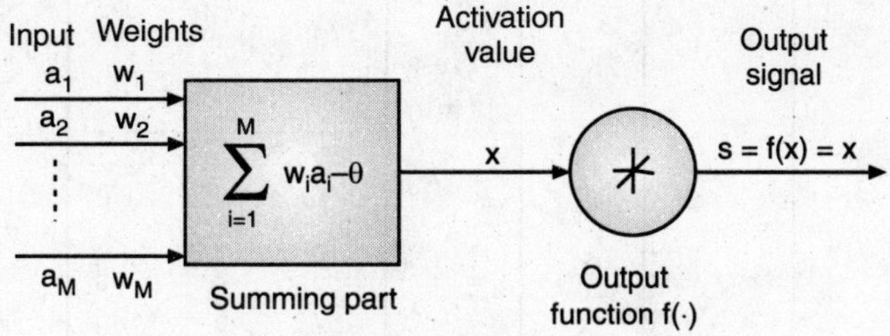

Fig. 1.18. Windrow's Adaline Model of Neuron

The following equations describe the operation of Adaline:

Activation, $x = \sum_{i=1}^{M} w_i a_i - \theta$

Output signal, $s = f(x) = x$

Error, $\delta = b - s = b - x$

Weight change: $\Delta w_i = \eta \delta a_i$, where η is learning rate constant.

This weight update rule minimizes the mean squared error δ^2, averaged over all inputs. Hence it is called Least Mean Squared (LMS) error learning law.

Algorithm:

1. Initialize weights (some non-zero random values) set learning rate a.
2. While stopping condition is false, do 3 – 7.
3. For each bipolar training pair, perform 4 – 6.
4. Set activation of input units for all inputs.
5. Compute net input to output unit $Y_{in} = b + S + w$.
6. Update weights and bias for all weights corresponding to activation inputs

   $w_{in}(n) = w_{i(o)}(n)\ \alpha\ (t - y_{in})\ x_i$ {No threshold, no output activation function}

   $b_n = b_o + \alpha\ (t - y_{in})$
7. Test for stopping condition.

	Inputs			Target	Net		Weight change			Weight			Error E
Epoch 1	x_1	x_2	(b) x_3	t	y_{in}	t-y_{in}	Δw_1	Δw_2	Δw_3	w_1	w_2	w_3	$(t - y_{in})^2$
	– 1	– 1	1	– 1	– 0.2	– 1.2	– 16			.04			
	– 1	+ 1	1	– 1	– 0.8	– .16							
	1	– 1	1	– 1									
	1	1	1	1									
Epoch 2	To be done by reader												

$$y_{in} = b + \Sigma wx$$
$$\Delta w_1 = \alpha(t - y_{in})\, x_i$$

MATLAB Programme for AND function using Adaline

```
Clean;
Clc;
disp ('Adaline networks for AND function');
x1 = [- 1 -1 1 1]                          Δω1 = del w1
x2 = [- 1 1 -1 1] ;                        Δω2 = del w2
x3 = [1 1 1 1];                            Δb = del b
w1 = w2 = b = alpha = 0.2;
% stopping criterior error = 2 or less
e = 2;
del w1 = del w2 = del b = 0;
epoch = 0
while e ≥ 2
    epoch = epoch + 1
    e = 0
    for i = 1: 4
    net y(i) = w1 * x1(i) + w2 * x2(i) + b;
    nt = [net y(i), t(i)];
```

```
    del w1 = alpha * (t(i)) – net y(i) * x1(i);
    del w2 = alpha * (t(i)) – net y(i) * x2(i);
    del b = alpha * (t(i)) – net y(i) * b(i);
    wc = [del w1 del w2 del b ];                      % weight change
    w1 = w1 + del w1 ;
    w2 = w2 + del w2 ;
    b = b + del b;
    w = [w1 w2 b ]
    x = [x1 (i) x2 (i) x3 (i)]                        % new weights
  [x          x             wc              w]
   ↓          ↓             ↓               ↓
 x1 x2 x3  net in target  del w1 del w2 del b  w1 w2 b
  end
  for i = 1: 4
    net y(i) = w1 *  x1(i) + w2 * x2(i) + b;
    e = e + (t(i) – net y(u) ∧ 2;
    end
  end
```

1.6. ACTIVATION FUNCTIONS

The activation function is used to calculate the output response of a neuron. The sum of the weighted input signal is applied with an activation to obtain the response. For neurons in same layer, same activation functions are used. There may be linear as well as nonlinear activation functions. The nonlinear activation functions are used in a multilayer network. A few linear and nonlinear activations are discussed here.

1.6.1. Linear Functions

1.6.1.1. Identity Function

The function is given by,

$$f(x) = x, \text{ for all } x \quad \text{...(2.1)}$$

This is shown in Fig. 1.19.

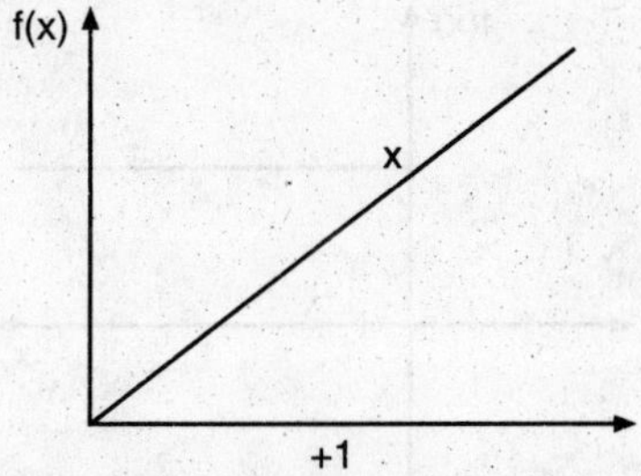

Fig. 1.19. Identity Function

1.6.1.2. Purelin function

The function is given by,

$$f(x) = x \text{ for } -\infty \leq x \leq \infty + \infty \qquad (2.2)$$

This is shown in Fig. 1.20.

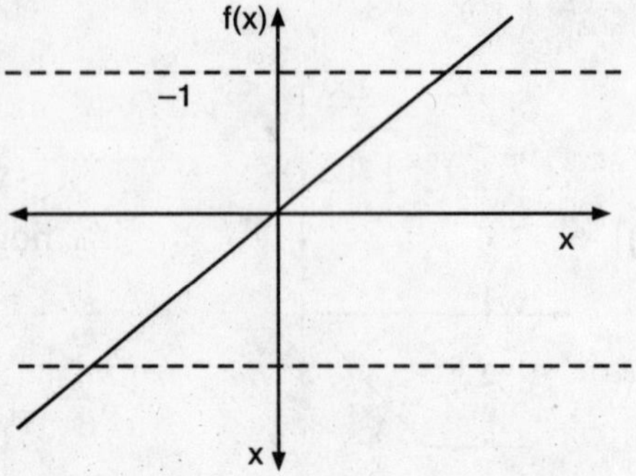

Fig. 1.20. Purelin Activation Function

Nonlinear activation functions:

1.6.2. Binary Step Function

The function is given by,

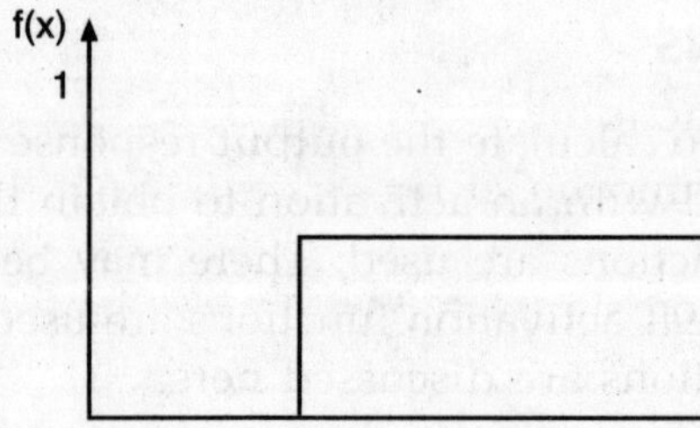

Fig. 1.21. Binary Step Function

1.6.3. Hardlim Function

The function is given by,

$$f(x) = 1 \text{ for } x \geq 0$$
$$= 0 \text{ for } x \quad 0$$

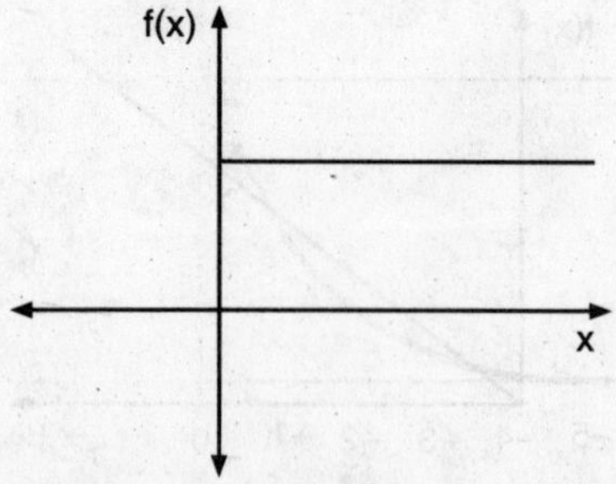

Fig. 1.22. Hardlim Function

1.6.4. Saturated Linear Activation Function

The function is given by,

$$f(x) = x \text{ for } -1 \le x \le +1$$
$$= 1 \text{ for } x > 1$$
$$= -1 \text{ for } x -1 \quad (2.5)$$

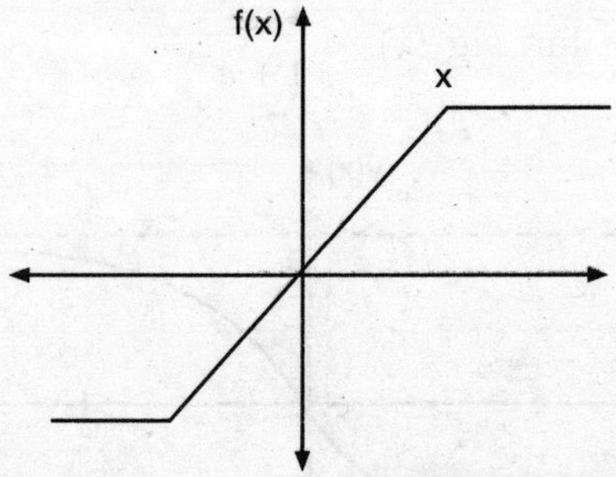

Fig. 1.23. Saturated Linear Function

1.6.5. Sigmoidal Functions

These functions are usually S-shaped curves. The hyperbolic and logistic functions are commonly used. These are used in multilayer nets. There are two main types of sigmoidal functions.

1.6.5.1. Binary Sigmoidal Function

This is also called logistic function. It ranges between 0 and 1.

$$f(x) = \log \text{sig}\ (x) = \frac{1}{1 + \exp^{-x}} \quad ...(2.6)$$

If $f(x)$ is differentiated we get,

$$f(x) = f(x)\ [1 - f(x)] \quad ...(2.7)$$

Figure 1.24. Shows the binary sigmoidal function.

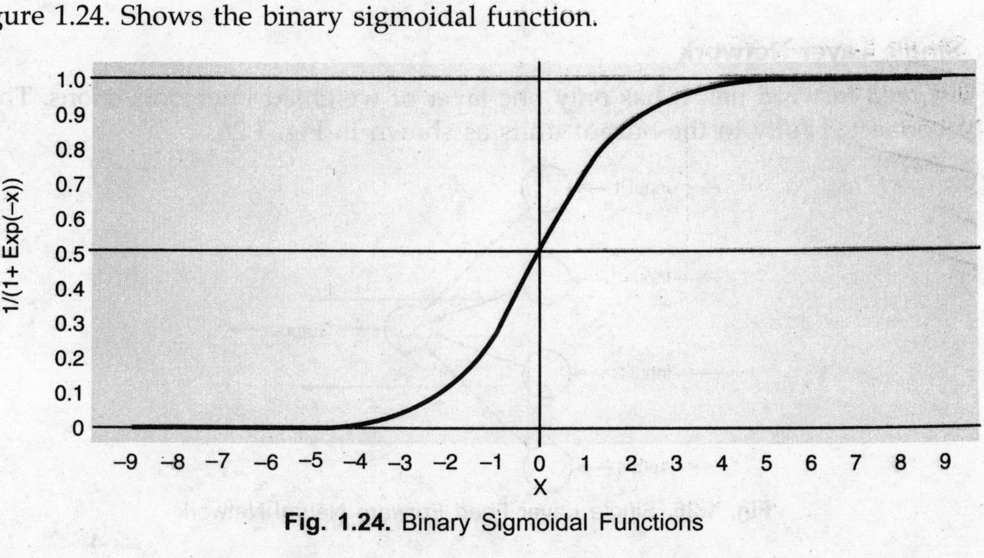

Fig. 1.24. Binary Sigmoidal Functions

1.6.5.2. Bipolar Sigmoidal Function

The desired range here is between + 1 and – 1.The function is related to the hyperbolic tangent function. The bipolar sigmoidal function is given as,

$$y(x) = 2f(x) - 1 \quad ...(2.8)$$

Substituting the value of f(x) from eq. (2.6) into the eq. (2.8) we get,

$$y(x) = \tan \text{sig } (x) = \frac{1 - e^{-x}}{1 + e^{-x}} \quad ...(2.9)$$

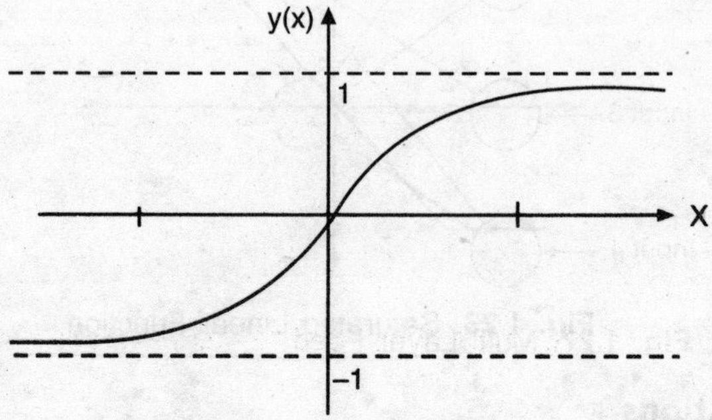

Fig. 1.25. Bipolar Sigmoidal Function

1.7. NEURAL NETWORK ARCHITECTURE

The arrangement of neurons into layers and pattern of connection within and in between layer are generally called as the architecture o the network or network topology. There are several network topologies. Each topology is designed to solve specific problem or to represent certain point of view. Each topology has some drawbacks and benefits. These topologies can be categorized in two main categories: Feed forward network and recurrent network.

Feed Forward Network: Feed forward network may have single layer of weights where the inputs are directly connected to the outputs or multilayer with intervening sets of hidden units.

1.7.1. Single Layer Network

It is a feed forward net. It has only one layer of weighted interconnections. The inputs may be connected fully to the output units as shown in Fig. 1.26.

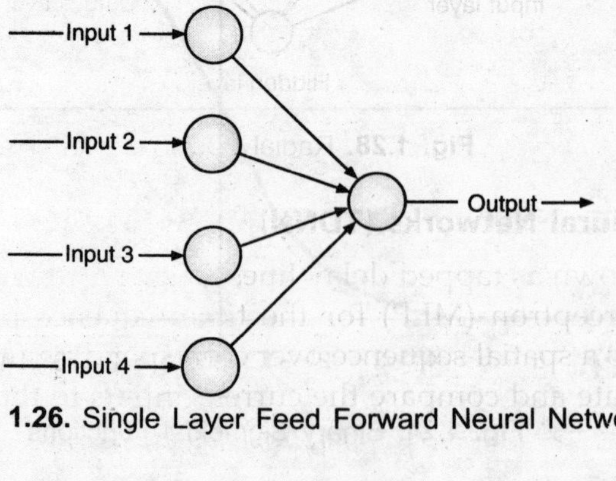

Fig. 1.26. Single Layer Feed Forward Neural Network

1.7.2. Multi Layers Feed Forward Neural Network

It is also a feed forward net i.e., the net where the signals flow from the inputs units to the output units in forward direction. The multi-layer net poses one or more layers of nodes between the inputs and output unit.

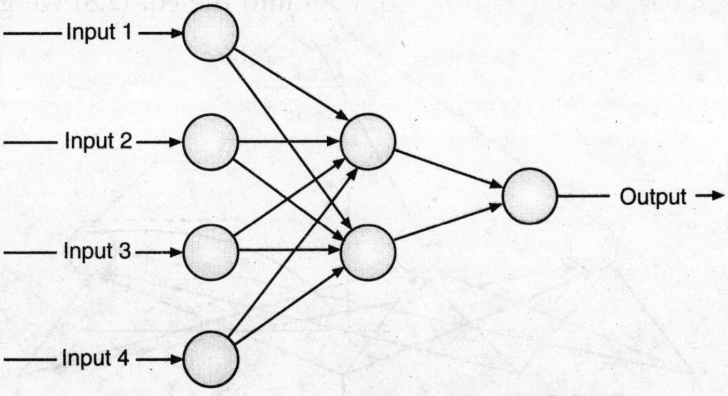

Fig. 1.27. Multi Layer Feed Forward Neural Network

1.7.3. Radial Basis Network

The basic topology of the RBF network consists of an input layer, one hidden layer, and an output layer as shown in Fig. 1.28. Additional ramifications in the basic architecture are determined by the particular learning rules used. With the availability of a large number of hidden layer units, the network can well approximate a broad class of mappings. There is also considerable flexibility in the choice of transfer characteristics of the hidden layer units. The RBF networks use Gaussian activation functions in addition to other proven possibilities.

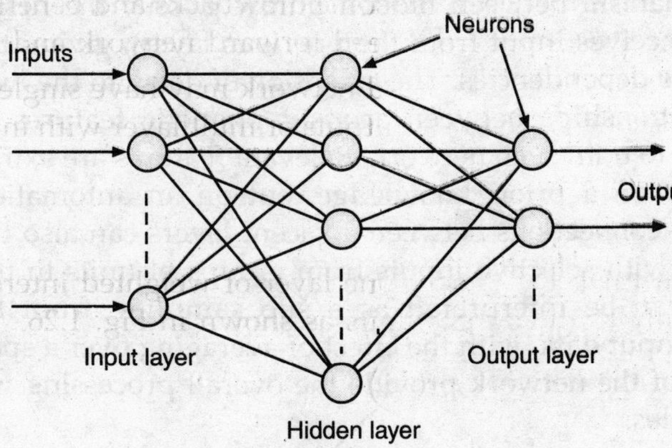

Fig. 1.28. Radial Basis Networks

1.7.4. Time Delay Neural Networks (TDNN)

TDNN are also known as tapped delay lines. They represent an effective attempt to train a static Multilayer Perceptron (MLP) for the time-sequence processing, by converting the temporal sequence into a spatial sequence over corresponding units of time. This modification gives the ability to relate and compare the current inputs to their past history. The idea was applied in a variety of Speech Recognition applications, mostly for phoneme recognition. An example of TDNN is shown in the Fig. 1.29 as:

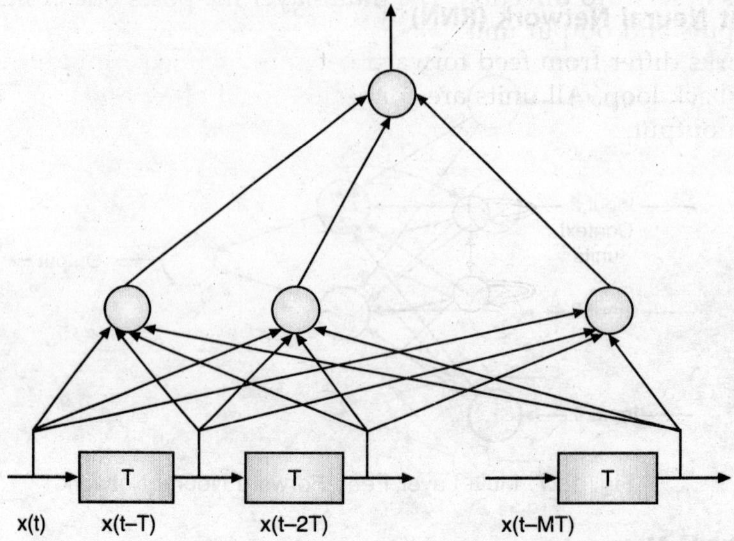

Fig. 1.29. Time Delay Neural Network

The input layer has been enlarged to accept as many input patterns as the fixed sequence length to be processed at each time step. Input vectors enter the network from the leftmost set of input units. At each step, inputs are shifted to the right through the time delay line that links each set of input units to the right-adjacent one, and the next input pattern is fed into the leftmost position. The same mechanism may be extended to subsequent layers, introducing a tapped- delay mechanism between hidden units(e.g. only the first block of units in the tapped line actually receives input from the previous layer), giving the ability to deal with more complicated time dependencies. The extension of delay to the hidden layers allows the discovery of the relationships between acoustic-phonetic features. The *Back Propagation* algorithm can be used to train such network. Relevant features are extracted from the original input, not based only on a priori knowledge, but on an automatic selection during the network training. The connections between adjacent layers can also be arranges in order to provide hidden units with selective inputs from groups of units in the previous layer. The resulting dynamics can be interpreted as a sub sampling from the original temporal representation of the input data, with the effect of averaging over a specific time slice. In this way, different layers of the network provide the overall processing with representations of different temporal scales.

Architectures for temporal pattern recognition tasks have evolved from the well-understood principles of multilayer feedforward and feedback neural networks. In order to use models based on these known architectures, it is necessary to represent the temporal pattern as a static spatial pattern. For this representation, delay units are used to store a fixed number of components belonging to the preceding instants. Thus a temporal pattern is represented using a tapped delay line as shown in the input layer in Fig. 1.29. The figure illustrates architecture for temporal pattern recognition using a multilayer feedforward neural network. The disadvantage of this approach is that the length of the sequence has to be fixed a priori. Also, a large number of training sample sequences are required for learning and generation. Moreover, the input signal must have precise time registration. Many natural signals like speech do not conform to these restrictions.

1.7.5. Recurrent Neural Network (RNN)

These networks differ from feed forward network architecture in the sense that there is at least one feedback loop. All units are connected to all other units and every unit is both an input and an output.

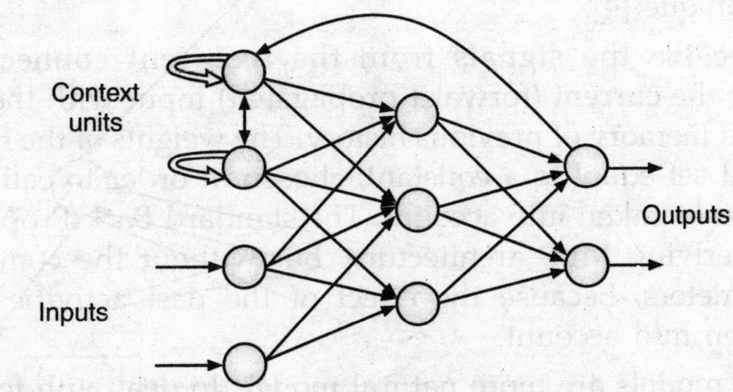

Fig. 1.30. Recurrent Neural Networks

This network provides a powerful extension of feed-forward connectionist model by allowing introducing connections between arbitrary pairs of units, independently from their position with the topology of network. Self recurrent loops of a unit onto itself, as well as back propagation connections to previous or lateral layers (Lateral links between the units belonging to the same layer) are allowed. A RNN is shown in the Fig. 1.31 as:

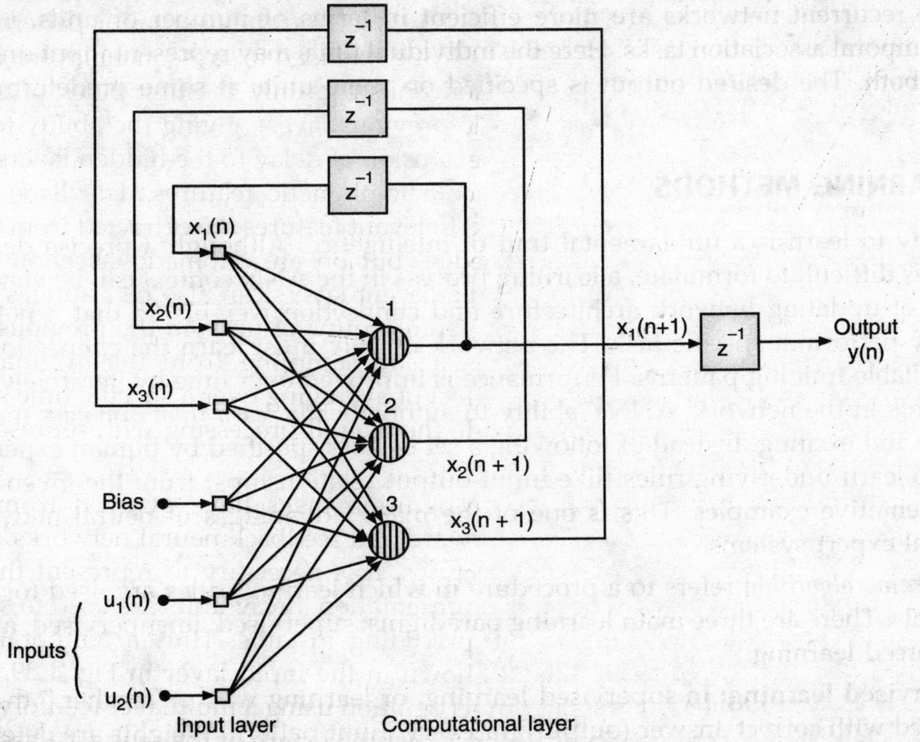

Fig. 1.31. Recurrent Neural Network

RNNs behave like dynamical systems. Once fed with input, the recurrent connections are responsible for an evolution in time of the internal state of the network. RNNs are particularly suited for sequence processing, due to their ability to keep an internal trace, or memory. This memory is combined with the current input to provide a context-independent output. Several RNN architectures are being tested with various training algorithm e.g. Recurrent Back Propagation, Back Propagation for Sequence, Real Time Recurrent Learning, Time Dependent Recurrent Back Propagation, Back Propagation through Time; all based on Gradient Descent Technique [4].

The units that receive the signals from the recurrent connections act either as preprocessors, filtering the current (forward propagated) input with the previous signal, or as a register that keeps a memory of previous history. The weights of the recurrent connections are generally fixed and set equal to a constant, chosen in order to calibrate the amount of previous information to be taken into account. The standard Back Propagation algorithm is used to train the underlying MLP architecture, but without the computation of the full gradient on the parameters, because the effect of the past activities through recurrent connections is not taken into account.

Recurrent network models are more natural models to deal with temporal pattern. But training will be a problem with these models. Several partially recurrent models were proposed in the literature (Elman, 1990; Jordan, 1986; Stornetta et al, 1988; Mozer, 1989). The connections are mostly feedforward with a few selected fixed feedback connections so as to keep the training within manageable complexity. Thus the recurrent part is realized with context units in different configurations. The context units received the feedback signals as shown in Fig. 1.7 for the configuration proposed by Jordan (1986).

Fully recurrent networks are more efficient in terms of number of units, in order to realize temporal association tasks. Here the individual units may represent input units, output units or both. The desired output is specified on some units at some predetermined time instants.

1.8. LEARNING METHODS

The ability to learn is a fundamental trait of intelligence. Although a precise definition of learning is difficult to formulate, a learning process in the ANN context can be viewed as the problem of updating network architecture and connection weights so that a network can efficiently perform a specific task. The network usually must learn the connection weights from available training patterns. Performance is improved over time by iteratively updating the weights in the network. ANNs' ability to automatically *learn from examples* makes them attractive and exciting. Instead of following a set of *rules* specified by human experts, ANNs appear to learn underlying rules (like input-output relationships) from the given collection of representative examples. This is one of the major advantages of neural networks over traditional expert systems.

A *learning algorithm* refers to a procedure in which learning rules are used for adjusting the weights. There are three main learning paradigms: supervised, unsupervised, and hybrid or Reinforced learning.

Supervised learning: In supervised learning, or learning with a "teacher," the network is provided with correct answer (output) for every input pattern. Weights are determined to allow the network to produce answers as close as possible to the known correct answers.

Unsupervised learning: In contrast, unsupervised learning, or learning without a teacher, does not require a correct answer associated with each input pattern in the training data set. It explores the underlying structure in the data, or correlations between patterns in the data, and organizes patterns into categories from these correlations.

Reinforced learning: In this learning, a teacher is assumed to be present, but the right answer is not presented to the network. Instead, the network is presented with an indication of whether the output is right or wrong. The network must then use this information to improve its performance. Reinforcement learning is a very general approach to learning that can be applied when the knowledge required to supervise learning is not available.

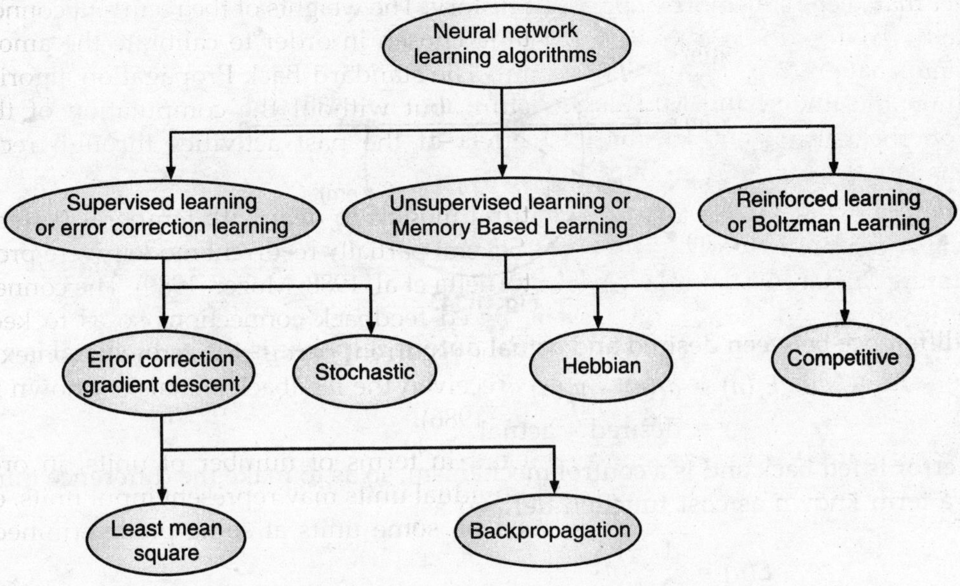

Fig. 1.32. Classification of Learning Algorithms

The property which distinguish the ANN from classical computer is its ability to learn from the environment and improve. This learning and adaption is applied to bias level and synaptic weights, thereby, making the network wiser, by each iteration.

As defined by Mendel and McClaren, learning is a process by which the free parameters of a neural network are adapted through a process of stimulation by the environment in which the network is embedded. The type of learing is determined by the manner in which the parameter changes take place.

These are the ways in which the synaptic weights of a neuron change. In detail:

(i) Supervised or Error correction learning

Let there be a NN which senses the lumens in a room and identifies whether a given light source is in working condition. Thus input to this network in lumensoty and output is yes/no *'y'* for working sources and *'N'* for non-working source. The block diagram of such a system may be

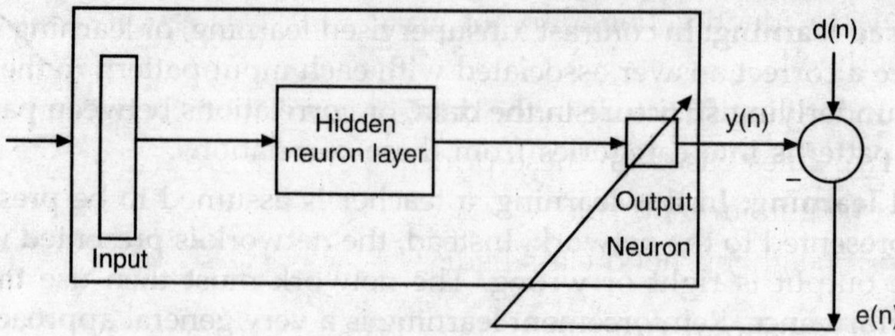

Fig. 1.33.

The error is fed back to all weights feeding to hidden neuron layer. Its signal flow graph will be:

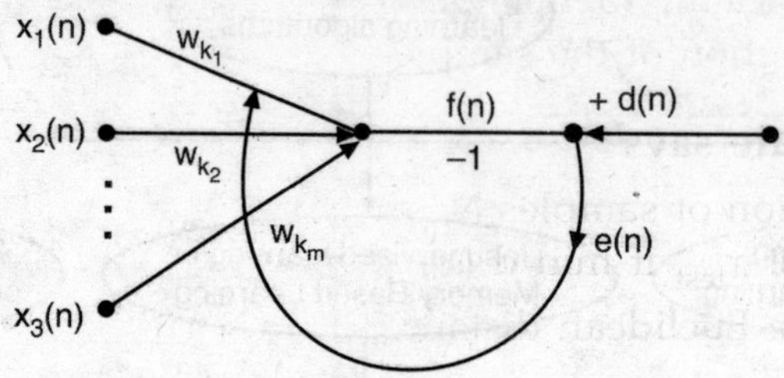

Fig. 1.34.

The difference between desired and actual output response is the error signal is given by

$$E_k(n) = d_k(n) - y_k(n)$$
$$= \text{desired} - \text{actual}$$

This error is fed back and is a control mechanism, so as to make the difference minimum. There is a term known as cost function defined as:

$$E(n) = \frac{1}{2} e_k^2(n)$$

$\frac{1}{2}$ is the normalization factor and maxima and minima are decided by second derivation. Thus cost function or index of performance is identified. The minimization leads to delta rule or windrow Hoff rule. Which states that $(n + 1)$ th value of w_{k_j} (weight from neuron j to k) is Dw more/less than its previous value (at nth state)

i.e., w_{k_j} (new) $= w_{k_j}$ (old) $+ \Delta\, \omega_{k_j}(n)$

V_k is post synaptic signal and x_k is presynaptic input signal. Thus with a feedback loop. The learning rate (*i.e.,* rate of change of weight w) must be selected carefully to ensure the stability accuring and convergence of sytem.

(ii) Unsupervised Memory-Based Learning

It operates by memorizing the training data explicitely in proper classification. The network is tested by analyzing and classifying the data in the neighborhood of set of testing data. All the algorithms based on memory learning have two components to be identified explicitily:

(*i*) The rule to identify boundary of class and centroid or identification of nearest neighborhood.

(*ii*) The rule to identify the distance between two classes/centhoid of classes.

There are two rules based on these principles which are widely studied.

(*a*) Nearest neighbourhood rule

(*b*) K-nearest neighbourhood classified

(*a*) Nearest neighborhood rule

The rule is based on two assumptions:

(*i*) The sample population is infinitely large.

(*ii*) The training samples are idenpendently and identically distributed as per the point probability distribution of the samples.

The nearest neighbor rule says

Let then be a population of samples *N*; so that x_N e $\{x_1, x_2, \ldots x_3\}$. This vector is said to be the nearest neighbor of x_{test} if min d $(x_i, x_{test}) = d\ (x_N, x_{test})$,

where $d(x_i, x_{test})$ is the Euclidean distance.

(*b*) k-nearest neighbor rule:

The rule says: Given a test pattern x_{text}

(*i*) Identify the *k* classified patterns that lie closest to the test vector x_{test}.

(*ii*) If most of the neighbors out of *k* patterns lie in on class *x*, *y* then x_{test} will be part of class *y* where most of the members of *k* patterns lie.

(*iii*) Reinforced or Boltzmann learning

It is a stochastic learning algorithm. The behave an binary units. The machine has an energy function E, given as

$$E = \frac{1}{2} \sum_j \sum_k w_{kj} . x_k \; x_j$$

The states x_k or x_j are denoted by eithers 1 or – 1 w_{kj} in synaptic weight of link between *j* to *k*. $j \neq k$ represents a no self feedback neuron.

(*i*) The machine choses a neuron at random.

(*ii*) It then flips the state of neuron *k* from x_k to $-x_k$ with probability.

$$P(x_k - x_k) = \frac{1}{1 + \exp\left(-\Delta \frac{E_k}{T}\right)}$$

Where $\Delta\ E_k$ is change in energy function of machine *T* is temperature.

(*iii*) This rule is applied repeatedly till a thermal equilithium is attained.

These are two types of neurons:

(*a*) Visible neurons are interface between network and operating environment.

(*b*) Hidden neurons operate freely.

There are two states of operation of neurons:

(*a*) **Clamped:** Where invisible neurons are all clamped into specific states (r_{kj}^+).

(*b*) **Free running condition:** Where all neuron (visible/hidden) freely (r_{kj}^-) .

According to Boltzmanu rule:

$$\Delta\omega_{kj} = \eta\,(\rho_{kj}^{+} - \rho_{kj}^{-}) \quad (j \neq k \text{ i.e., no self loop})$$

Where, ρ_{kj}^+ be correlation between states of neurons *j* and *k* in the free-running condition of network.

ρ_{kj}^+ and ρ_{kj}^- vary between – to + 1.

And η is learning rate.

1.9. VARIOUS LEARNING TECHNIQUES

1.9.1. Supervised Learning Algorithms

The ability to learn is a fundamental trait of intelligence. Although a precise definition of learning is difficult to formulate, a learning process in the ANN context can be viewed as the problem of updating network architecture and connection weights so that a network can efficiently perform a specific task. The network usually must learn the connection weights from available training patterns. Performance is improved over time by iteratively updating the weights in the network. ANNs' ability to automatically *learn from examples* makes them attractive and exciting. Instead of following a set of *rules* specified by human experts, ANNs appear to learn underlying rules (like input-output relationships) from the given collection of representative examples. This is one of the major advantages of neural networks over traditional expert systems. Common learning rules are described below [9]:

1.9.1.1. Perceptron Learning Rule

For the perceptron learning rule, the learning signal is the difference between the desired and actual neuron's response. This type of learning is supervised.

$$\Delta\omega_i = c\delta x_i \qquad \text{...(3.1)}$$

Where *c* is a small learning constant.

$$\delta = d - 0 \qquad \text{...(3.2)}$$

$$\Delta\omega_i = cx_i\,(d - \text{sign (net)}) \qquad \text{...(3.3)}$$

$$\text{net} = \sum_{i=1}^{n} w_i x_i \qquad \text{...(3.4)}$$

1.9.1.2. Delta Learning Rule or Convergence Rule

The LMS method assumes linear activation function ***net* = *o***, and the obtained solution is sometimes far from optimum. If error is defined as,

$$\text{Error}_j = \sum_{p=1}^{p} (o_{jp} - d_{jp})^2 \qquad \text{...(3.7)}$$

Then the derivative of the error with respect to the weight w_{ij} is

$$\frac{\partial \text{Error}_j}{\partial w_{ij}} = 2\sum_{p=1}^{P} (o_{jp} - d_{jp}) \frac{\partial f(\text{net}_{jp})}{\partial \text{net}_{jp}} x_i \quad ...(3.8)$$

Note, that this derivative is proportional to the derivative of the activation function *f′(net)*.

Using the cumulative approach, the neuron weight w_{ij} should be changed with a direction of gradient,

$$\Delta w_{ij} = cx_i \sum_{p=1}^{P} (d_{jp} - o_{jp}) f'_{jp} \quad ...(3.9)$$

in case of the incremental training for each applied pattern,

$$\Delta w_{ij} = cx_i \sum_{p=1}^{P} (d_{jp} - o_{jp}) f'_{jp} \quad ...(3.10)$$

the weight change should be proportional to input signal x_i , to the difference between desired and actual outputs $d_{jp} - o_{jp}$, and to the derivative of the activation function f'_{jp}. Similar to the LMS rule, weights can be updated in both the incremental and the cumulative methods. In comparison to the LMS rule, the delta rule always leads to a solution close to the optimum

1.9.1.3. Gradient Descent (Error-BackPropagation) Learning Rule

The delta learning rule can be generalized for multilayer networks. Using a similar approach, as it is described for the delta rule; the gradient of the global error can be computed in respect to each weight in the network. The back propagation algorithm has many disadvantages which lead to very slow convergence. One of the most painful disadvantage is that in the backpropagation algorithm the learning process almost perishes for neurons responding with the maximally wrong answer.

Momentum term: The backpropagation algorithm has a tendency for oscillation. In order to smooth up the process, the weights increment "w_{ij} can be modified according to Rumelhart, Hinton, and Wiliams (1986)[9],

$$w_{ij}(n) = w_{ij}(n-1) + \Delta w_{ij}(n-1) + \beta \Delta w_{ij}(n-2) \quad ...(3.11)$$

Or according to Sejnowski and Rosenberg (1987) [9],

$$w_{ij}(n) = w_{ij}(n-1) + (1-\beta) \Delta w_{ij}(n-1) + \beta \Delta w_{ij}(n-2)... \quad (3.12)$$

Where β is a momentum factor.

1.9.1.4. Widrow-Hoff (LMS) Learning Rule

Widrow and Hoff (1962) developed a supervised training algorithm which allows to train a neuron for the desired response. This rule was derived so the square of the difference between *net* and output value is minimized.

$$\text{Error}_j = \sum_{p=1}^{P} (\text{net}_{jp} - d_{jp})^2 \quad ...(3.5)$$

Where $Error_j$ is the error for *j-th* neuron, *P* is the number of applied patterns, d_{jp} is the desired output for *j-th* neuron when *pth* pattern is applied, and *net* is given by equation (3.4). This rule is also known as the LMS (Least Mean Square) rule. The formula for the weight change is,

$$\Delta w_{ij} = cx_{ij}\, p = \sum_{p=1}^{p} (d_{jp} - \text{net}_{jp}) \qquad ...(3.6)$$

Note that weight change is w_{ij} is a sum of the changes from each of the individual applied patterns. Therefore, it is possible to correct weight after each individual pattern was applied. This is *incremental updating*, while *cumulative updating* is when weights are changed after all patterns were applied. The incremental updating usually leads to a solution faster, but it is sensitive to the order in which patterns are applied. If the learning constant c is chosen to be small, then both methods give the same result. The LMS rule works well for all type of activation functions.

1.9.1.5. Levenberg-Marqardt (LM) Algorithm

It combines the speed of the Newton algorithm with the stability of the steepest descent method. Weights are calculated using the following formula,

$$W_{k+1} = W_k - (J_k^T J_k + \mu I)^{-1} J_k^T E \qquad ...(3.13)$$

Where, J = Jacobian matrix

I = Identity matrix

μ = Learning parameter

E = Cumulative error vector for all patterns.

For $\mu = 0$, it becomes the Gauss-Newton method. For very large μ the LM algorithm becomes the steepest descent or the error backpropogation algorithm. The μ parameter is automatically adjusted at each iteration in order to secure convergence.

Table 1.3. Summaries of Various Learning Algorithms and their Associated Network Architectures

Paradigm	S. No.	Learning Rule Rule	Network Architecture	Weight Change
Supervised	1	Perceptron Learning Rule	Single and multi layer feed forward	$\Delta w_i = cx_i\,(d - \text{sign (net)})$
	2	Delta Learning Rule	Multilayer feed forward	$\Delta w_i = cx_i \sum_{p=1}^{p} (d_{jp} - o_{jp}) f'_{jp}$
	3	Gradient Descent Learning Rule	Multilayer feed forward	$w_{ij}(n) = w_{ij}(n-1) + (1-\beta)\,\Delta w_{ij}(n-1)(n-2)$
	4	Widrow-Hoff Learning Rule	Multilayer feed forward	$\Delta w_i = cx_{ij}\, p = \sum_{p=1}^{p} (d_{jp} - \text{net}_{jp})$
	5	Levenbreg-Marquard Learning Rule	Multilayer feed forward	$w_{k+1} = w_k - (J^T{}_K J_k + \mu I)^{-1} J^T{}_k E$

1.10. PERCEPTRON AND CONVERGENCE RULE

This learning suggest that N.N has no knowledge of environment and teacher has all the knowledge of environment. Thus, the difference is desired and actual response is fed to the learning neural network; which will gradually becomes as competent as the teacher.

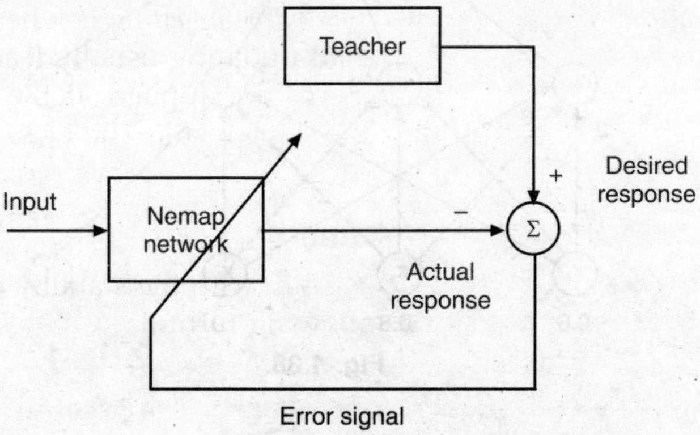

Fig. 1.35.

Supervised learning may be correction or stochastic. The error correction can very well be implemented by back propagation or least mean square algorithm. Here, a concept of error surface must be introduced. The error is a function of $E = \Sigma E^p (V, W, I)$ when V is input wt, W is output wt. and I is new neural information.

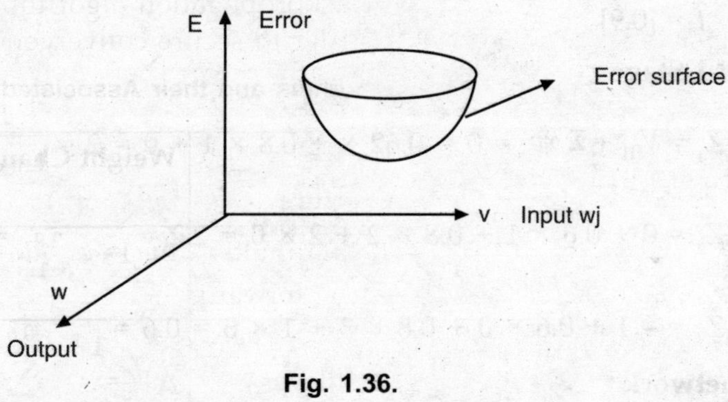

Fig. 1.36.

Effect of learning rate

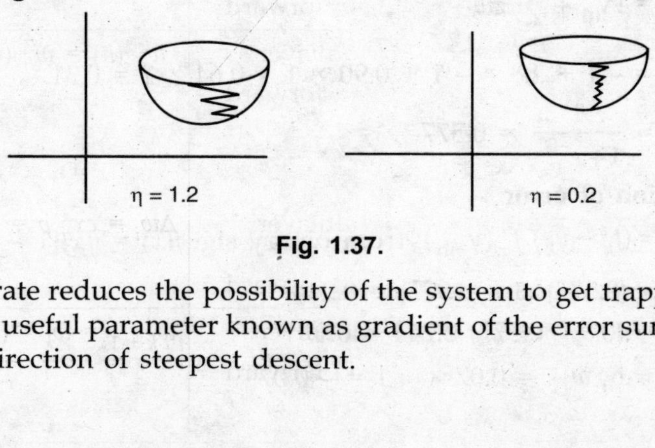

Fig. 1.37.

A low learning rate reduces the possibility of the system to get trapped to local minima. Thius gives rise to a useful parameter known as gradient of the error surface. This is a vector that points in the direction of steepest descent.

Example 2

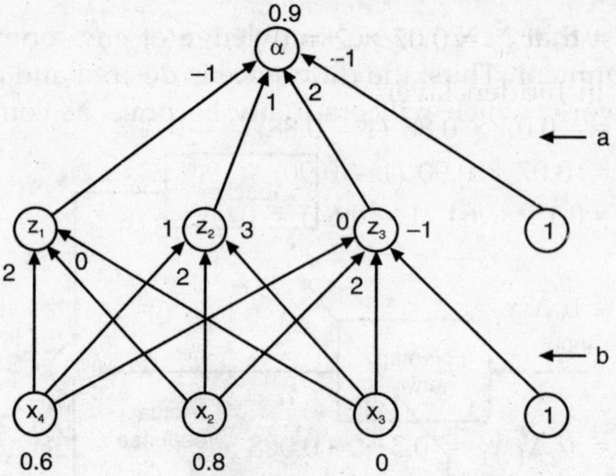

Fig. 1.38.

Initialization

$W = [-1 \;\; 1 \;\; 2]$ $\qquad W_0 = [-1]$ Bias $\qquad \leftarrow$ a

$$V = \begin{bmatrix} 2 & 1 & 0 \\ 1 & 2 & 2 \\ 0 & 3 & 1 \end{bmatrix} \qquad v_0 = [0 - 0 - 1] \text{ (Bias)} \qquad \leftarrow b$$

$x = [0.6 \quad 0.8 \quad 0]$

$t = [0.9]$

Feed Forward Stage

$$Z_1 = V_{01} + \Sigma\, xv = 0 = 0.62 \times + 0.8 \times 1 + 0 = 2 = \frac{1}{1+e^{-2}} = 0.88$$

$$Z_2 = 0 + 0.6 \times 1 + 0.8 \times 2 + 2 \times 0 = 2.2 = \frac{1}{1+e^{-2.2}} = 0.90$$

$$Z_3 = -1 + 0.6 \times 0 + 0.8 \times 3 + 1 \times 0 = 0.6 = \frac{1}{1+e^{-0.6}} = 0.64$$

Output of network

$$Y_{\text{in}} = W_{01} + \sum_{j=1}^{3} zw$$

$$= -1 + .88 \times -1 + 0.90 \times 1 + 0.64 \times 2 = 0.31$$

$$Y_1 = \frac{1}{1+e^{-31}} = 0.577$$

Back Propagation of error

$$\delta_k = (t_k - y_k)\, f'(y_{-\text{in}}) \text{ from binary sig. } f(x) = f(x)[1 - f(x)]$$

So, $\quad f(y_1) = (0.577)\,(1 - 0.577) = 0.244$

$\delta_1 = (0.9 - 0.577)\; 0.244 = 0.07$

So, $\quad \delta_{\text{in}\,1} = \delta_1\, w_{11} = 0.07 \times -1 = -0.07$

$$\delta_{\text{in }2} = \delta_1\, w_{21} = 0.07 \times 1 = 0.07$$

$$\delta_{\text{in }3} = \delta_1\, w_{31} = 0.07 \times 2 = 0.15$$

For error term in hidden layer:

$$\Delta_1 = -\ 0.07 \times 0.88\ (1 - 0.88) = -\ 0.008$$
$$\Delta_2 = 0.07 \times 0.90\ (1 - 0.90) = .06$$
$$\Delta_3 = 0.15 \times .64\ (1 - 0.64) = 0.03$$

Wt updation:

$$\Delta V_{ij} = \alpha\, \Delta_j x_j \qquad x = [0.6\ 0.8\ 0]$$
$$\Delta = [-0.008,\quad .09,\quad .03]$$
$$\alpha = 0.3$$

$$\Delta V_{11} = \alpha\, \Delta_1\, x_1 = 0.3 \times -\ 0.008 \times 0.6 = -\ .0015$$

$$V_{11(n)} = V_{11(0)} + \Delta V_{11} = 2 - .0015 = 1.9985$$

So, $$V = \begin{bmatrix} 1.99 & 1.00 & .054 \\ 0.99 & 2.00 & 2.00 \\ 0 & 3 & 1 \end{bmatrix}$$

Then $$\Delta_{ok} = \alpha\, \delta_k\, Z_z$$
$$W_N = W_0 + \Delta_{W11} = -\ 1 + 0.3 \times 0.07\ 8 \times .8808 = -\ 1 + 0.02 = 0.9792$$
$$\Delta W_{12} = 1 + 0.3 \times .078 \times 0.900 = = 1 + 0.02 = 1.02$$
$$\Delta W_{13} = 2 + 0.3 \times 0.078 \times 0.646 = +\ 2 + 0.015 = 2.01$$
$$\Delta W_o = \alpha\, \delta_1 = 0.3 \times 0.078 = 0.023$$

Thus the weights one calculated.

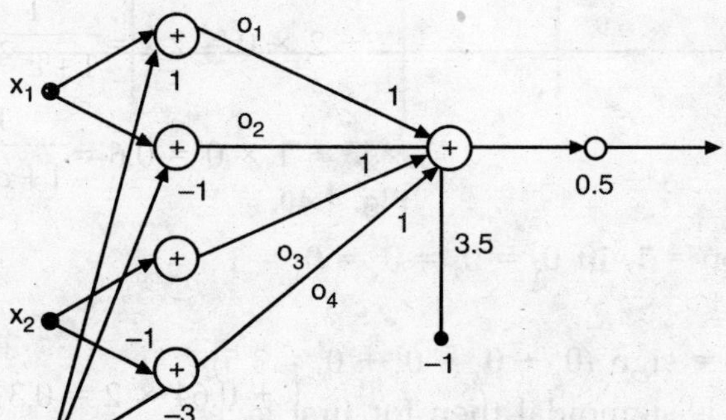

Fig. 1.39.

For Ist layer:

Input vector $[x_1\quad x_2\ -\ 1]$

Output $[0_1\quad 0_2\quad 0_3\quad 0_4]$

$$W_1 = \begin{bmatrix} w_{11} & w_{12} & w_{13} \\ w_{21} & w_{22} & w_{23} \\ w_{31} & w_{32} & w_{33} \\ w_{41} & w_{42} & w_{43} \end{bmatrix} = \begin{bmatrix} 1 & 0 & 1 \\ -1 & 0 & -2 \\ 0 & 1 & 0 \\ 0 & -1 & -3 \end{bmatrix} \begin{bmatrix} x_1 \\ x_2 \\ -1 \end{bmatrix}$$

$$\begin{bmatrix} 1 & 0 & 1 \\ -1 & 0 & -2 \\ 0 & 1 & 0 \\ 0 & -1 & -3 \end{bmatrix} \begin{bmatrix} x_1 \\ x_2 \\ -1 \end{bmatrix}$$

For II layer

Input vector $[0_1\ 0_2\ 0_3\ 0_4]$

Output vector $[0_5]$

$$w_2 = [w_{11}\ w_{12}\ w_{13}\ w_{14}\ w_{15}]$$
$$= [1\ 1\ 1\ 1\ 3.5]$$

Response of Ist layer

$$0 = [\text{sign}\ (x_1 - 1) + \text{sign}\ (x_1 + 2)]\ \text{sign}\ (x_2) + \text{sign}\ (-x_2 + 3)$$

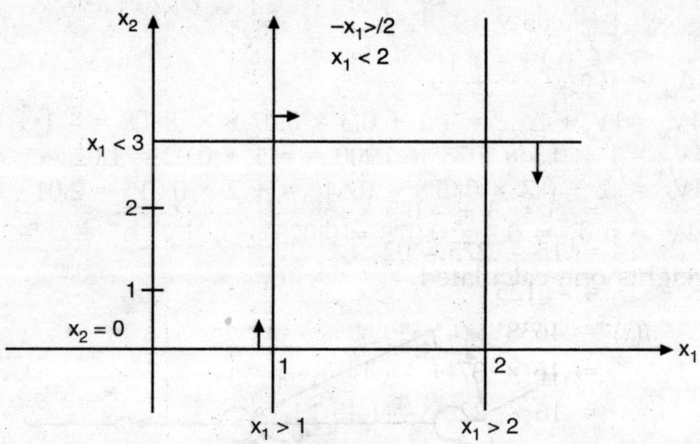

Fig. 1.40.

So, the fifth neuron = 1, iff $0_1 = 0_2 = 0_3 = 0_4 = 1$

Response of II layer

$$0_5 = \text{sign}\ (0_1 + 0_2 + 0_3 + 0_4 - 3.5)$$

If the activation f_x is sigmoidal then for first layer

$$0 = \begin{bmatrix} \dfrac{2}{1+\exp(1-x_1)\lambda} - 1 \\ \dfrac{2}{1+\exp(x_1-2)\lambda} - 1 \\ \dfrac{2}{1+\exp(-x_2)\lambda} - 1 \\ \dfrac{2}{1+\exp(x_2-3)\lambda} - 1 \end{bmatrix}$$

And for second layer

$$0_5 = \frac{2}{i.e., \exp(3.5 - 0_1 - 0_2 - 0_3 - 0_4)\lambda} - 1$$

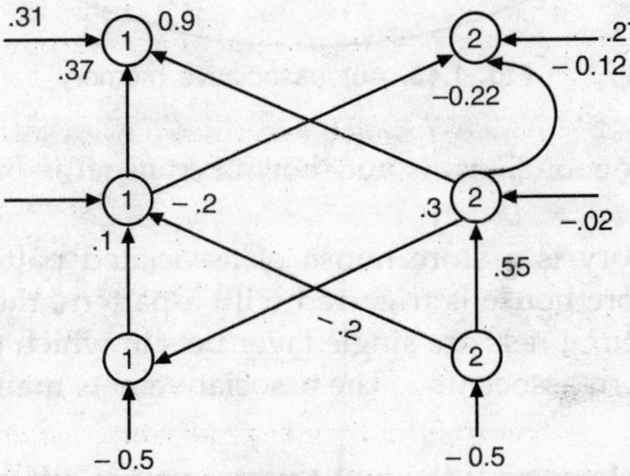

Fig. 1.41.

For [0.5, – 0.5]

$$Z_1 = 0.5 \times 0.1 + - 0.5 \times - .2$$
$$= 0.05 + 10$$
$$= 0.15 + 0.1$$
$$= 16 - .275$$
$$f(\overset{.16}{x}) = .5399$$
$$Z_2 = .5 \times .3 + - 0.5 \times .55 - 0.2$$
$$= .15 - .275 - 02$$
$$= - .125$$
$$f(x) = .4638$$
$$= .16 \times .37 + - (.145)\ 0.9 + .31$$
$$= .16 \times .22 + - (1.45)\ .12 + .27$$

1.11. AUTO ASSOCIATIVE AND HETRO ASSOCIATIVE MEMORY

Let then be two layers of neurons, namely L_1 and L_2. When all connection are made between them, we get a memory.

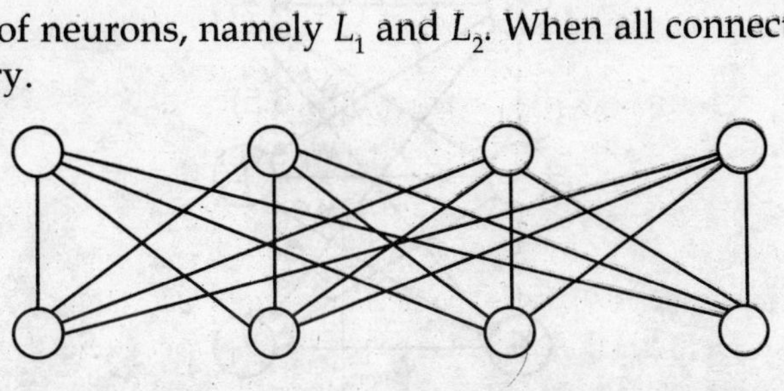

Fig. 1.42. Bidirectional Heteroassociatve memory

If all these connections are two way, we get a Bidirectional Associate Memory.

If the two layers coincide and the weights and symmetric between two nodes or neurons i and j such that $w_{ij} = w_{ji}$ and $i \neq j$, then we get an Autoassociative Memory. Here, all the units are connected to each other and to it self as shown below

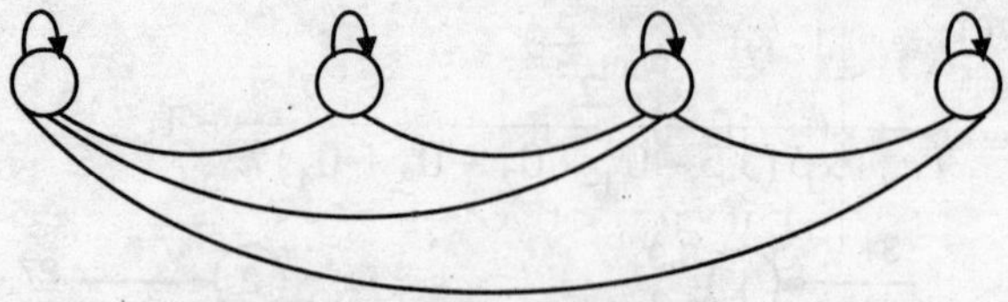

Fig. 1.43. Autoassociative memory

NOTE: Note that the flow of signal is not shown because it is Bidirectional in both cases.

An associative memory is a store house of associated pattern, which are encoded in some form. When the store house is trigerred with a pattern, the associated pair is recalled or output. Associative neural nets are single layer nets in which the weights are determined to store as assest of pattern association. The associative nets mainly consists of two types of nets.

Hetero Associative Memory: If the input vector pair is different from that of the output vector pair, then it forms a hetero associative net. It mear the associative pattern pairs (*X*, *Y*) are different and if the model recalls a *Y* for given *X* or vice-versa.

Architecture of Heteroassociative Memory: The architecture resembles a single layer feed for forward network. There exist *n* = number of input neurons and '*m*' number of inputs neurons and *m* number of output neurons. The training process is based on Hebb Learning Rule.

Application Algorith:

(*i*) Weights aer initialised using Hebb or Delta Rule.

(*ii*) For each input vector do step 3 to 5

(*iii*) Set the activation for input x_i.

(*iv*) Compute net input to output units

$$Y_{inj} = \sum_{\rho=1}^{n} x_i W_{ej}$$

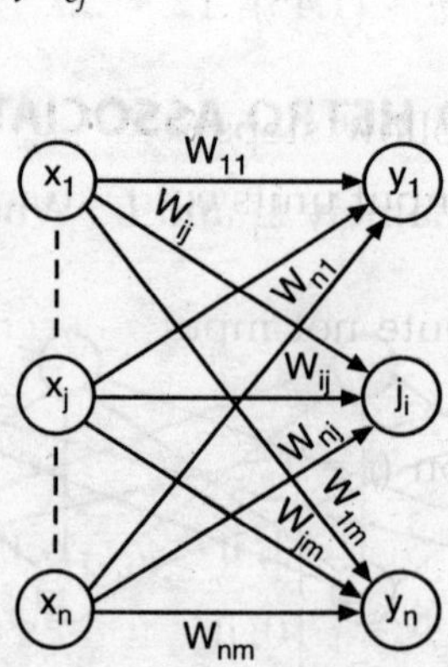

Fig. 1.44. Architecture of a Hetero Associative Neural Net

Determine the activation of output unit

$$y_j = \begin{cases} 1, & \text{if } \; y_{inj} > 0 \\ 0, & \text{if } \; y_{inj} = 0 \\ -1, & \text{if } \; y_{inj} < 0 \end{cases}$$

If the target response of the net is binary a suitable activated is

$$f(x) = \begin{cases} 1, & \text{if } \; x > 0 \\ 0, & \text{if } \; x \le 0 \end{cases}$$

Autoassociative Memory: In autoassociative net the training input Σ_j target output vector should be identical or same. The auto associative net training is often called storing the vectors, which may be binary or bipola. In autoassociative net the weights on the diagonal are set to 20. The diagonal component are those, whcih connect an input pattern component to the corresponding component in the output pattern. Setting the weights to zero improves the net ability to generalise or increase the plausability of the net.

Architecture of Auto Associative Memory: This architecture resemoies a single consists of only one layer of weighted interconnections. There exists 'n' number of input neurons and 'n' number of output layers in which n input neurons are there.

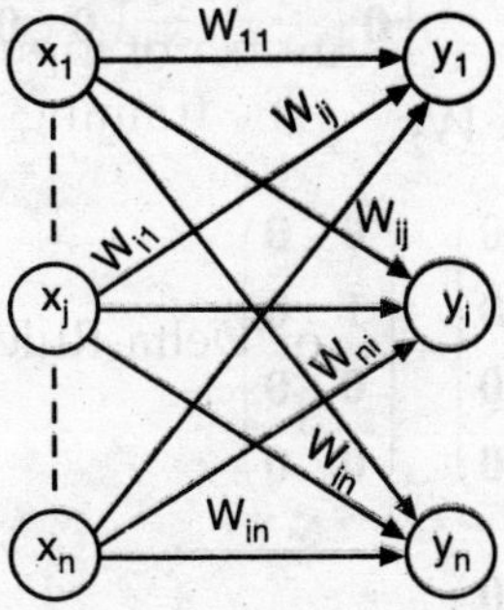

Fig. 1.45. Architecture of an auto associative net

Application Algorithm:

Step 1: Initialize the weights

Step 2: For each testing input follow steps 3 to 5

Step 3: Set the activation of the input units equal to input vector. Architecture of an auto associative net

Step 4: Set activation and compute not input to each output unit

$J = 1, \ldots n \quad y_{inj} = \Sigma X_i W_{ij}$

Step 5: Apply activation function ($j = 1, \ldots n$)

$$y = \text{f}(y_{-\text{inj}}) = \begin{cases} 1, & \text{if } \; y_{\text{in}\,j} > 0 \\ 0, & \text{if } \; y_{-\text{in}\,j} \le 0 \end{cases}$$

SHORT QUESTIONS WITH ANSWERS

Q.1. A hetero associative memory is given as:

$$S_1 = (0,1,1,0) \quad t_1(1,0)$$
$$S_2 = (0,1,0,0) \quad t_2(1,0)$$

Find the weight matrix.

Ans.
$$W_1 = S_1^T t_1 = \begin{bmatrix} 0 \\ 1 \\ 1 \\ 0 \end{bmatrix} [1\ 0] = \begin{pmatrix} 0 & 0 \\ 1 & 0 \\ 1 & 0 \\ 0 & 0 \end{pmatrix}$$

$$W_2 = S_2^T t_2 = \begin{bmatrix} 0 \\ 1 \\ 0 \\ 0 \end{bmatrix} [1\ 0] = \begin{pmatrix} 0 & 0 \\ 1 & 0 \\ 0 & 0 \\ 0 & 0 \end{pmatrix}$$

$$W = W_1 + W_2$$

$$W = \begin{pmatrix} 0 & 0 \\ 1 & 0 \\ 1 & 0 \\ 0 & 0 \end{pmatrix} + \begin{pmatrix} 0 & 0 \\ 1 & 0 \\ 0 & 0 \\ 0 & 0 \end{pmatrix}$$

$$W = \begin{pmatrix} 0 & 0 \\ 2 & 0 \\ 1 & 0 \\ 0 & 0 \end{pmatrix}$$

Q.2. What are the different types of activation functions are there?

Ans. 1. Heaviside function

2. Identify function
3. Sigmoidal function
4. Signum function

Q.3. What is a neural network (NN)?

Ans. An ANN is a network of many very simple processors ("units"), each possibly having a (small amount of) local memory. The units are connected by unidirectional communication channels ("connections"), which carry numeric (as opposed to symbolic) data. The units operate only on their local data and on the inputs they receive via the connections.

Q.4. What can you do with a Neural Network and what not?

Ans. In principle, NNs can compute any computable function, *i.e.* they can do everything a normal digital computer can do. Especially anything that can be represented as a mapping between vector spaces be approximated to arbitrary precision by feedforward NNs (which is the most often used type).

In practice, NNs are especially useful for mapping problems which are tolerant of a high error rate, have lots of example data available, but to which hard and fast rules can not easily be applied. NNs are, at least today, difficult to apply successfully to problems that concern manipulation of symbols and memory.

Q.5. Who is conce.ned with Neural Networks?

Ans. Neural Networks are interesting for quite a lot of very dissimilar people:

- Computer scientists want to find out about the properties of non-symbolic information processing with neural nets and about learning systems in general.

Q.6. What are the types of neural network architecture?

Ans. 1. Single layer feedforward networks

2. Multi layer feedforward networks

3. Recurrent networks

- Engineers of many kinds want to exploit the capabilities of neural networks on many areas (*e.g.* signal processing) to solve their application problems.
- Cognitive scientists view neural networks as a possible apparatus to describe models of thinking and conscience (High-level brain function).
- Neuro-physiologists use neural networks to describe and explore medium-level brain function (e.g. memory, sensory system, motorics).
- Physicists use neural networks to model phenomena in statistical mechanics and for a lot of other tasks.
- Biologists use Neural Networks to interpret nucleotide sequences.
- Philosophers and some other people may also be interested in Neural Networks for various reasons.

LONG QUESTIONS WITH ANSWER

Q.1. Define Neural Network Architecture.

Ans. Actificia Neural Network is composed of a number of an interconnects local computation or function. The output of any unit is determined its input-output characteristics, its interconnection with other units and possible external inputs.

A typical neural network is fully connected which means that each neuron is any given layer is connected with all the neuron in next layer.

An artificial neuron network is defined as a data processing input system consisting of a large number of simply highly interconnected processing elements (artificial neurons) in an architecture inspired by the structure of the neuron contex of the brain. Generally, an ANN structure can be represented using a directed graph. A graph in an ordered 2-tuple (V, ε) consisting of a set of edges. When each edge is assigned an arienation the graph is directed graph or a diagraph.

Digraphs assume significance in Neural Network theory since signals in the Network systems are restricted to flow in sepcial directions.

→ The vertex of the graph may represent neurons (input/output) and the edges, the synaptic links.

→ The edges labelled by the weights attached to the synaptic link.

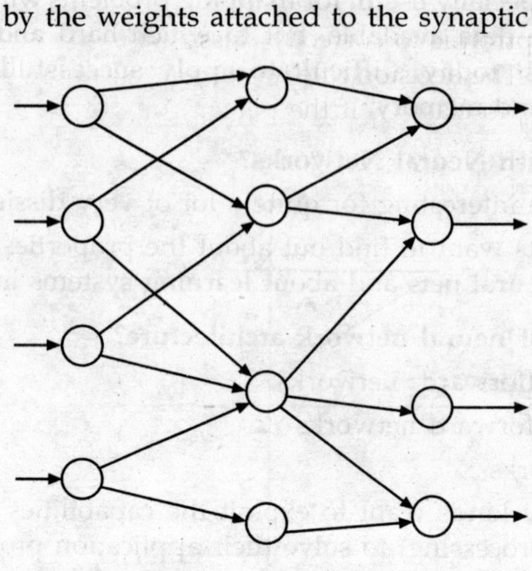

Fig. 1.46. Neural Network Architecture

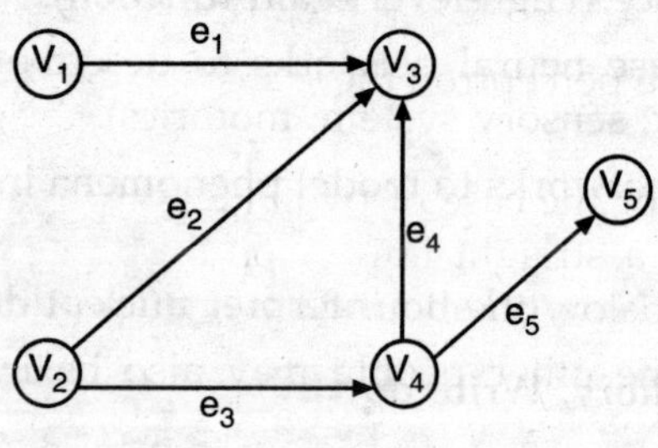

Fig. 1.47. An Example Diagraph

Q.2. What is Rosenblatt's perceptron?

Ans. For the perception learning rule, the learning signal is the difference between the desired and actual neurons response [Rosenblatt, 1958]. This neural model consist of a linear combination followed by a signum functin.

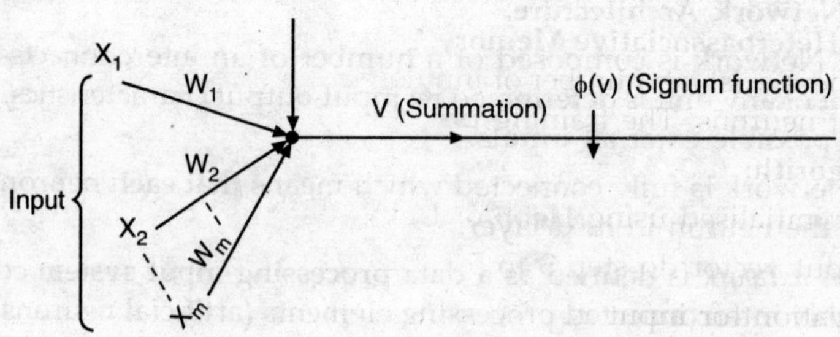

Fig. 1.48.

The synaptic weights of the perceptron aer denoted by W_1, W_2 ... W_m corresponding the inputs applied to the pereceptron are denoted by X_1, X_2 .. X_m. The externally applied bias is denoted by b.

$$V = \sum_{i=1}^{m} W_i \, x_i + b$$

The goal of the perceptron is to correctly classify the set of externally applied $X_1, X_2, \ldots X_m$ into two classes C_1 and C_2. The decision rule for the classification is to assign the ponit represented by the inputs $X_1, X_2 \ldots X_m$ to class C_1 if the perceptron output Y is + 1 and to class C_2 if it is – 1.

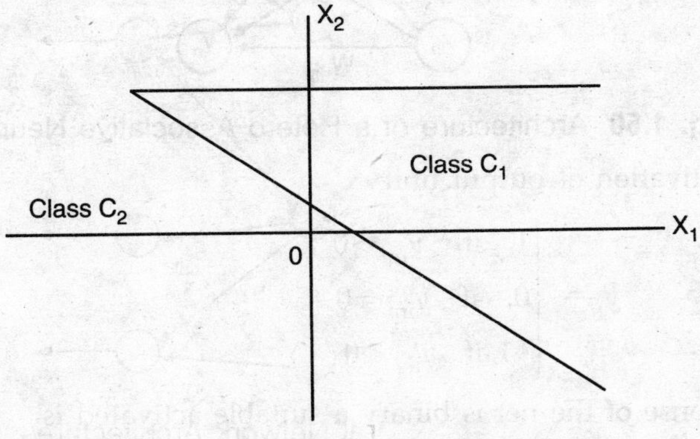

Fig. 1.49

In the simplest form of the perceptron there are two decision regions separated by hyperplane defined by.

This is shown in figure for the case of two input variables X_1 and X_2 for which the decision boundary takes the form of a straight line. A point (x_1, X_2) that lies above the boundary line belongs to the class C_1 and below the boundary line belongs to the class C_2.

Q.3. Define associative memory. Write down its application.

Ans. An associative memory is a store house of associated pattern, which are encoded in some form. When the store house is trigerred with a pattern, the associated pair is recalled or output. Associative neural nets are single layer nets in which the weights are determined to store as assest of pattern association. The associative nets mainly consists of two types of nets.

Hetero Associative Memory: If the input vector pair is different from that of the output vector pair, then it forms a hetero associative net. It mear the associative pattern pairs (X, Y) are different and if the model recalls a Y for given X or vice-versa.

Architecture of Heteroassociative Memory: The architecture resembles a single layer feed for forward network. There exist n = number of input neurons and 'm' number of inputs neurons and m number of output neurons. The training process is based on Hebb Learning Rule.

Application Algorith:

(*i*) Weights aer initialised using Hebb or Delta Rule.

(*ii*) For each input vector do step 3 to 5

(*iii*) Set the activation for input x_i.

(*iv*) Compute net input to output units

$$Y_{\text{inj}} = \sum_{\rho=1}^{n} x_i W_{ej}$$

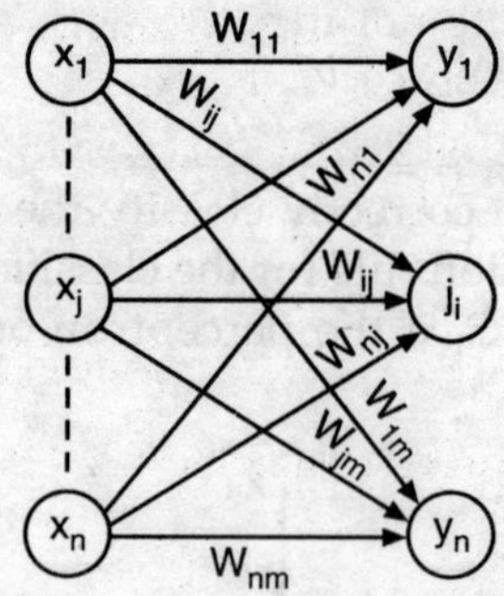

Fig. 1.50. Architecture of a Hetero Associative Neural Net

Determine the activation of output unit

$$y_j = \begin{cases} 1, & \text{if } \; y_{inj} > 0 \\ 0, & \text{if } \; y_{inj} = 0 \\ -1, & \text{if } \; y_{inj} < 0 \end{cases}$$

If the target response of the net is binary a suitable activated is

$$f(x) = \begin{cases} 1, & \text{if } \; x > 0 \\ 0, & \text{if } \; x \le 0 \end{cases}$$

Autoassociative Memory: In autoassociative net the training input Σ_j target output vector should be identical or same. The auto associative net training is often called storing the vectors, which may be binary or bipola. In autoassociative net the weights on the diagonal are set to 20. The diagonal component are those, whcih connect an input pattern component to the corresponding component in the output pattern. Setting the weights to zero improves the net ability to generalise or increase the plausability of the net.

Architecture of Auto Associative Memory: This architecture resemoies a single consists of only one layer of weighted interconnections. There exists *'n'* number of input neurons and *'n'* number of output layers in which *n* input neurons are there.

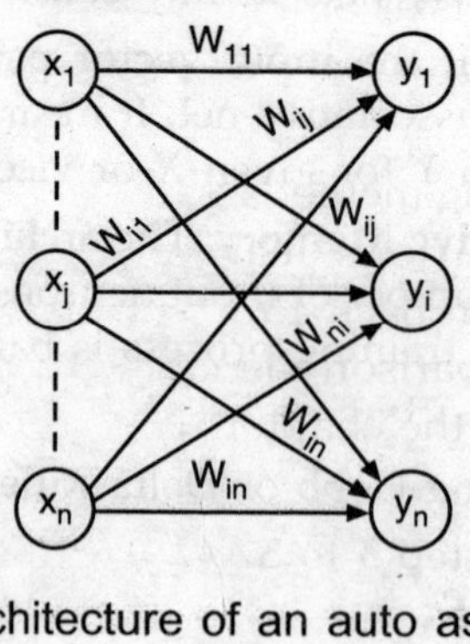

Fig. 1.51. Architecture of an auto associative net

Application Algorithm:

Step 1: Initialize the weights

Step 2: For each testing input follow steps 3 to 5

Step 3: Set the activation of the input units equal to input vector. Architecture of an auto associative net

Step 4: Set activation and compute not input to each output unit

$J = 1, \ldots n \quad y_{inj} = \Sigma X_i W_{ij}$

Step 5: Apply activation function ($j = 1, \ldots n$)

$$y = f(y_{-inj}) = \begin{cases} 1, \text{ if } & y_{in\,j} > 0 \\ 0, \text{ if } & y_{-in\,j} \leq 0 \end{cases}$$

SUMMARY

In this chapter, We have given the Introduction of Neural network from a layman point of view. We sailed through he Definition of Neural Network and its features like parallel structure with distributed storages and distributed processing of information and its learning ability by adjusting weights even with erroneous data. With missing neurons, there is only a marginal and graceful degradation of the performance of the network. These motivations are found from Biological Neural Network, so the composition of a neuron, properties of brain, generation and propagation of action potential and information flow in Nervous system can be discussed. The same connection pattern that lies within brain's neural network; that was tried with Artificial Neurons to form Artificial Neural Network. Thus, an analogy and comparison between Biological network and Artificial network has been created.

Next, We take up the three mathematical models of Neuron, *i.e.*, McCulloch and Pitts Model, Rosenblatt's Perceptron Model and Adaline Model. The examples to create gates out of them with MATLAB implementation has also been discussed. Since we needed to implement the algorithm of Least Mean Squared error Learning Law, we discussed that in brief here. Later in Chapter 2, all the training algorithms will be discussed in complete details. Thereafter, we have discussed different activation function for the nonlinear behavior of the neuron.

With all the fundamentals of neuron, learning and architecture in place, next, we discuss the Neural Network Architecture. There are basically two main categories: Feed Forward and Recurrent. Feed forward has variants as Single Layer Network, Multi Layer Network, Radial Basis Network and Time delay Neural Network. Recurrent Networks are the networks with at least one feedback loop.

Next, we discuss Various learning methods, *e.g.*, Supervised, Unsupervised, and Reinforced Learning. Also, we discussed various learning rules likePerceptron Learning, Delta LearningGradient Descent ot Error Back Propagation Learning, Windrow Hoff Learning, Levenberg Learning and their comparison therein. Perceptron Convergence rule is discussed with examples thereby completing the chapter with Auto associative and Hetero associative memory.

REVIEW QUESTIONS

1. Compare and contrast between Digital Computer and Artificial Neural Network? Prove the features which make Biological Neural Network superior to most advanced Artificial Intelligence Computer.

2. Describe the biological Neuron and working of Biological Neural Network.
3. How can you prove that the Biological Neural Network does the tasks of pattern recognition better than any AI computer?
4. Discuss McCulloch and Pitts model and show that ex-or gate is its limitation.
5. Adaline Model is an improvement to Perceptron model – prove this statement by a suitable example.
6. What are the main difference between the three models of Artificial Neurons?
7. Write a short note on Activation functions.
8. Compare Single Layer Network with Multilayer feed forward network.
9. What are the basic learning laws?
10. Write a short note on unsupervised neighborhood rule.
11. Compare between Delta Learning Rule and Windrow Hoff's Least Mean Square learning rule.
12. Explain the significance of the initial values of weights and the learning rate parameter in the seven basic learning laws.
13. Discuss the effect of learning rate over Perceptron convergence rule.
14. Compare LMS, Perceptron and Delta Learning Laws.
15. Compare and contrast between Auto Associative and Hetero Associative Memory

NUMERICAL PROBLEMS

1. Explain the logic functions performed by the following networks with McCulloch Pitts model of neurons

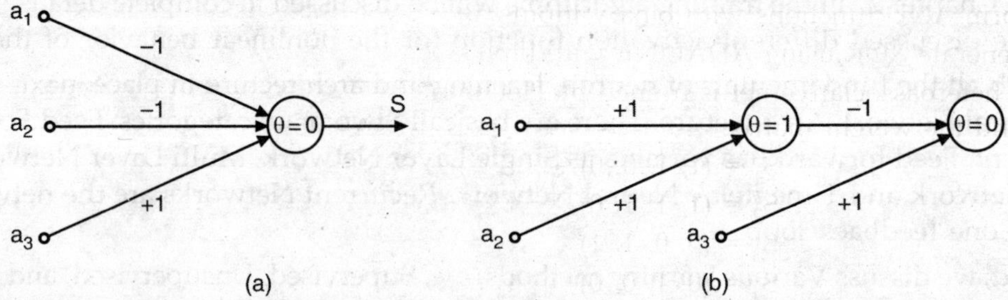

Fig. 1.52.

2. Design networks for the following functions using ± 1 as weights. Use McCulloch Pitts model:
   (a) $s(a_1, a_2, a_3) = a_1a_3 + a_2a_3 + a_1a_2$
   (b) $s(a_1, a_2, a_3) = a_1\ a_2\ a_3$
   (c) $s(a_1, a_2, a_3) = a_1 + a_2 + a_3$
   (d) $s(a_1, a_2, a_3) = \bar{a}_1\ a_2\ \bar{a}_3$
3. Realise the Hebb net for the AND function with bipolar inputs and targets.
4. Use Hebb net to the training pattern that define the XOR function with bipolar inputs and targets. [hint: you might have to prove that Hebb net cannot be used for this problem]
5. Classify the input pattern I and J using a 3 × 3 matrix using Hebb rule.

6. Find out the output for the following network for the input $[111]^T$

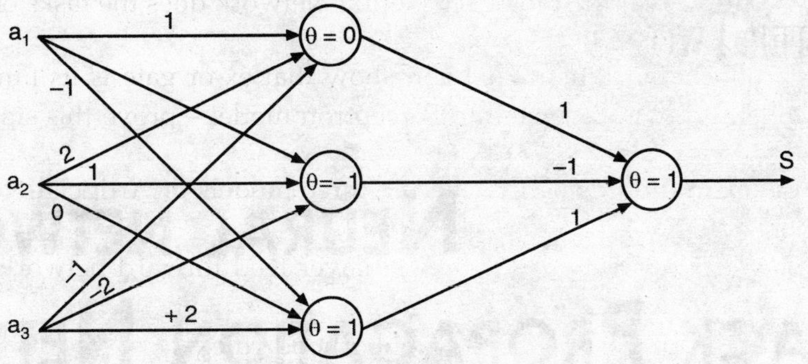

Fig. 1.53.

7. Find out the McCulloch Pitts model for NAND and NOR gate.
8. Realize a ANDNOT function using Hebb net. Draw the decision boundary also.
9. Realize a Hebb net to classify between A and E in a 5 × 5 matrix.
10. Make the letters (u, v) and (s, t) using 5 × 5 matrix. Use Hebb net to classify these patterns.
11. Using unit step function, determine a set of weights for the truth table given below. Use McCulloch Pitts model.

X1	X2	Output
– 0.8	– 0.38	1
0.5	– 0.5	1
0.4	0.3	0
0.2	0.5	0

12. Form AND functions with binary inputs and outputs using Adaline net.
13. Generate XOR using ADALINE with bipolar inputs and targets.
14. A heteroassociative net is trained by Hebb rule for the following input output vector pairs. Find the weight matrix:

$S_1 = (1, 1, 0, 0) \quad t_1 = (1, 0)$
$S_2 = (1, 1, 1, 0) \quad t_2 = (0, 1)$
$S_3 = (0, 0, 1, 1) \quad t_3 = (1, 0)$
$S_4 = (0, 1, 0, 0) \quad t_4 = (1, 0)$

15. Use the bipolar Hebb rule to find the weight matrix for the associative memory network based on the following binary input output vector pairs:

$S_1 = (1, 1, 0, 0) \quad t_1 = (1, 0)$
$S_2 = (1, 1, 1, 0) \quad t_2 = (0, 1)$
$S_3 = (0, 0, 1, 1) \quad t_3 = (1, 0)$
$S_4 = (0, 1, 0, 0) \quad t_4 = (1, 0)$

16. In an auto associative network, use Hebb rule to store the vector (11 –1 –1)
    (*a*) Find the weight matrix
    (*b*) Test the input vector x = (11 –1 –1)
    (*c*) Test the net for tolerance for one mistake/missing component in the vector.
    (*d*) Test the net for tolerance for one mistake/missing component in the vector.

□□□

2 CHAPTER TWO

NEURAL NETWORKS-II (BACK PROPAGATION NETWORKS)

CHAPTER OUTLINE

2.1 The Perceptron
2.2 Solution to the Neuron Model
2.3. Multi Layer Perceptron (MLP)
2.4 Back Propagation Learning Methods
2.5 Back Propagation Algorithm
2.6 Factors Affecting Back Propagation Learning
2.7 Some Application Domains

2.1. THE PERCEPTRON

The perceptron is a binary classifier which maps its input x (a real-valued vector) to an output value $f(x)$ (a single binary value) across the matrix.

$$f(x) = \begin{cases} 1 \text{ if } w \,.\, x + b > 0 \\ 0 \text{ else} \end{cases}$$

where w is vector of real-valued weights and $w.x$ is the dot product (which computes a weighted sum). b is the 'bias', constant term that does not depend on any input value.

The value of $f(x)$ (0 ro 1) is used to classify x as either a positive or a negative instance, in the case of a binary classification problem. If b is negative, then the weighted combination of inputs must produce a positive value greater than $|b|$ in order to push the classifier neuron over the 0 threshold. Spatially, the bias alters the position (though not the orientation) of the decision boundary. The perceptron learning algorithm does not terminate if the learning set is not linearly separable.

The perceptron is considered the simplest kind of feed-forward neural network.

2.2. SOLUTION TO THE NEURON MODEL

If we smoothen the threshold function so that it more or less turns on or off as before but has a sloping region in the middle that will give us some information on the inputs, we will be able to determine when we need to strengthen or weaken the relevant weights. Now, the network will be able to learn as required. A couple of possibilities for the new thresholding function are given following table which will be described later in this chapter in detail. Even now, the value of the outputs will practically be one if the input exceeds the value of the threshold a lot and will be practically zero if the input is far less than the threshold. However, in cases when input and threshold are almost same, the output of the neuron will have a value between zero and one, meaning that the output to the neurons can be related to input in a more informative way.

An artificial neuron is developed to mimic the characteristics and functions of a biological neuron. Analogous to a biological neuron, an artificial neuron receives much input representing the output of the other neurons. Each input is multiplied by the corresponding weights analogous to synaptic strengths. All of these weighed inputs are then summed up and passed though an activation function to determine the neuron output. The artificial neural model is shown in Fig 1.9.

$$\mu\ (t) = W_1 l_1 + W_2 l_2 + ... + W_n l_n$$

$$\mu = \Sigma Wl$$

Table 2.1. (a) Typical nonlinear activation functions

Type	Equation	Functional form
Linear	$O = \text{gl}$ $g = \tan\phi$	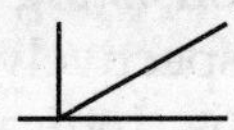
Piecewise Linear	$O = \begin{cases} 1 & \text{if ml} > 1 \\ gl & \text{if ml} < 1 \\ -1 & \text{if ml} > -1 \end{cases}$	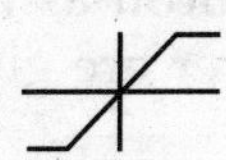
Hard Limiter	$O = \text{sgn}\ [l]$	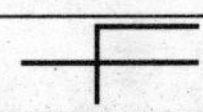
Unipolar Sigmoidal	$O = \dfrac{1}{(1 + \exp(-\lambda I))}$	
Bipolar Sigmoidal	$O = \tan\text{h}\ [\lambda l]$	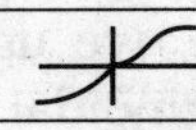
Unipolar Multimodal	$O = \dfrac{1}{2}\left[1 + \dfrac{1}{M}\Sigma_{m=1}^{M} \tanh h\ (g^m (I - W_o^m))\right]$	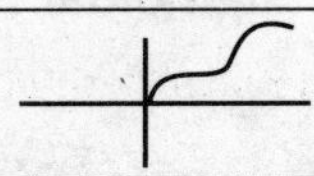
Radial Basis Function (RBF)	$O = \exp\ (I)$ $I = \left[-\dfrac{\sum_{i=1}^{N} (W_i(t) - x_i(t))^2}{2\sigma^2} \right]$	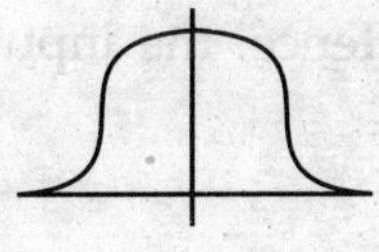

Considering threshold θ, the relative input to the neuron is given by

$$u(t) = W_1 I_1 + W_2 I_2 + \ldots + W_n I_n - \theta$$

$$= \sum_{i=0}^{n} W_i I_i \text{ where, } W_0 = -\theta_i\, I_o = 1$$

The output using the nonlinear transfer function 'f' is given by

$$O = f(u)$$

The activation function $f(u)$ is chosen as a nonlinear function to emulate the nonlinear behaviour of conduction current mechanism in a biological neuron. However, as the artificial neuron is not intended to be a xerox copy of the biological neuron, many forms of nonlinear functions have been suggested and used in various engineering applications. Some of the activation functions along with their mathematical descriptions given it Table 3.1 are most commonly used activation functions in multilayered static neural networks. It is also seen that for sigmoidal functions, the output of a neuron varies continuously but not linearly with the input. Neurons with sigmoidal functions bear a greater resemblance to biological neurons than with other activation functions. Even if the sigmoidal function is differentiated, it gives continuous values of the output. Hard limiter and radial basis functions are also equally popular.

2.2.1. Single Layer Artificial Neural Network

In the preceding section, we have seen the mathematical details of a neuron at a single level. Although a single neuron can perform certain simple pattern detection problem, we need larger networks to offer greater computational capabilities. In order to mimic the layered structure of certain portions of the brain. Let consider a single layer feedforward neural network consisting of an input layer to receiver the inputs and an output layer to output the vectors respectively. The input layer consists of 'n' neurons and the output layer consists of 'm' neurons. Indicate the weight of the synapse connectiong i th input neuron to the jth output neuron as W_{ij}. The inputs of the input layer and the corresponding output of the output layer are given as

$$I_i = \begin{Bmatrix} I_{I1} \\ I_{I2} \\ \vdots \\ I_{I3} \end{Bmatrix} ; O_o = \begin{Bmatrix} O_{O1} \\ O_{O2} \\ \vdots \\ O_{O3} \end{Bmatrix}$$

Assume, we use linear transfer function for the neurons in the input layer and the unipolar sigmoidal function for the neurons in the output layer (refer Table 3.1).

$$(O_I) = (I_I) \text{ (linear transfer function)}$$

$$= n \times 1 \quad m \times 1$$

$$I_{oj} = W_{Ij} I_n + W_{2j}\, I_{I2} + \ldots + W_{nj} I_n$$

Hence, the input to the output layer can be given as

$$\{I_O\} = [W]^T \{O_I\} = [W]^T \{I_I\}$$

$$= m \times 1 \quad m \times n \quad n \times 1$$

Using the unipolar sigmoidal or squashed-S function for neurons in the output layer, the output is given by-

$$O_{ok} = \frac{1}{(1 + e^{-\lambda I_{O^z}})}$$

Or $$\{O_O\} = f[WI]$$

The nonlinear activation function *operates component wise on the activation values 'I' of* each neuron. Each activation value is in turn a scalar product of the input with respect to weight vectors. The sigmoidal function is given as

$$f(I) = \frac{1}{(1 = e^{-\lambda I})}$$

And $$f'(I) = \lambda f(I)\,(1 - f(I))$$

2.2.2. Model for Multilayer Perceptron

The adapted perceptrons are arranged in layers and so the model is termed as multilayer perceptron. This model has three layers; an input layer, and output layer, and a layer in between not connected directly to the input or the output and hence, called the hidden layer. For the perceptrons in the input layer, we use linear transfer function, and for the perceptrons in the hidden layer and the output layer, we use sigmoidal or squashed-S functions. The input layer serves to distribute the value they receive to the next layer and so, does not perform a weighted sum of threshold. Because we have modified the single layer perceptron by changing the nonlinearity form a step function to a sigmoidal function and added a hidden layer, we are forced to alter the learning rules as well. So now, we have a network that should be able to learn to recognize more complex things. The input-Output mapping of multilayer perceptron is shown as–

$$O = N_3[N_2\,[N_1[I]]]$$

N_1, N_2 and N_3 represent nonlinear mapping provided by input, hidden and output layers respectively. Multilayer perceptron provides no increase in computational power over a single layer neural network unless there is a nonlinear activation function between layers. Many capabilities of neural networks, such as nonlinear functional approximation, learning, generalization etc. are in fact due to nonlinear activation function of each neuron.

The three-layer network show that the activity of neurons in the input layers represent the raw information that is fed into the network. The activity of neurons in the hidden layer is determined by the activities of the neurons in the input layer and the connecting weights between input and hidden units. Similarly, the activity of the output units depends on the activity of neurons in the hidden layer and the weight between the hidden and output layers. This structure is interesting because neurons in the hidden layeres are free to construct their own representations or the input.

2.2.3. Backpropagation Learning

Consider the network where the subscripts *I,H,O* denote input, hidden and ouput neurons.

Consider a problem in which an "nset" of "*I*" inputs and the corresponding "nset" of "*n*" output data is given as shown in Table 2.2.

Table 2.2. "nset" of input and output data

No.	Input		Output	
	I_1	$I_2 \ldots I_2$	O_1	$O_2 \ldots O_n$
1.	0.3	0.4 ... 0.8	0.1	0.56 ... 0.82
2.				
⋮				
Nset				

Input Layer Computation

Consider linear activation function the output of the input layer is input of input layer (considering $g = \tan\theta = 1$). Taking one set ot data

$$\{O\}_I = \{I\}_I$$
$$= I \times 1 \; I \times 1$$

The hidden neurons are connected by synapses to input neurons and (let us denote) v_{ij} is the weight of the arc between ith Input neuron to jth hidden neuron. The input to the hidden neuron is the weighted sum of the outputs of the input neurons to get I_{Hp} (*i.e.* Input to the *p*th hidden neuron) as

$$I_{Hp} = V_{1p}\, O_{I1} + V_{2p}\, O_{12} + \cdots V_{1p}\, O_{I1}$$

(p =1, 2, 3, ... m)

Denoting weight matrix or connectivity matrix between input neurons and hidden neurons as [V], we can get an input to the hidden neuron as $I \times m$

$$\{I\}_H = [V]^T \{O\}I = m \times 1 \; m \times I \quad I \times 1$$

Hidden Layer Computation

Considering sigmoidal function or squashed -S function, the output of the *p*th hidden neuron is given by

$$O_{HP} = \frac{\mathbf{1}}{(\mathbf{1} + e^{-\lambda(I\,H_p - O\,H_p)})}$$

Where O_{Hp} is the is the output of the *p*th hidden neuron, I_{Hp} is the input of the *p*th hidden neuron, and θ_{HP} is the threshold of the *p*th neuron. A non-zero threshold neuron is computationally equivalent to an input that is always held at –1 and the non-zero threshold becomes the connecting weight values.

But in our derivations we will not treat threshold as shown in Fig. 3.8. Now, output to the hidden neuron is given by

$$\{O\}_H = \left\{ \frac{\mathbf{1}}{(\mathbf{1} + e^{-\lambda(I_{Hp} - O_{Hp})})} \right\}$$

Treating each component of the input of the hidden neuron separately, we get the outputs of the hidden neuron as given by

The input to the output neurons is the weighted sum of the outputs of the hidden neurons. To get I_{Oq} (*i.e.* the input to the *q*th ouput neuron)

$$I_{Oq} = W_{Iq}\ O_{H1} + W_{zq}\ OH_2 + \ldots + W_{mq}\ O_{Hm}$$

$(q = 1, 2, 3, \ldots, n)$

Denoting weight matrix or connectivity matrix between hidden neurons and output neurons as $[W]$, we can get input to the output neuron as

$$\{I\}_O = [W]^T \{O\}_H = n \times 1\ n \times m\ m \times 1$$

Output Layer Computation

Considering sigmoidal function, the output of the qth output neuron is given by

$$O_{Oq} = \left\{ \frac{1}{(1 + e^{-\lambda(I_{Oq} - \theta_{Oq})})} \right\}$$

Where, O_{Oq} is the output of the qth output neuron, I_{Oq} is the input to the qth output neuron, and θ_{Oq} is the threshold of the qqth neuron. This threshold may also be tackled again by considering extra Oth neuron in the hidden layer with output of –1 and the threshold value Q_{Oq} becomes the connecting weight value as shown in

Hence, the outputs of output neurons are given by

$$\{O\}_O = \left\{ \frac{1}{(1 + e^{-\lambda(I_{Oq} - \theta_{Oq})})} \right\}$$

Calculation of Error

Considering any rth output neuron and for the training example we have calculated the output 'O' for which the target output 'T' is given

Hence, the error norm in output for the rth output neuron is given by

$$E_r^1 = \frac{1}{2} e_r^2 = \frac{1}{2} (T - O)^2$$

Where $E_r^1 = \frac{1}{2}$ second norm of the error in the rth neuron ('e_r') for the given training pattern. The square of the error is considered since irrespective of whether error is positive or negative, we consider only absolute values. The Euclidean norm of error E^1 for the first training pattern is given by–

$$E^1 = \frac{1}{2} \sum_{r=1}^{n} (T_{Or} - O_{Or})^2$$

This is the error function in one training pattern. If we use the same technique for all the training patterns, we get

$$E(V, W) = \sum_{j=1}^{n\,zet} E^j(V, W, I)$$

Where E is the error function depending on the m $(I = n)$ weights of $[W]$ and $[V]$. This is a classic type of optimization problem. For such problems, an objective function or cost function is usually defined to be maximized or minimized with respect to a set of parameters.

In this case, the network parameters that optimize the error function E over the 'nset' of pattern sets $[I^{nset}, t^{nset}]$ are synaptic weight values$[V]$ and $[W]$ whose sizes are

$$= [V] \text{ and } [W]$$
$$= I \times m \; m \times n$$

Below is an example of a learning algorithm for a single-layer (no hidden-layer) perceptron. For multilayer perceptrons, more complicated algorithms such as backpropagation must be used. Or methods such as the delta rule can be used if the function is non-linear and differentiable, although, the one below will work as well.

The learning algorithm we demonstrate is the same across all the output neurons, therefore everything that follows is applied to a single neuron in isolation. We first define some variables:

- $D = \{(x_1, d_1), \dots (x_m, d_m)\}$ is the training set of m samples, where X_j is the n-dimensional input vector with $x_{j,0} = 1$ and d_i is the desired output value of the perceptron for that input.
- $x_{j,i}$ is the value of the ith node of the jth training input vector.
- w_i is the ith value in the weight vector, corresponding to the ith input node.
- $y = f(z)$ denotes the output from the perceptron.
- a is the learning rate where $0 < \alpha \leq 1$
- b is the bias term which for conveniene we take to be 0.

An extra dimension $n + 1$ can be added to all input vectors $x_{j,n+1} = 1$, in which case w_{n+1} replaces the bias term.

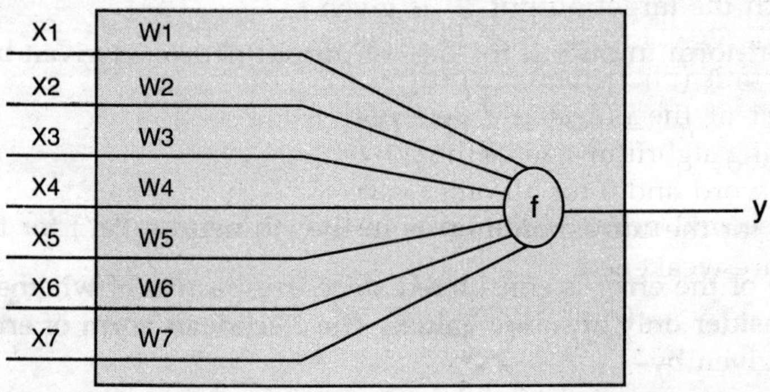

Fig. 2.1. Output as function of weighted summation of inputs

The appropriate weights are applied to the inputs, and the resulting weighted sum passed to a function which produces the output y.

Learning algorithm steps:

1. initialise weights and threshold:
   - Set $w_i(t)$ to be the weight i at time t for all $1 \leq i \leq n$ input nodes.
   - Set $w(0)$ to be 0 and all x_i inputs in this initial case to be 1.
   - Set $w_i(1)$ to small random values (which do not necessarily have to be normalised), thus initialising the weights. We take a firing threshold at $y = 0$.

2. Present input and desired output:
    - Present from our training samples D the input X_j and desired output d_j for this training set.
3. Calculate the actual output:
    - $y(t) = f[w(t) \cdot x_j] = f[w_0(t) + w_1(t)\, x_{i,1} + w_2(t)\, x_{i,2} + \ldots + w_n(t)\, x_{j,n}]$
4. Adapt weights:
    - $w_i(t+1) = w_i(t) + a(d_j - y(t))\, x_{j,i}$ for all nodes $0 < i \leq n$.

Steps 3 and 4 are repeated until the iteration error $d_j - y(t)$ is less than a user-specified error threshold γ, or a predetermined number of iterations have been completed.

2.3. MULTI LAYER PERCEPTRON (MLP)

Minsky and Papert had proposed a solution to the problem posed by functions such as the Ex-OR. They suggested that an extra layer of nodes with non-linear activation functions could be introduced. The output would now be a non-linear combination of the inputs so more complicated decision surfaces could be represented. The problem that remained was that no training algorithm was available to train such a network of perceptrons at the time.

During the 1970s more research turned towards the representation of knowledge and away from learning and many new ideas were developed. Then, in the 1980s, there was a resurgence of interest in neural networks and it was during this time that an effective algorithm, called back propagation, for the training of multi-layer perceptron (MLP) structures was developed [48]. Later, these algorithms for constraint descent method [36], for on-line training [24] were also studied.

The most obvious way to use MLP for Speech Recognition is to present all acoustic vectors of a speech unit (phoneme of word) at once at the input and to detect the most probable speech unit at the output by determining the output neuron with the highest activation. The learning algorithm can be the conventional back propagation with the desired output 1 for correct word and 0 for all other speech units. In this way, not only the correct output is reinforced for the corresponding sequence of acoustic vector, but simultaneously the wrong outputs are weakened.

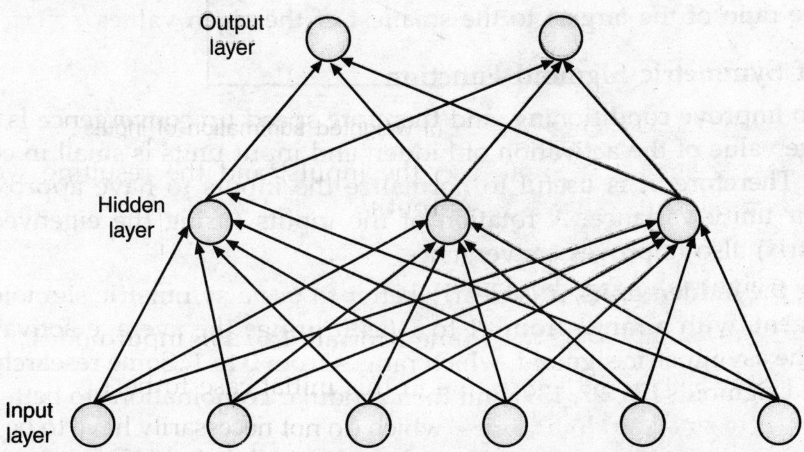

Fig. 2.2. Fully Connected Multi Layer Perceptron (6-3-2)

2.3.1. Problems with MLP

The problems associated with this approach are multiple.

1. For word recognition, a huge number of input units must be used. This means that even larger number of parameters to be determined by learning and consequently the necessity to dispose a large database.
2. The method is appropriate for phoneme recognition. However, in this case, a phoneme segmented database has to be available for learning; which generally is not available or partly available.
3. The time variability of speech in general presents a problem for the MLP. The same word pronounced by different speakers or even be the same speaker has different duration for different utterances. Therefore, with a fixed number of inputs to network, either some acoustic vectors have to be cut if the word pronounced is too long, or some inputs have to be set to some arbitrary values of the word pronunciation is too short. This implies that a given neuron has not always an input that corresponds to the same part of word pronunciation. Several methods are used to align the speech signal to a fixed length, before feeding it to the MLP. One such method suggested by Esprit project was to decompose the speech signal to small stationary frames of, say, 10ms each. Once the stationary parts are discovered, a classical feature extraction is applied to each stationary part. A time warping (DTW or TS)of the speech signal is then easier to perform. The training algorithm is gradient descent method.

2.3.2. Important Heuristics for Training MLP

Although the implementation of MLP and of most of ANN is generally straight forward and this is why MLP are so widely used. There are several simple tricks, generally not mentioned in the literature, which should be used to obtain speedy convergence of the training algorithm and good generalization.

Many of these tricks are motivated by the idea that the speed of convergence of gradient directly depends on the conditioning of the Hessian matrix (of second derivatives of the training criterion with respect to the parameters). Although it is not generally practical to compute, and invert, this matrix (which would allow one to have perfect conditioning), some of these tricks are computationally simple ways to improve the conditioning. The latter depends on the ratio of the largest to the smallest of the eigen values.

2.3.2.1. Use of Symmetric Sigmoid Function

One way to improve conditioning, and therefore speed up convergence is to make sure that the average value of the activation of hidden and input units is small in comparison to their variance. Therefore, it is useful to normalize the inputs to have approximately zero mean, and near unity variance. A rotation of the inputs (using the eigenvectors of their covariance matrix) also improves convergence.

Concerning the hidden units, it is clearly better to use a symmetric sigmoid such as the hyperbolic tangent, with a range from –1 to 1 (as it brings the average activation closer to 0), rather that the asymmetric sigmoid, which ranges from 0 to 1. Some researchers have also tried multi rated sigmoids [26, 27, 139] and their modular combinations to better develop the network.

2.3.2.2. Choice of Targets

For classification problem, two target output values are usually defined, a 'low' and a 'high' target, for each output unit. For these tasks, it is advisable to use a saturating non-linearity on the output units, such as sigmoid or hyperbolic tangent. The main advantage of this non-linearity is that it penalizes the network less for producing outputs above the 'high' and below the 'low' target. On the other hand, these values should not be chosen too close to asymptotic values of the non-linearity (i.e. 0 or 1 for the sigmoid), since they will yield very large weights on the output layer, and will stall or halt training because of the near zero derivatives through the output non-linearity. It is advisable to choose target outputs near the inflection points of the non-linearity *i.e.* 0.2 and 0.8 for the sigmoid [107].

2.3.2.3. Use of Batch Gradient Descent

Stochastic Gradient Descent updates the ANN parameters after each pattern presentation, whereas, batch gradient descent updates the parameters after having processed the whole training set (summing all the gradient contributions). For large training sets (several hundred or thousand), stochastic gradient descent usually gives significantly faster convergence than batch gradient descent (sometimes hundreds of times faster). The main reason for faster convergence is the redundancy present in training sets. Furthermore, stochastic gradient descent appears to be more robust to local minima and to generalize better. The resistance to shallow local minima is due to the 'noise' in the gradient, since only one or few patterns are used to 'estimate' it [108]. This often causes small (temporary) uphill jumps in the error function, thus allowing it to escape shallow (but not wide) minima. Finally, better generalization is generally observed because stochastic gradient descent tends to find solutions in wide basins of the error function. In these wide basins, many similar values of the parameters yield performance, therefore the solution obtained is robust to small changes in the parameters (which could occur when a different training set is used). The set of solutions within wide basins require less precision for the parameters and therefore correspond to a smaller effective capacity (hence better control of the over fitting problem).

When stochastic gradient descent is used, convergence can be further accelerated by presenting more 'surprising' examples more often (where surprise can be defined by the error performed by the network on these patterns). The first application of this simple idea is to randomly 'shuffle' all the examples of the training set. A very bad training set for stochastic gradient descent would have grouped all the examples of the same class together. Shuffling guarantees that such dependencies on the sequence of patterns are removed. A more general application of this idea is to present examples that have generated a large error in the past more often, therefore focusing on the more ambiguous patterns near the decision surface, and yielding faster and more discriminate training. However, one has to be careful with outliers (wrongly labelled data), and not concentrate the training efforts on these cases. Many studies were done for equalized, on-line [24] and multi-level threshold [26].

2.3.2.4. Early Stopping

When an iterative training algorithm (such as gradient descent) is used, the effective capacity of the learning machine increases as training machine increases as training proceeds. The effective capacity is smaller than the actual capacity, and takes into account the constraints (and the limitation of resources) of the training algorithm. Before the training iterations start, the *size* of the set of solution explored by the learning algorithm is very small (e.g. 1). As training proceeds, the algorithm is allowed to explore a large set of solutions, and the effective capacity increases.

By estimating after each training epoch the generalization performance on a validation set (or test set, not used for training), training can be stopped early. When the estimated generalization does not improve any more (or even worsens), the effective capacity is too large and the machine starts to over fit [10].

2.3.2.5. Use of a Priori Knowledge

Learning can be seen as combining two different types of information: Prior knowledge or hints as suggested by Abu-Mostafa in 1990 and training data. Prior knowledge can be any information about the problem that can be used to impose constraints on the learning process or help it to reach a 'better' solution. Constraints can be used to reduce the size of the Euclidian space. However, the fact that the 'prior knowledge' can be partially incorrect may induce some undesirable bias. Three categories of constraints can be considered, namely: architectural constraints, constraints that guide training, and constraints used for initialization. Architectural constraints can be introduced in several ways. The network architecture itself can be constrained by reducing connectivity, using shared weights, or more generally using a weight space transformation. Input pre-processing and coding, as well as output representation and coding are also important. Finally, the system can be partitioned into modules by decomposing the problem into sub-problem introducing intermediate state variables. Prior knowledge can be used to decide on a representation for these intermediate states.

An important training constraint is in the choice of cost function (say desired – actual response) and possible penalty terms (may be *I* terms of weight decay). A particularly interesting type of training constraint imposes some invariance on the function computed by the network [110].

2.3.2.6. Input Representation

Similar to other learning algorithms, it is very important for ANNs to appropriately select pre-processing schemes and the input representation, In the case of Automatic Speech Recognition (ASR) there is evidence of important type of signal processing used. Cosi, showed in 1990 that auditory model performed better as a preprocessor for a feed forward neural network than a mel-scale filter bank. Thus, it is clear that the choice of good features help to train the MPL better.

2.3.3.7. Output Representation

Most ANNs designed for phoneme recognition have a simple output coding scheme consisting of one input per phoneme, with a high desired output for the output unit corresponding to the target phoneme and a low desired output for the other units. In these cases, one can interpret the output activation of the network as representing a degree of evidence. Using phonetic knowledge, coding schemes based on articulatory features have been considered.

Another idea is again to use a priori knowledge for the design of output representations is to use 'hint' outputs units(again Cosi, 1990, vowel were supported by extra units of phonetic features). These additional constraints on the network guided it to take these phonetic features into account, with an improved generalization.

2.3.2.8. Learning Rate

The learning rate h determines the rate of convergence of the algorithm, i.e. the length of the step away from the current point in the weight space. Larger values of h allows faster convergence. However, too large a h can diverge the algorithm. The optimal value of h depends on the largest eigenvalue of the Hessian matrix of second order derivatives of the cost function with respect to the weights.

2.4. BACK PROPAGATION LEARNING METHODS

Error back propagation is a gradient descent algorithm where weights and biases are adjusted to minimize a cost function equal to the mean square error in the network.

The training patterns are applied in some random order one by one, and the weights are adjusted using the back propagation learning law. Each application of the training set patterns is called a cycle. The patterns may have to be applied for several training cycles to obtain the output error to an acceptable low value. Once the network is trained, it can be used to recall the appropriate pattern (in this case some interpolated output pattern) for a new input pattern. The computation for recall is straightforward, in the sense that the weights and the output functions of the units in different layers are used to compute the activation values and the output signals. The signals from the output layer correspond to the output.

Back propagation learning emerged as the most significant result in the field of artificial neural networks. In fact it is this learning law that led to the resurgence of interest in neural networks, nearly after 15 years period of lull due to exposition of limitations of the perception learning by Minsky and Papert (1969).

As noted earlier, the backpropagation learning involves propagation of the error backwards from the output layer to the hidden layers in order to determine the update for the weights leading to the units in a hidden layer. The error at the output layer itself is computed using the difference between the desired output and the actual output at each of the output units. The actual output for a given input training pattern is determined by computing the outputs of units for each hidden layer in the forward pass of the input data. The error in the output is propagated backwards only to determine the weight updates. There is no feedback of the signal itself at any stage, as it is a feedforward neural network.

Since the back propagation learning is exactly the same as the delta learning, at the output layer and is similar to the delta learning with the propagated error at the hidden layers, it is also called *generalized delta rule*. The term *generalized* is used because the delta learning could be extended to the hidden layer units. Backpropagation or error is possible only if the output functions of the nonlinear processing units are differentiable. Note that if these output functions are linear, then the advantage of a multilayer network to generate complex decision boundaries can not be explored for nonlinearly separable (hard) classification problems. In fact a multilayer feedforward network with linear processing units is equivalent to a linear associative network. On the other hand, hard-limiting output function as in a multilayer perceptron cannot be used for learning the weights. A common differentiable output function used in the backpropagation learning is one which possesses a sigmoid nonlinearity. Two examples of sigmoidal nonlinear function are the logistic function and hyperbolic tangent function. The hyperbolic tangent function can be viewed as a biased and scaled version of the logistic function. The asymmetry of the hyperbolic tangent function makes the learning faster by reducing the number of iterations required for training.

The back propagation learning is based on the gradient descent along the error surface. That is, the weight adjustment is proportional to the negative gradient of the error with respect to the weight. The error is the instantaneous error between the desired and the actual values of the output of the network. This instantaneous error is due to a given training pattern, which can be assumed to be a sample function of a random process. Thus the error can be assumed to be a random variable. Therefore this gradient descent method is a path to the minimum of the error surface will be zigzag. The error surface itself will be an approximation to the true error surface determined by the entire training set of patterns. Moreover, even the true error surface is not a smooth quadratic surface as in the case of the Adaline. In fact, the error surface may contain several local minima besides the global minimum. Hence the stochastic approximation of the gradient decent used in the backpropagation learning need not converge. There is no proof of convergence even in the mean as in the case of the LMS algorithm. Since there is no proof of convergence, heuristic criteria are used to stop the process of learning. There are based on the values of the gradient and the error in successive iterations and also on the total number of iterations. The average gradient value over each training cycle (presentation of all the training patterns once) is observed, and if this average value is below a preset threshold value for successive cycles, then the training process may be stopped. Likewise, the training process may be stopped using a threshold for average error and observing the average error in successive cycles.

Sometimes both the average gradient as well as the average error may be used in the stopping criterion. But the main objective is to capture the implicit pattern behaviour in the training set data so that adequate generalization takes place in the network. The generalization feature is verified by testing the performance of the network for several new (test) patterns.

2.5. BACK PROPAGATION ALGORITHM

The step of BPN are as follows:

Step 1: Normalized the inputs and outputs with respect to their maximum values. It is proved that the neutral networks work better if input dnd outputs lie between 0-1. For each training pair, assume there are 'l' inputs given by $\underset{l\times 1}{\{r\}_r}$ and 'n' outputs $\underset{n\times 1}{\{O\}_o}$ in a normalized form.

Step 2: Assume the number of neurons in the hidden layer to lie between $l < m < 2l$.

Step 3: [V] represents the weights of synapses connecting input neurons and hidden neurons and [W] represents weights of synapses connecting hidden neurons and output neurons. Initialize the weights to small random values usually from – 1 to 1. For general problems, λ can be assumed as 1 and the threshold values can be taken as zero.

$$[V]^0 = [\text{random weights}]$$
$$[W]^0 = [\text{random weights}]$$
$$[\Delta V]^0 = [\Delta W]^0 = [0]$$

Step 4: For the training data, present one set of inputs and outputs. Present the pattern to the input $(I)_I$ as inputs to the input layer. By using linear activation function, the output of the input layer may be evaluated as

$$\underset{l\times r}{\{O\}_I} = \underset{1\times 1}{\{I\}_r}$$

Step 5: Compute the inputs to the hidden layer by multiplying corresponding weights of synapses as

$$\underset{m\times 1}{\{I\}_H} = \underset{m\times l}{[V]^T}\ \underset{l\times 1}{\{O\}_I}$$

Step 6: Let the hidden layer units evaluate the output using the sigmoidal function as

$$\{O\}_H = \underset{m\times 1}{\begin{Bmatrix} \cdot \\ \cdot \\ \dfrac{1}{(1+e^{-I_{Hi}})} \\ \cdot \\ \cdot \end{Bmatrix}}$$

Step 7: Compute the inputs to the output layer by multiplying corresponding weights off synapses as

$$\underset{n\times 1}{\{I\}_o} = \underset{n\times m}{[W]^T}\ \underset{m\times 1}{\{O\}_H}$$

Step 8: Let the output layer units evaluate the output using sigmoidal function as

$$\{O\}_o = \begin{Bmatrix} \cdot \\ \cdot \\ \dfrac{1}{(1+e^{-I_{oj}})} \\ \cdot \\ \cdot \end{Bmatrix}$$

The above is the network output.

Step 9: Calculate the error and the difference between the network output and the desired output as for the *i*th training set as

$$E^P = \frac{\sqrt{\Sigma(T_j - O_{oj})^2}}{n}$$

Step 10: Find $\{d\}$ as

$$\{d\} = \underset{n\times 1}{\begin{Bmatrix} \cdot \\ (T_k - O_{ok})\,O_{ok}\,(1-O_{ok}) \\ \cdot \\ \cdot \end{Bmatrix}}$$

Step 11: Find $[Y]$ matrix as

$$\underset{m\times n}{[Y]} = \underset{m\times n}{\{O\}_H}\,\underset{1\times n}{\langle d\rangle}$$

Step 12: Find $[\Delta W]^{t+1} = \alpha\,\underset{m\times n}{[\Delta W]^t} + \eta\,\underset{m\times n}{[Y]}$

Step 13: Find $\underset{m\times 1}{[e]} = \underset{m\times n}{[W]}\,\underset{n\times 1}{\{d\}}$

$$\underset{m\times 1}{\{d^*\}} = \underset{m\times 1}{\begin{Bmatrix} \cdot \\ \cdot \\ e_i(O_{Hi})(1-O_{Hi}) \\ \cdot \\ \cdot \end{Bmatrix}}$$

Find [X] matrix as

$$\underset{1\times m}{[X]} = \underset{1\times 1}{\{O\}_r}\,\underset{1\times m}{\langle d^*\rangle} = \underset{1\times 1}{\{I\}_I}\,\underset{1\times m}{\langle d^*\rangle}$$

Step 14: Find $[\Delta V]^{t+1} = \underset{1\times m}{\alpha[\Delta V]^t} + \underset{1\times m}{\eta[X]}$

Step 15: Find $[V]^{t+1} = [V]^t + [\Delta V]^{t+1}$

$$[W]^{t+1} = [W]^t + [\Delta W]^{t+1}$$

Step 16: Find error rate as

$$\text{error rate} = \frac{\Sigma E_p}{\text{nset}}$$

Step 17: Repeat steps 4–16 until the convergence in the error rate is less than the tolerance value.

2.6. FACTORS AFFECTING BACK PROPAGATION LEARNING

The performance of the back propagation learning law depends on the initial setting of the weights, learning rate parameter, output functions of the units, presentation of the training data, ad already discussed in previous sections. It is important to initialize the weight value properly before applying the learning law for a given training set. Initial weights correspond to a priori knowledge in the form of initial weights. The overall performance of the resulting trained network in terms of speed of learning and generalization would improve significantly. In general it is not known how to collect the relevant knowledge a priori. The more difficult part is to know how to include it in the form of weights. Therefore all the weights in the network are initialized to random numbers that are uniformly distributed in a small range of values.

The range is typically $\left[-\frac{a}{\sqrt{N_i}}, +\frac{a}{\sqrt{N_i}}\right]$ where N, is the number of inputs to the ith unit.

Thus the range can be different for each unit. The value of ± is typically in the range (1 to

3) [1]. On the other hand, large initial weight values may result in the network output values in the saturation region of the output function. In the saturation region the gradient value is small. If the saturation is at the incorrect level, it may result in slow learning due to small changes made in the weights in each iteration. Incorrect saturation rarely occurs if the unit operates in the linear range of the output function.

Adjustment of the weights using backpropagation learning law is done by presenting the given set of training patterns several times. Randomizing the presentation of these patterns tends to make the search in the weight space stochastic, and thus reduces the possibility of 'limit cycles' in the trajectory in the weight space during learning [4]. Presentation of the training data pattern by pattern for adjustment of the weights makes it possible to have the learning online. This pattern mode also reduces the problem of local minima. But to speed up the learning process it is preferable to update the weights in a batch mode, in which the gradient of the error, computed over all the training patterns, is used. The batch mode gives a better estimation of the gradient of the overall error surface.

Effect of Learning Rate Coefficient

Learning rate parameter plays a crucial role in the backpropagation learning. The order of values for n depends on the variance of the input data. For the case of Adaline, the learning rate parameter $\frac{\eta < 1}{(2\lambda_{max})}$, where λ_{max} is the largest eigen value of the autocorrelation matrix of the input data. This gives an indication for the choice of ·, since the derivation in the back-propagation does not suggest any clue for this choice. Since it is stochastic gradient descent learning, too small an · will result in a smooth trajectory in the weight space, but takes long time to converge. On the other hand, too large an · may increase the speed of learning, but will result in large random fluctuations in the weight space, which in turn may lead to an unstable situation in the sense that the network weights may not converge. Learning rate coefficient determines the size of weight adjustments made at each iteration and hence affects the convergence. Too large a learning rate will make the search path oscillate and slow convergence whereas, too small a learning rate will increase the convergence time.

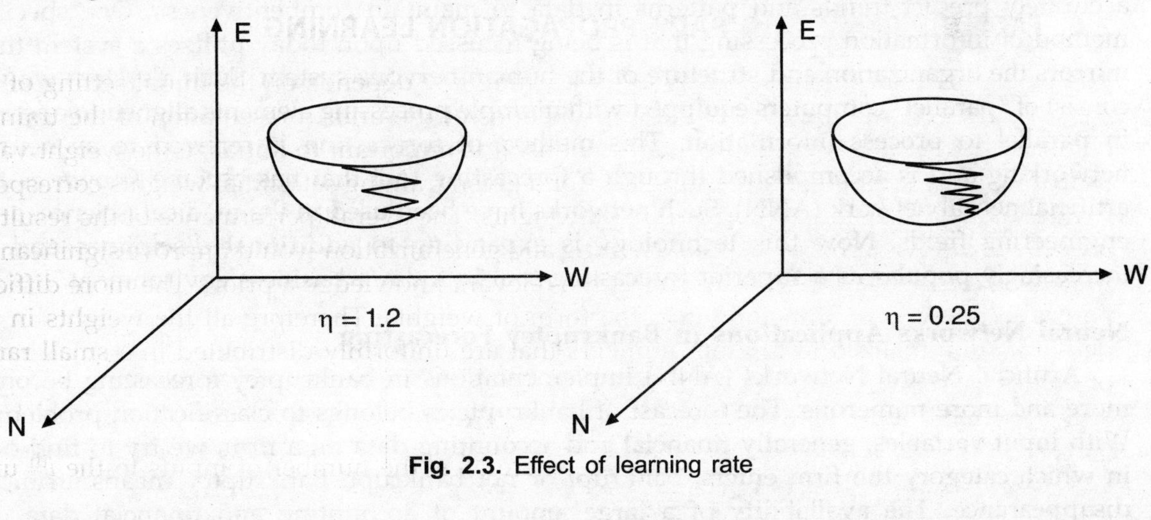

Fig. 2.3. Effect of learning rate

2.7. SOME APPLICATION DOMAINS

Neural networks, inspired by the information-processing strategies of the brain, are proving to be useful in a variety of applications, and their full potential is far from realization. However, some of the common application domains have been listed below:

Applications of Neural Networks in Bioinformatics

Due to advancement in genetic engineering especially the advent of the human genome project, the pressure for molecular computation and sequencing technologies have been immensely increased. Since the human genome project is aiming to map and sequence the entire human genome, it is expected to produce data beyond our current abilities to sequence and interpret, given our current computing ways. There are many sequencing problems in the area of Bioinformatics for machines to solve and rationalize. A problem, many scientists believe to the most significant problem remaining in genetic engineering is *Protein Folding*. Protein folding refers to the problem of predicting a protein's three-dimensional structure from a one-dimensional amino-acid sequence. So far, neural networks have shown a lot of promise and initial experimental success towards the protein folding problem. The protein folding problem is the a key initiative in the fields of protein engineering and 'rational' drug design. Traditionally, the protein structure has been predicted using methods such as X-ray crystallography, which involves manual and experimental procedures. However due to the human genome project, it is becoming impossible for these approaches to keep up with the number of protein sequences needed to be mapped. As a result the existence of an artificially intelligent process of sequencing is becoming imminent necessity.

Neural Network in Forecasting

Most forecasting today is based upon traditional linear and non-linear statistical models, such as regression analysis. Although these models are useful and have been utilized for many years to predict, the models are somewhat limited in their ability to forecast in certain situations. Given the changing nature of technology and the globalization of business and financial markets, it is becoming increasingly important to be able to mire quickly and accurately predict trends and patterns in data to maintain competitiveness. One specific method of information processing that is being focussed upon today utilizes a system that mirrors the organization and structure of the human nervous system. Such a system would consist of "parallel" computers equipped with multiple processing elements aligned to operate in parallel to process information. This method of forecasting is referred to as neural networking and is accomplished through a forecasting tool that has become known as an artificial neural network (ANN). Such networks have been used in the medical, science, and engineering fields. Now this technology is expanding to additional disciplines and is increasingly popular as a superior forecasting tool in today's business environment.

Neural Networks Applications in Bankruptcy Forecasting

Artificial Neural Networks (ANN) implementations in bankruptcy forecasting become more and more numerous. The forecast of bankruptcies belongs to classification problems. With input variables, generally financial and accounting data on a firm, we try to find out in which category the firm enters, bankrupt or not bankrupt. Bankruptcy means juridical disappearance. The availability of a large amount of accounting and financial data on computerise databases, facilitates the use of artificial neural networks with quantitative data.

They are tested as substitutes of traditional statistical tools such as multivariate discriminate analysis.

Neural Network in Healthcare

The advantage of the neural networks over conventional programming lies in their ability to solve problems that do not have an algorithmic solution or the available solution is too complex to be found. Neural networks have been applied within the medical domain for–

- Clinical diagnosis
- Image analysis and interpretation
- Signal analysis and interpretation
- Drug developments

Application of Neural Networks to Intrusion Detection

Intrusion Detection System (IDS) are now mainly employed to secure company networks. Ideally, an IDS has the capacity to detect in real-time all (attempted) intrusions, and to execute work to stop the attack (for example, modifying firewall rules). We present here state of art Intrusion Detection System, and a new way to improve false-alarm detection using Neural Network approach. This approach is still in development; nevertheless it seems to be very promising in future.

Neural Networks in Pattern Recognition

Pattern Recognition is the association of an observation to the past experience or knowledge. Humans continuously perform perceptual pattern recognition, and the amazing complexity of the cognitive processes involved has, made pattern recognition an active area of search in psychology and neurophysiology for many decades. Pattern Recognition is the science and art of giving names to the natural objects in the real world. Neural networks have shown remarkable progress in the recognition of visual images, handwritten characters, printed characters, speech and other PR based tasks.

Application of Neural Network in Optimization/Constraint Satisfaction

A wide variety of problems in mathematics, statistics, engineering, science, medicine, and economics can be posed as optimization problems. The goal of an optimization algorithm is to find a solution satisfying a set of constrains such that an objective function is maximized or minimized. Examples of such problems include manufacturing scheduling, finding the shortest possible tour given a set of cities, etc.

Applications of Neural Networks in Control systems

The applications included are:

- Neural networks based controller for induction motor: In this application the neural networks are used as controller to control the speed of the induction motor.
- Neural network control of the neutralization systems: In this application the pH value is used to control the neutralization of the system. The pH value is controlled by controlling the flow rate of the base.
- Neural network based controller for broom stick balancer: In this application the position and the angle of the broom stick is used to control the balance of the broom stick.

SOLVED EXAMPLES

EXAMPLE 2.1.

Find weight matrix

Test network with training input vector.

$$X_1 \qquad W$$

$$[1\ 1\ 0\ 0]_{1\times 4} \begin{bmatrix} 1 & 0 \\ 2 & 0 \\ 0 & 2 \\ 0 & 1 \end{bmatrix}_{4\times 2} = [3 \quad 0]$$

$$Y_{inj} = \sum_{P=1}^{4} X_i W_{Pj}$$

$$Y_{in1} = X_1 W_{11} + X_2 W_{21} + X_3 W_{31} + X_4 W_{41}$$

$$= 3$$

$$Y_{in2} = X_1 W_{12} + X_2 W_{22} + X_3 W_{32} + X_4 W_{42}$$

$$= 0$$

$$Y_1 = 1;\ Y_2 = 0$$

$$X_2 = [0\ 1\ 0\ 0]$$

$$[0\ \ 1\ \ 0\ \ 0] \begin{bmatrix} 1 & 0 \\ 2 & 0 \\ 0 & 2 \\ 0 & 1 \end{bmatrix} = [2\ \ 0]$$

$$Y_1 = 1; \quad Y_2 = 0$$

$$Y_3 = [0\ \ 0\ \ 1\ \ 1] \qquad Y_4 = [0\ \ 0\ \ 1\ \ 0]$$

Similarly other two responses can be shown correctly.

EXAMPLE 2.2.

Test the (1110) as training input

$$[1\ \ 1\ \ 1\ \ 0] \begin{bmatrix} 1 & 0 \\ 2 & 0 \\ 0 & 2 \\ 0 & 1 \end{bmatrix} = [3\ \ 2]$$

$$Y_1 = 1; \quad Y_2 = 1$$

$$X_2 = [0\ \ 1\ \ 0\ \ 0]$$

EXAMPLE 2.3.

A hetero-associative network is given as s_1 = (0 1 1 0) t_1 (1 0) s_2 = (0 1 0 0) t_2 (1 0)

Find the weight matrix. Also test the (1 1 1 0) as training input.

Solution:

$$W_1 = \begin{bmatrix} 0 \\ 1 \\ 1 \\ 0 \end{bmatrix} [1\ 0] = \begin{bmatrix} 0 & 1 \\ 1 & 0 \\ 1 & 0 \\ 0 & 0 \end{bmatrix}$$

$$W_2 = \begin{bmatrix} 0 \\ 1 \\ 0 \\ 0 \end{bmatrix} [1\ 0] = \begin{bmatrix} 0 & 0 \\ 1 & 0 \\ 0 & 0 \\ 0 & 0 \end{bmatrix}$$

$$W = \begin{bmatrix} 0 & 0 \\ 2 & 0 \\ 1 & 0 \\ 0 & 0 \end{bmatrix}$$

$$Y_{in} = [1\,1\,1\,0] \begin{bmatrix} 0 & 0 \\ 2 & 0 \\ 1 & 0 \\ 0 & 0 \end{bmatrix} = [3\ 0]$$

$$Y_1 = 1; \quad Y_2 = 0$$

SHORT QUESTIONS WITH ANSWERS

Q.1. What does 'backprop' mean?

Ans. It is an abbreviation for 'backpropagation of error' which is the most widely used learning method for neural networks today.

Q.2. How many learning methods for NNs exist? Which?

Ans. The main categorization of these methods is the distiction of supervised from unsupervised learning:

- In supervised learning, there is a "teacher" who in the learning phase "tells" the net how well it performs ("reinforcement learning") or what the correct behavior would have been ("fully supervised learning").
- In unsupervised learning the net is autonomous: it just looks at the data it is presented with, finds out about some of the properties of the data set and learns to reflect these properties in its output. What exactly these properties are, that the network can learn to recognise, depends on the particular network model and learning method.

Q.3. What are the types of perceptrons?

Ans. Single layer perceptron

Multilayer perceptron

LONG QUESTIONS WITH ANSWER

Q.1. Give a comparison between single layer feed forward network multilayer feed forward network and recurrent network.

Ans.

Single layer feed	Multilayer feed forward forward Neural network	Recurrent network Neural network
1. This type of network sets of two layers namely the input layer.	1. This network is made up of multiple layers.	1. This network is also made up of multiple layers.
2. It does not have any hidden layer.	2. It does have hidden layer.	2. It has hidden layer.
3. There is no feed back loop hence called feed forward network.	3. It also does not have any feedback loop.	3. There exist atleast on feedback loop from out layer to input layer.
4. Computation task is performed by output layer itself.	4. Hidden layer isued to provide connection between external input and network output and compotational task.	4. Computation task is performed by hidden layer.
5. Synaptic links carrying the weights connect every input neuron to output neuron. It is called acylic in nature.	5. Let *l* = input neurons *m* = neurons in hidden layer. *n* = output neurons configuration = *l* – *m* – *n*	5. There are neurons with self-feedback links the output of a neuron is feedback into itself as input.

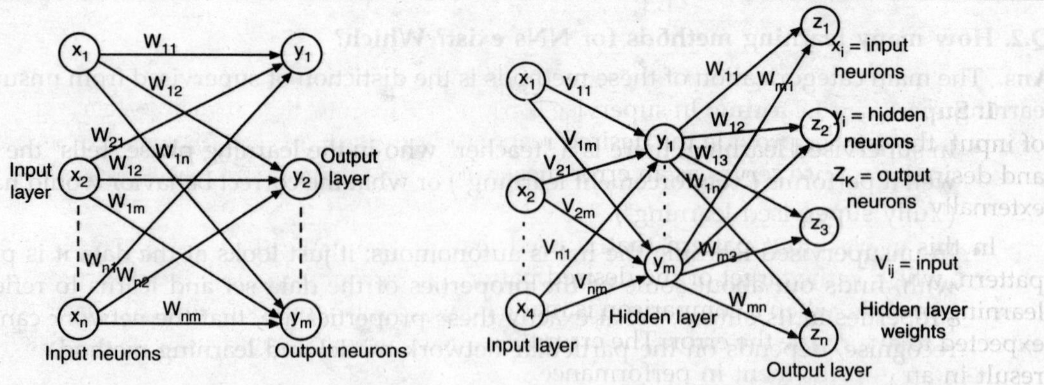

(a) Single Layer Feed Forward Neural Network

(b) Multilayer Feed Forward network (l – m – n configuration)

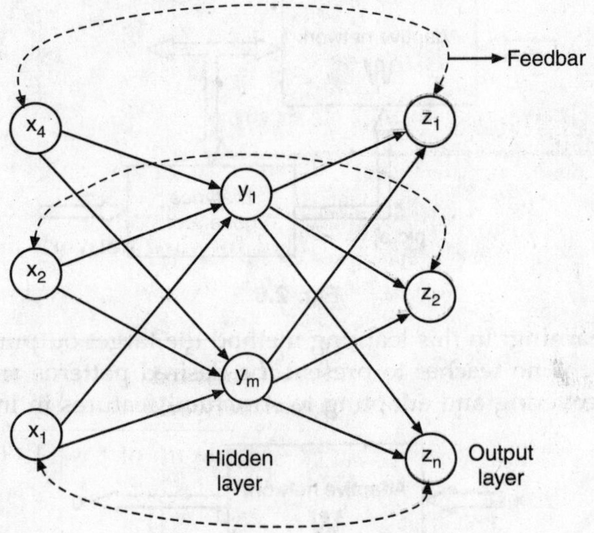

(c) Recurrent Network

Fig. 2.4.

Q.2. Discuss any three learning methods.

Ans. Learning methods in neural networks can be broadly classified

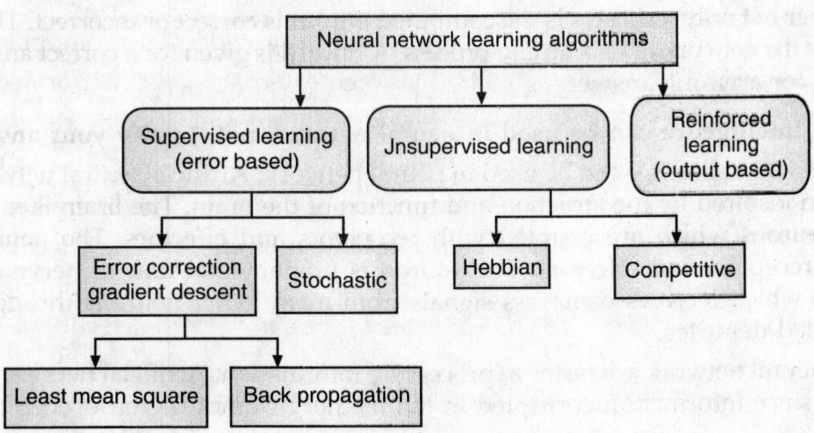

Fig. 2.5.

1. Supervised Learning: In supervise learning we assume that at each instance of application of input, the teacher provide the desired response of the system. The distance between the actual and desired response serves as an error measured and is used to correct the network parameters externally.

In this every input pattern that is used to train the network is associated with an output pattern, which is the target or the desired pattern. A teacher is assumed to be present during the learning process, when a comparison is made between the networks comuted output, and correct expected to determine the error. The error then can be used to change network parameters, which result in an improvement in performance.

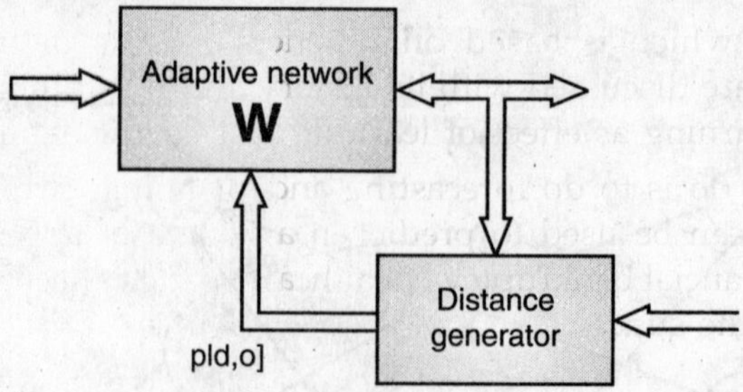

Fig. 2.6

2. Unsupervised Learning: In this learning method the target output is not presented to the network. It is as if there is no teacher to present the desired patterns and therefore the system learn of its own by discovering and adapting to structural features in input patterns.

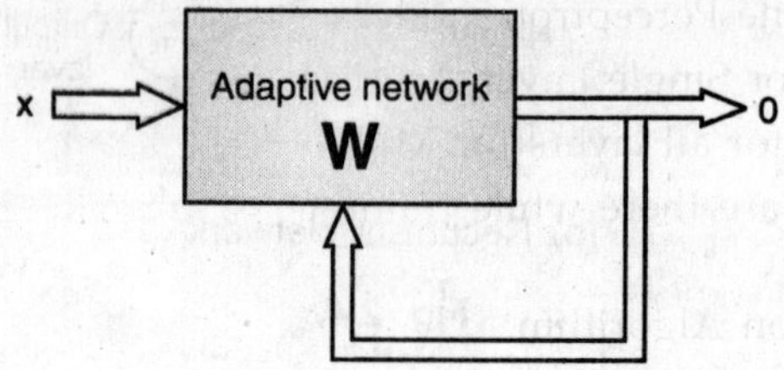

Fig. 2.7

3. Reinforced Learning: In this method a teacher through available, does not present the expected answer but only indicates is the computed output is correct or incorrect. The information provided helps the network in its learning process. A reward is given for a correct answer computer and a penalry for a wrong answer.

Q.3. Artificial intelligence can be used in neural network not. Justify your answer.

Ans. Yes Artificial intelligence can be used in neural network. Artificial neural network we a form of computation inspired by the struction and function of the brain. The brain is made up of vast network of neurons which are coupled with recepetors and effectors. The neuron which are coupled with receptors and effectors. The neuron is fundamental unit of nervous system. The neuron is that which receives compress signals from many other neurons through filamentary input path called dentrites.

Artificial neural network are faster in processing information. Artificial nets are inherently not fault tolerant, since information corrupted in teh memory cannot be retrieved.

SUMMARY

This chapter is actually the detailing of the previous one. The chapter deals in with the concept of The perceptron and its solution using weight correction. Also, different ANN models, as Single Layer and Multiple Layer perceptron and its weight correction methods are discussed. Back Propagation of Error is an algorithm of network training, which is backbone of all learning processes. This is discussed in detail with calculation of error in each of its layers i.e., input, hidden and output layers. A detail discussion about the heuristics for training Multi Layer

Perceptron is discussed which is based on practical aspects of training MLPs. Next, Back Propagation Algorithms are discussed with its history and Algorithm. Different factors affecting the Back Propagation Learning as effect of learning rate coefficient are also discussed.

All statistics we ever do is to do forecasting and planning. Artificial Neural Networks is a forecasting technique. It can be used to predict in any area of application, *e.g.*, bioinformatics, radar signal modeling, financial bankruptcy , healthcare, pattern Recognition, control systems and many more areas under the sun.

REVIEW QUESTIONS

1. Derive the solution to the Perceptron model of the artificial neuron.
2. Derive the expression for Single Layer Artificial Neural Network.
3. Derive the expressions for all layers for Multilayer Perceptron.
4. Many practical aspects are there while training MLP; mathematical model of which is not very clear. Discuss them.
5. Discuss Back propagation Algorithm.
6. Discuss the affect of Learning rate coefficient on Back propagation algorithm.
7. Discuss the role of Artificial Neural Network in our daily lives. Try to save yourself from fantasy.

NUMERICAL PROBLEMS

1. Make a Perceptron net for OR function using binary input and output. Compare the results if you use bipolar input and target.
2. Use Perceptron learning rule and bipolar data, let the learning rates be 1 in the first calculation and 0.5 in the second calculation. Compare the ANDNOT function generated.
3. Use bipolar inputs and targets. Compare the operation of AND function using Hebb net and Perceptron net.
4. Do the same exercise as that of previous question for input pattern (1, –1) and target out of 0.7. Again use binary sigmoidal activation function.
5. Use the backpropagation network given below to find new weights, when the input pattern presented is [0.6 0.8 0] and target output is 0.9. Try learning rate as 0.2 and use the binary sigmoidal activation function. (Fig. of Problem 6)
6. For the back propagation network given below find out the new weights when the input is (1, 0) and target output is 1. Use learning rate of 0.4 and binary sigmoidal activation function as output.

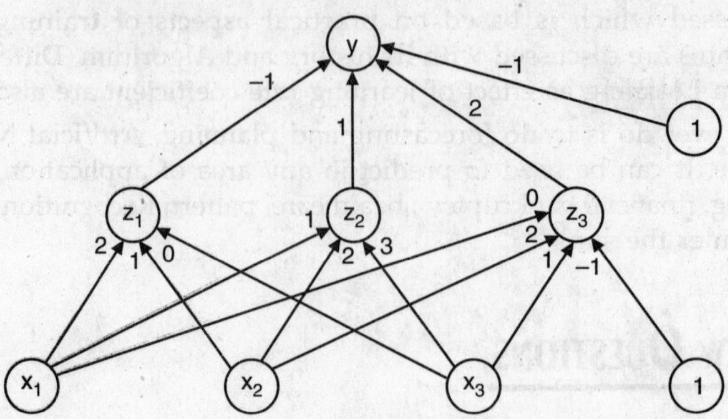

Fig. 2.8.

7. Realise a neural network using BPN algorithm for XOR gate. Use the initial weights and biases as given in the architecture.

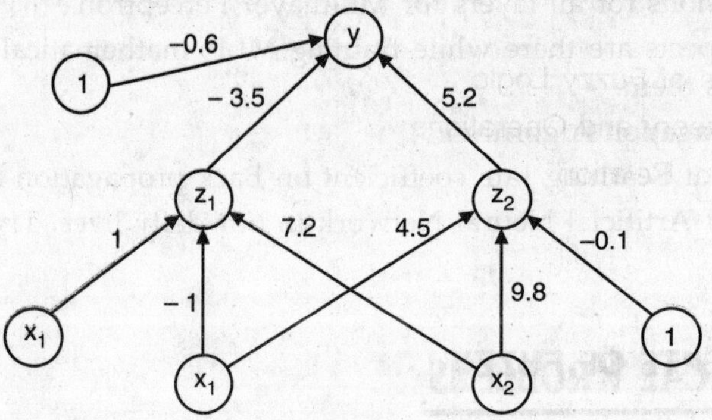

Fig. 2.9.

3 CHAPTER THREE

FUZZY LOGIC-I (INTRODUCTION)

CHAPTER OUTLINE

3.1. Basic Concepts of Fuzzy Logic
3.2. Crisp Sets
3.3. Fuzzy Sets: Theory and Operations
3.4. Properties of Fuzzy Sets
3.5. Crisp and Fuzzy Relation
3.6. Fuzzy to Crisp Conversion

3.1. BASIC CONCEPTS OF FUZZY LOGIC

In 1979, in a football ground, our coach shouted "Don't Fuzz". None of us could understand the word at that time; but understood that, whatever we were doing, should not be repeated. In fact, it was a derogatory slang term for the police and he was asking us not to be one after one player, so as to stop movement of the player each player was covering. Fuzzy logic means inexact reasoning with inexact data granulating, computing with words and so on.

It refers to imperfection or haziness. Haziness is always present in any realistic process. This vagueness may come up from the analysis of the data inputs and in the directives used to describe the relationships between the informative attributes. Very number of times, we wonder, what to do if a student earns 59.92 percent marks. It is not a first class, nevertheless, it is never a second class either. Fuzzy logic grants an inference structure that allows the human reasoning capacities to be applied to artificial knowledge-based structures. Fuzzy logic gives a means for adapting linguistic strategy into control actions and thus offers a high-level computation.

No computer can match the cognitive abilities of human beings. No computer can recognize the faces, voices better than any super computer in the world. The science of neutral networks provides new computing tool with learning and adaptation capabilities. The theory of fuzzy logic gives an inference method under cognitive uncertainty. Computational neutral networks and fuzzy logic offer exciting benefits such as learning, adaptation, fault tolerance, parallelism, and generalization.

We can find different kinds of explanation for fuzzy systems theory. Human knowledge has become more and more important. We increase it from living through the world a, which we live and use our capability to reason to create an order in the mass of information (i.e., to formulate human knowledge in a organized manner). The million dollar question here tis to store and preserve that 'Tacit knowledge'. We are all limited in our ability to perceive the world and to deep reasoning, we find ourselves everywhere faced by uncertainty that is a result of lack of information, in particular, miscalculation of measurements or our own biases.

The other restriction factor in our quest for precision is a natural language used for describing/sharing knowledge, communication, etc. Every word has a spirit. We know the difference between listening and hearing. Difference between watching and seeing and many more. While all of know the word 'walk', though saunter, stroll, amble, meander, ambulate, etc. all have different spirits. We understand core meanings of word and are able to communicate without error to an adequate degree, but generally we cannot precisely agree among ourselves on the singular word or terms of communal sense meaning. In other words, ordinary languages are unclear.

NOTE

Our impression of the real world is pervaded by concepts, which do not have severely defined boundaries – for example, many, tall, much larger than, young, etc. just relative terms. A 5 feet 5 is average in Indian context but short in American/Irish/Afghani context. These concepts can be called fuzzy or vague concepts – a human brain works with them, while computers may not do it. The computers work in black and white, whereas there is a thousand greys between them in real life. Natural languages, which are much higher in level than programming languages, are fuzzy.

The door to the development of fuzzy computers was opened in 1985 by the design of the first logic chip by Masaki Togai and Hiroyuki Watanabe at Bell Telephone Laboratories. in the years to come fuzzy computers will empty both fuzzy hardware and fuzzy software, and they will be much closer in structure to the human brain than the present-day computers are.

The Fuzzy Logic tool was introduced in 1965 by Lotfi Zadeh. It is a mathematical tool for dealing with uncertainty. It offers to a software computing system which is the important concept of computing with words. It grants a technique to deal with vagueness. The fuzzy theory provides a mechanism for representing linguistic constructs such as "many", "low", "medium", "often", "few", "seldom", "high" and many more qualitative/linguistic adjectives.

The traditional binary set theory describes crisp events, events that have two options; one or zero. It uses probability theory to explain whether an event in about to happen, measuring the chance with which is a given event is expected to come about. The domain of fuzzy logic is based leading the concept of relative graded connection and so are the functions of awareness and cognitive processes.

It is important to keep in mind that there is a close connection between *Fuzziness* and *Complexity*. As the complexity of a task, or of a system for performing that task, exceeds a certain limit, the system must unavoidably become fuzzy in nature. Zadeh, initially an engineer and systems scientist, was concentrated with the fast turn down in information afforded by common mathematical models as the complexity of the target system increased. Though he was proved wrong shortly by Japanese as more than 2000 patents were filed and acquired.

With the increasing of complexity our skill to make accurate and yet major statements about its behaviour diminishes. Real world problems are too complex, and the *difficulty involves the degree of vagueness* – as uncertainty increases, or the data are erroneous, the complexity of event increases to an extent that it cannot be handled by human brain alone. Habitual system modeling and analysis techniques are too precise for such problems. Nevertheless, appropriate simplifications, assumptions, etc. are made to accomplish a reasonable compromise between the information we have and the amount of uncertainty we are willing to accept. In this aspect, fuzzy systems theory is similar to other engineering theories, because almost all of them describe the real world in an approximate manner.

NOTE: Engineers define fuzzy logic as a useful tool with diverse applications; especially on solving problems that linear computing is not able to do. A good example of this functionally is the capability for image recognition, since human brain is capable to distinguish an object from an image even if is blur, meanwhile linear computing is able to real just pixels as set of colour. There is an analog factor: vagueness, images may have fuzzy maps. All this must be kept in mind, while we tour through our journey to fuzziness.

Since our matriculation, we are learning about Boolean algebra dealing with 'ones' and 'zeros' and 'Venn' diagram thereof. The crisp sets lead to classical or Boolean logic, whereas, Fuzzy Sets lead to Fuzzy Logic. In the following section, we have a review of crisp sets and classical logics. Thereafter, we move towards Fuzzy Logic and Fuzzy Sets.

3.2. CRISP SETS

3.2.1. Characteristics of Crisp Sets

Characteristics of Crisp sets are defined as follows:

Universe of Discourse or Universal Set

The universe of discourse or universal set is the set which, with reference to a particular context, contains all possible elements having the same characteristics and from which sets can be formed. The universal set is denoted bE.

Example:

(*i*) The universal set of all numbers in Euclidean space.

(*ii*) The universal set of all population of India (Census).

Set

A set is a well defined collection of objects. Here, well defined means the object either belongs to or does not belong to the set (observe the "crispness" in the definition).

A set in certain context may be associated with its universal set from which it is derived.

Given a set A whose objects are $a_1, a_2, a_3, ..., a_n$ we wrote A as $A = \{a_1, a_2, ..., a_n\}$. Here, $a_1, a_2, a_3, ..., a_n$ are called the members of the set. Such a form of representing a set is known as list form.

Example: A = {Chanda Kochar, Kalpana Morparia}

B = {Swan, Peacock, Dove}

A set may also be defined based on the properties the members have to satisfy. In such a case, a set A is defined as:

$$A = \{x \mid P(x)\}$$

Here, $P(x)$ stands for the property P to be satisfied b the member x. This is read as 'A is the set of all X such that $P(x)$ is satisfied.'

Example: $A = \{x \mid x \text{ is an even number}\}$
$B = \{y \mid y > 0 \text{ and } y \bmod 5 = 0\}$

Venn Diagram

Venn diagrams are pictorial representations to denote a set. Give a set A defined over a universal set E, the Venn diagram for A and E is as shown in Fig. 3.1.

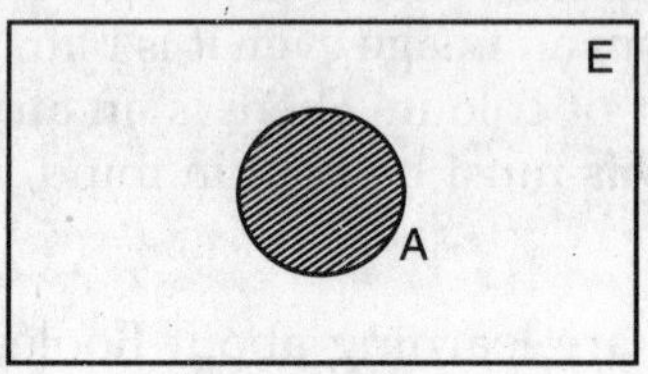

Fig. 3.1. Venn diagram of a set A

Example: In Fig. 3.1, if E represents the set of university students then A may represents the set of female students.

Membership

An element x is said to be a member of a set A if x belongs to the set A. The membership is indicated by "Î" and is pronounced "belongs to". Thus, xÎ A means x belongs to A and x Ï A means x does not belong to A.

Example: Given $A = \{3, 4, 5, 6, 7, 8, 10\}$, for $x = 2$ and $y = 4$, we have x Ï A and y Î A.

Here, observe that each element either belongs to or does not belong to a set. The concept of membership is definite and therefore crisp (1-belongs to, 0-does not belongs to). In contrast, as we shall see later, a fuzzy set accommodates membership values which are not only 0 to 1 but anything between 0 and 1. 'Life is not in zeros and ones – there are grey shades as well.'

Cardinality

The number of elements in a set is called its cardinality. Cardinality of a set A is denoted as $n(A)$ or $|A|$ or $\#A$.

Example: If $A = \{4, 5, 6, 7, 8\}$ than $|A| = 5$
If $B = \{2, 3, 4\}$ then $|B| = 3$

Family of Sets

A set whose members are sets themselves, is referred to as a family of sets.

Example: $A = \{\{1, 2, 5\}, \{2, 4, 6\}, \{7, 10\}\}$ is a set whose members are the sets $\{1, 2, 5\}$, $\{2, 4, 6\}$ and $\{7, 10\}$.

Null Set / Empty Set

A set is said to be a *null set or empty set* if has no members. A null set is indicated as ϕ or { } and indicates an impossible event. Also $|\phi| = 0$. Please note that a set having a member zero is not a null set. There is a difference in having a zero or having a null.

Example: The set of all presidents who are below 15 years of age.

Singleton Set

A set with a single element is called a *singleton set*. A singleton set has cardinality of 1.

Example: If $a = \{a\}$, then $|A| = a$

Subset

Given sets A and B defined over E the universal set, A is said to be a *subset* of B if A is fully contained in B, that is, every element of A is in B.

Denoted as $A \subset B$, we say that A is a subset of B, A is a *proper subset* of B. On the other hand, if A is contained in or equivalent to that of B then we denote the subset relation as $A \subseteq B$. In such a case A is called the *improper* subset of B.

Super set

Given sets A and B on E the universal set, A is said to be a *super set* of B if every elements of B is contained in A.

Denoted as $A \supset B$, we say A is a super set of B or A contains B. If A contains B and is equivalent to B, then we denote it as $A \supseteq B$.

Example: Let $A = \{3, 4\}$, $B = \{3, 4, 5\}$ and $C = \{4, 5, 6\}$

Here, $A \subset B$ and $B \supset A$

$C \subseteq B$ and $B \supseteq C$

Power Set

A power set of a set A is the set of all possible subsets that are derivable from A including null set. A power set is indicated as $P(A)$ and has cardinality of $|P(A)| = 2^{|A|}$

Example: Let $A = \{3, 4, 6, 7\}$

$P(A) = \{\{3\}, \{4\}, \{6\}, \{7\}, \{3, 4\}, \{4, 6\} \{6, 7\}, \{3, 7\}, \{3, 6\}, \{4, 7\}, \{3, 4, 6\}, \{4, 6, 7\}, \{3, 6, 7\}, \{3, 4, 7\}, \{3, 4, 6, 7\}, \varnothing\}$

Here, $|A| = 4$ and $|P(A)| = 2^4 = 16$

3.2.2. Operations of Crisp Sets

Four operations are possible on crisp set:

1. Union
2. Intersection
3. Complement
4. Difference

Union (∪)

The union of two sets A and B ($A \cup B$) is the set of all elements that belong to A or B or both

$$A \cup B = \{x/x \in A \text{ or } x \in B\}$$

Example: Given $A = \{a, b, c, 1, 2, 4\}$ and $B = \{1, 2, 3, a, b, c\}$ we get $A \cup B = \{a, b, c, 1, 2, 3, 4\}$

Fig. 3.2 illustrates the Venn diagram representation for $A \cup B$

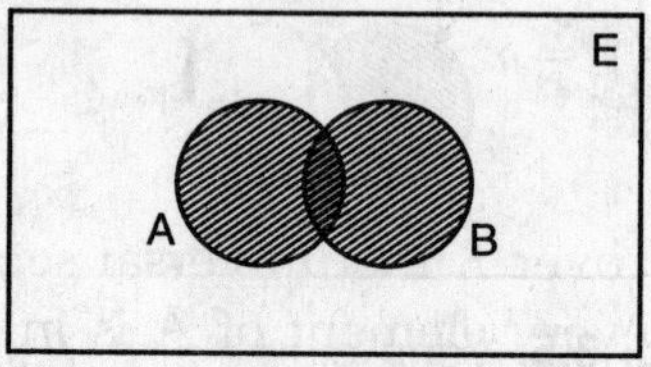

Fig. 3.2. Venn diagram for $A \cup B$

Intersection (∩)

The intersection of two sets A and B ($A \cap B$) is the set of all elements that belong to A and B

$$A \cap B = \{x/x \in A \text{ and } x \in B\}$$

Any two sets which have $A \cap B = \phi$ are called *Disjoint Sets.*

Example: Given $A = \{a, b, c, 1, 2, 4\}$ and $B = \{1, 2, 3, 1, b, c\}$ we get $A \cap B = \{b, c, 1, 2\}$

Figure 3.3 illustrates the Venn diagram for $A \cap B$

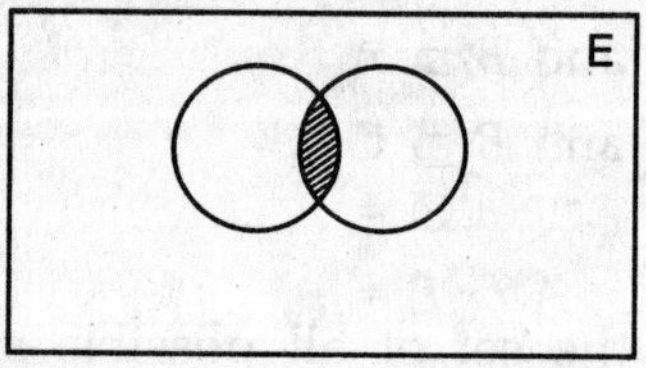

Fig. 3.3. Venn diagram for $A \cap B$

Complement (^{c})

The complement of a set A ($A \backslash A^c$) is the set of all elements which are in E but not in A.

$$A^c = \{x \ / \ x \notin E\}$$

Example: Given $X = \{1, 2, 3, 4, 5, 6, 7\}$ and $A = \{5, 4, 3\}$, we get $A^c = \{1, 2, 6, 7\}$

Figure 3.4 illustrate as the Venn diagram for A^c

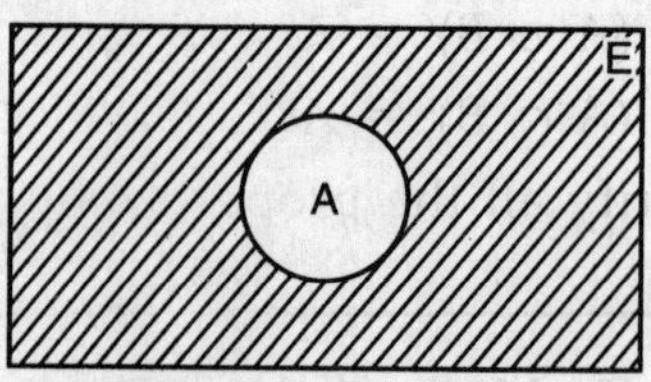

Fig. 3.4. Venn diagram for A^c

Difference (-)

The difference of the set A and B is $A - B$, the set of all elements which are in A but not in B.

$$A - B = \{x / x \in A \text{ and } x \notin B\}$$

Example: Given $A = \{a, b, c, d, e\}$ and $B = \{b, d\}$ we get $A - B \{a, c, e\}$.

Figure 3.5 illustrates the Venn diagram for $A - B$.

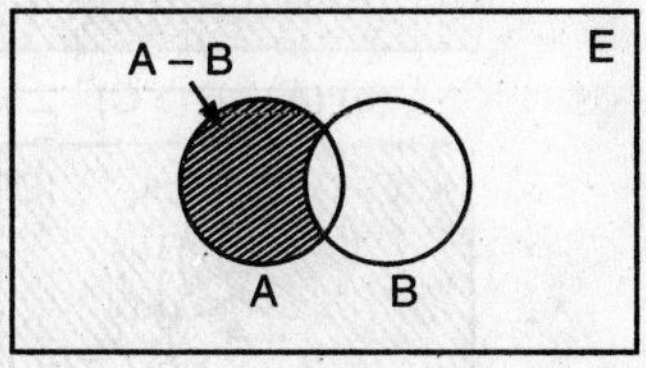

Fig. 3.5. Venn diagram for $A - B$

3.2.3. Properties of Crisp Sets

The following properties of sets are important for further manipulation of sets.

Commutativity: $A \cup B = B \cup A$

$A \cap B = B \cap A$

Associativity: $(A \cup B) \cup C = A \cup (B \cup C)$

$(A \cap B) \cap C = A \cap (B \cap C)$

Distributivity: $A \cup (B \cap C) = (A \cup B) \cap (A \cup C)$

$A \cap (B \cup C) = (A \cap B) \cup (A \cap C)$

Idempotence: $A \cup A = A$

$A \cap A = A$

Identify: $A \cup \varnothing = A$

$A \cap E = A$

$A \cap \varnothing = \varnothing$

$A \cup E = E$

Law of Absorption: $A \cup (A \cap B) = A$

$A \cap (A \cup B) = A$

Transitivity: If $A \subseteq B$, $B \subseteq C$ then $A \subseteq C$

Involution: $(A^c) = A$

Law of the Excluded Middle: $A \cup A^c = E$

Law of Contradiction: $A \cap A^c = \phi$

De Morgan's laws: $(A \cup B)^c = A^c \cap B^c$

$(A \cap B)^c = A^c \cup B^c$

I leave it to the readers to verify all the properties by means of Venn diagrams.

EXAMPLE 3.1.

Give three sets *A, B* and *C*. Prove De Morgan's laws Venn diagrams.

Solution: To prove De Morgan's laws, we need to show that

(*i*) $A \cup B \cup C)^c = A^c \cap B^c \cap C^c$

(*ii*) $A \cap B \cap C)^c = A^c \cap B^c \cap C^c$

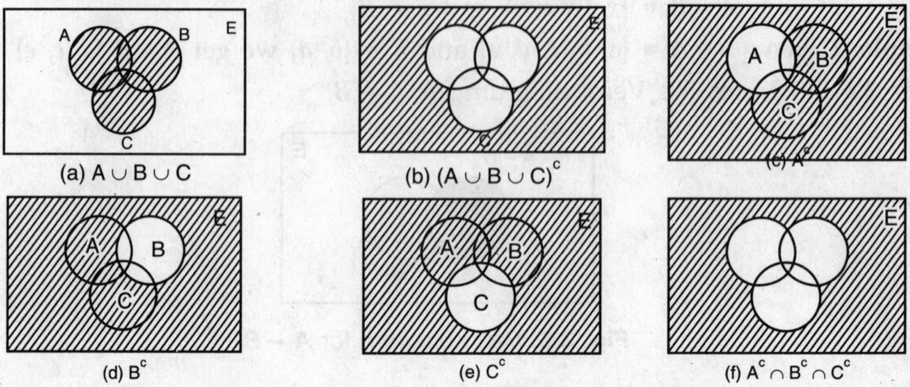

(a) $A \cup B \cup C$ (b) $(A \cup B \cup C)^c$ (c) A^c

(d) B^c (e) C^c (f) $A^c \cap B^c \cap C^c$

Fig. 3.6. Shows the Venn diagram for De Morgan's Law i.e., $(A \cup B \cup C)^c = A^c \cap B^c \cap C^c$

(a) $(A \cap B \cap C)^c$ (b) $A^c \cup B^c \cup C^c$

Fig. 3.7. Shows that $(A \cap B \cap C)^c = A^c \cup B^c \cup C^c$

EXAMPLE 3.2.

Let the sets, *A*, *B*, *C* and *E* be given as follows:

E = all students enrolled in the university cricket club.

A = male students,

B = offenders,

and C = defenders.

Draw individual Venn diagrams to illustrate (*a*) female student (*b*) offenders who cannot defend (*c*) female students who can both offened and defend.

Solution:

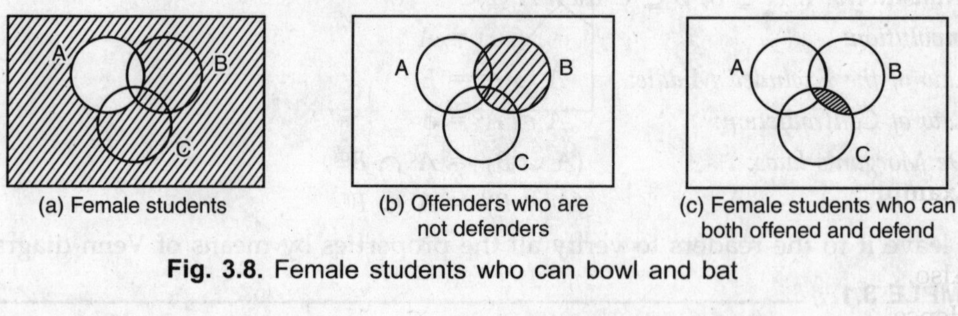

(a) Female students (b) Offenders who are not defenders (c) Female students who can both offened and defend

Fig. 3.8. Female students who can bowl and bat

EXAMPLE 3.3.

In Example 3.2, assume that $|E|$ - 600, $|A|$ = 300, $|B|$ = 225, $|C|$ = 160. Also, let the number of male students who are affenders ($A \cap B$) be 100, 25 of whom are defenders too ($A \cap B \cap C$), and the total number of male students who are batsmen ($A \cap C$) be 85.

Determine the number of students who are: (*i*) Females, (*ii*) Not bowlers (*iii*) Not batsmen, (*iv*) Female students who can bowl.

Solution: From the given data, the Venn diagram obtained is as follows:

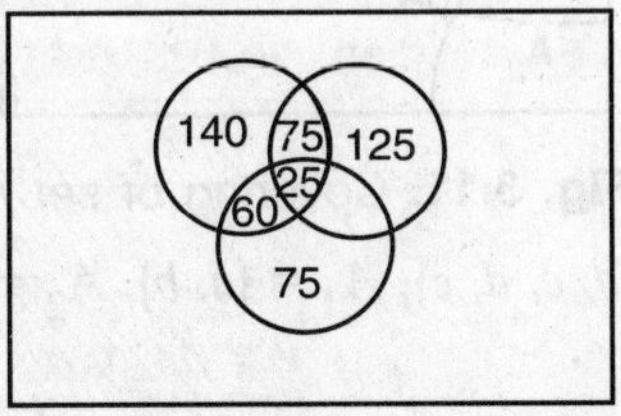

Fig. 3.9. Showing the Distribution of data

(*i*) No. of female students $|A^c| = |E| - |A| = 600 - 300 = 300$

(*ii*) No. of students who are no between $|B^c| = |E| - |B| = 600 - 225 = 375$

(*iii*) No. of students who are not batsmen $|C^c| = |E| - |C| = 600 - 160 = 440$

(*iv*) No. of female students who can bowl $|A^c \cap B| = 125$ (from the Venn diagram given in 3.9)

3.2.4. Partition and Covering

Partition

A *partition* on *A*is defined to be a set of non-empty subsets A_i, each of which is pair wise disjoint and whose union yields the original set *A*.

Partition on *A* indicated as $\pi(A)$, is therefore

(*i*) $A_i \cap A_j = \phi$ for each pair $(l, j) \in /, i \neq j$

(*ii*) $\cup_{\, i\, \hat{\imath}\, l} A_i = A$

The members A_i of the portion are known as block (refer Fig. 3.10).

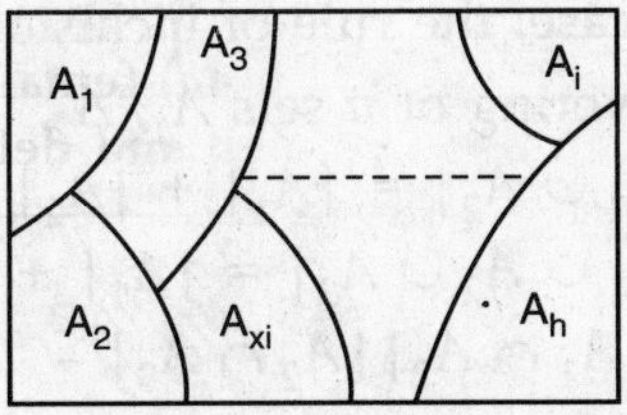

Fig. 3.10. Partition of set *A*

Example: Given, $A = \{a, b, c, d, e\}$, $A_1 = \{a, b\}$, $A_2 = \{c, d\}$ and $A_3 = \{e\}$, which gives

$$A_1 \cap A_2 = \phi,\ A_1 \cap A_3 = \phi,\ A_2 \cap A_3 = \phi$$

Also, $A_1 \cup A_2 \cup A_3 = A = \{a, b, c, d, e\}$

Hence, $\{A_1, A_2, A_3\}$, is a portion of *A*.

Covering

A covering on A is defined to be a set of non-empty subsets A_i, whose union yields the original set A. The non-empty subsets need not be disjoint (Fig. 3.11).

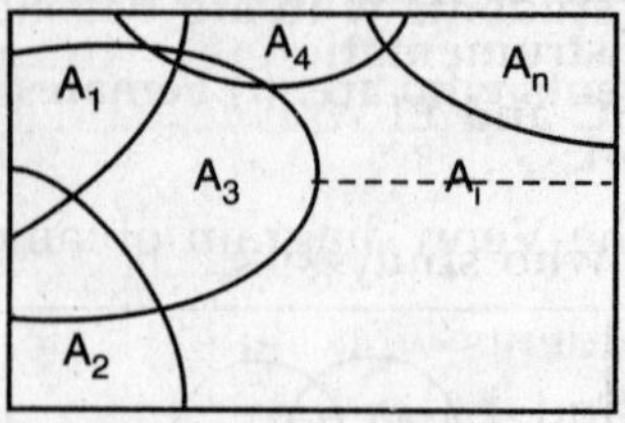

Fig. 3.11. Covering of set A

Example: Given, $A = \{a, b, c, d, e\}$, $A_1 = \{a, b\}$, $A_2 = \{b, c, d\}$ and $A_3 = \{d, e\}$. This gives

$$A_1 \cap A_2 = \{b\}$$
$$A_1 \cap A_3 = \phi$$
$$A_2 \cap A_3 = \{d\}$$

Also, $A_1 \cup A_2 \cup A_3 = \{a, b, c, d, e\} = A$

Hence, $\{A_1, A_2, A_3\}$, is a covering of A.

Rule of Addition

Given a portion on A where A_i, $l = 1, 2, ..., n$ are its non-empty subsets then,

$$|A| = \left|\bigcup_{i=1}^{n} A_i\right| = \sum\nolimits_{i=1}^{n} |A_i|$$

Example: Given $A = \{a, b, c, d, e\}$, $A_1 = \{a, b\}$, $A_2 = \{c, d\}$ and $A_3 = \{e\}$, $|A| = 5$,

and $\sum\nolimits_{i=1}^{n} |A_i| = 2 + 2 + 1 = 5$

Rule of Inclusion and Exclusion

Rule of addition is not applicable on the covering of set A, especially if the subsets are not pair wise disjoint. In such a case, the rule of inclusion and exclusion is applied.

Example: Given A to be a covering of n sets $A_1, A_2, ..., A_n$.

For $n = 2$, $|A| = |A_1 \cup A_2| = |A_1| + |A_2| - |A_1 \cap A_2|$

For $n = 3$, $|A| = |A_1 \cup A_2 \cup A_3| = |A_1| + |A_2| + |A_3|$
$- |A_1 \cap A_2| |A_2 \cap A_3| - |A_1 \cap A_3| + |A_1 \cap A_2 \cap A_3|$

Generalizing,
$$|A| = \left|\bigcup_{i=1}^{n} A_i\right| = \sum_{i=1}^{n} |A_i| - \sum_{i=1_{i \neq j}}^{n} \sum_{j=1}^{n} |A_i \cap A_j|$$

$$+ \sum_{i=1_{i \neq j \neq k}}^{n} \sum_{j=1}^{n} \sum_{k=1}^{n} |A_i \cap A_j \cap A_k| \; ... \; (-i)^{n+1} \left|\bigcup_{i=1}^{n} A_i\right|$$

EXAMPLE 3.4.

Give $|E|$ =100, where E indicates a set of students who have chosen subjects from different streams in the computer science discipline, if is found that 32 study subjects chosen from the Electronics Engineering (EE) stream, 20 from the Electronics and Communication (EC) stream, and 45 from the Electronics and instrumentation (EI) stream. Also, 15 study subjects from both EE and EI streams, 7 from both EC and EI streams, and 30 do not study any subjects chosen from either of the three streams.

Find the number of students who study subjects belonging to all three streams.

Solution: Let A, B, C indicate students who study subjects chosen from EE, EC, EI streams respectively. The problem is to find $|A \cap B \cap C|$.

The no. of students who do not study any subject chosen from either of the three streams = 30.

i.e., $|A^c \cap B^c \cap C^c| = 30$

$\Rightarrow$ $|\overline{A \cup B \cup C}| = 30$ (using De Morgan's laws)

$\Rightarrow$ $|E| - |A \cup B \cup C| = 30$

$\Rightarrow$ $|A \cup B \cup C| = |E| - 30$

$= 100 - 30$

$= 70$

From the principle of inclusion and exclusion.

$$|A \cup B \cup C| = |A| + |B| + |C| - |A \cap B| - |B \cap C| - |A \cap C| + |A \cap B \cap C|$$

$$\Rightarrow |A \cap B \cap C| = |A \cup B \cup C| - |A| - |B| - |C| + |A \cap B| + |B \cap C| + |A \cap C|$$

$$= 70 - 32 - 20 - 45 + 15 + 7 + 10$$

$$= 5$$

Hence, the no. of students who study subjects' chosen from all the three streams is 5.

3.3. FUZZY SETS: THEORY AND OPERATIONS

Fuzzy sets support a flexible sense of membership of elements to a set. While in crisp set theory, an element either belongs to or does not belong to a set, in fuzzy set theory many degrees of membership (between 0 and 1) are allowed. Thus, a membership function $m_A(x)$ is associated with a fuzzy set A such that the function maps every element of the universe of discourse X (or the reference set) to the interval [0, 1].

Formally, the mapping is written as $\mu_A(x)$: $X \rightarrow [0, 1]$

A fuzzy set is defined as follows:

If X is universe of discourse and x is a particular element of X, then a fuzzy set A defined on X may be written as a collection of ordered pairs.

$$A = \{(x, \mu_A(x)), x \hat{I} X\}$$

Where each pair $(x, \mu_A(x))$ is called a singleton. In crisp sets, $\mu_A(x)$ is dropped.

An alternative definition which indicates a fuzzy set as a union of all $\mu_A(x)/x$ singletons is given by $A = \sum \mu_k \frac{(x)}{x_i}$ in the discrete case.

And $A = \int \mu_A \frac{(x)}{x_i}$ in the continuous case.

Here, the summation and integration signs indicate the union of all $\mu_A(x)/x$ singletons.

Example: Let $X = \{g_1, g_2, g_3, g_4, g_5\}$ be the reference set of students. Let A be the fuzzy set of "smart" students where "smart" is a fuzzy linguistic term.

$$A = \{g_1, 0.4\} \{g_2, 0.5\} \{g_3, 1\} \{g_4, 0.9\} \{g_5, 0.8\}$$

Here A indicates that the smartness of g_1 is 0.4, g_2 is 0.5 and so on when graded over a scale of 0-1.

Though fuzzy sets model vagueness, it needs to be realize that the definition of the sets varies according to the context in which it is used. Thus, the fuzzy linguistic term "tall" could have one kind of fuzzy set while refering to the height of a building and another kind of fuzzy set while referring to the height of human beings.

3.3.1. Membership Function

The membership function values need not always be described by discrete values. Quite often, these turn out to be as described by a continuous function.

The fuzzy membership function for the fuzzy linguistic term "cool" relating to temperature may turn out to be as illustrated in Fig. 3.12.

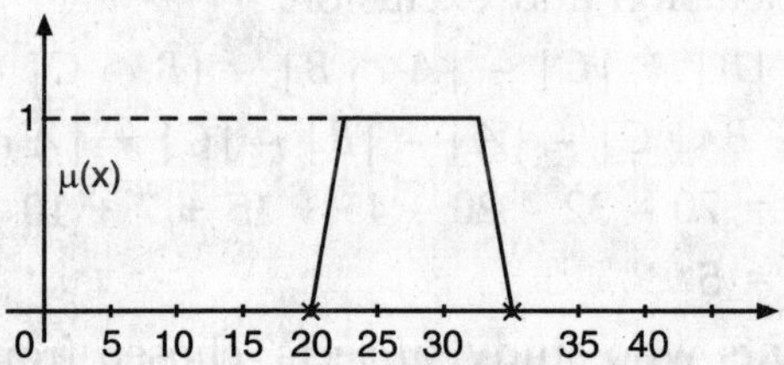

Fig. 3.12. Continuous membership functions for "cool"

A membership function can also be given mathematically as

$$\mu_A(x) = \frac{1}{(1+x)^2}$$

The graph is as shown in Fig. 3.13.

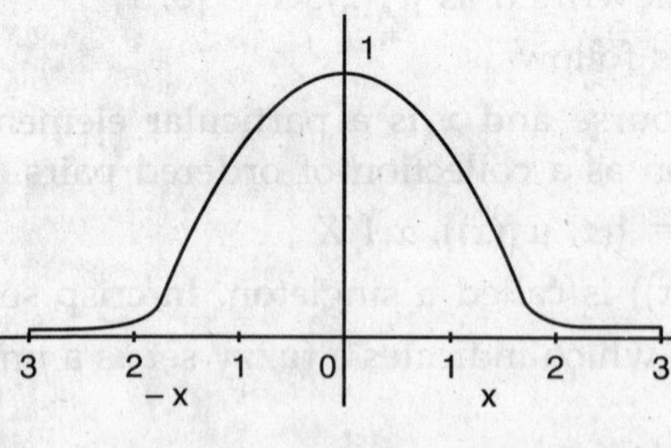

Fig. 3.13. Continuous Gaussian membership function dictated by a mathematical function

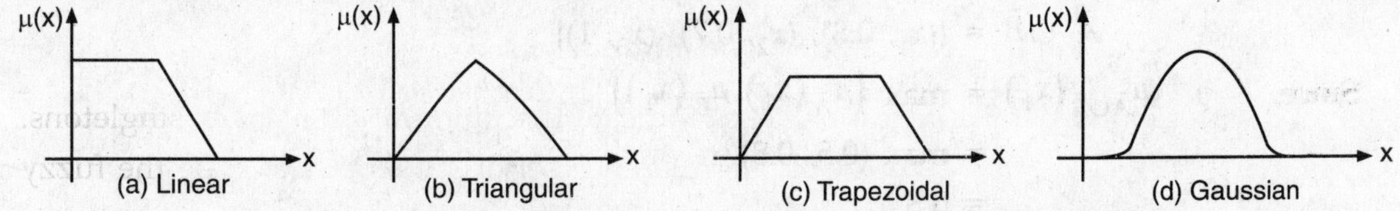

Fig. 3.14. Different shapes of membership function graphs

Example: Consider the set of people in the following age groups:

0-10 40-50
10-20 50-60
20-30 60-70
30-40 70 and above.

The fuzzy sets "young", "middle-aged", and "old" are represented by the membership function graphs as illustrated in Fig. 3.15.

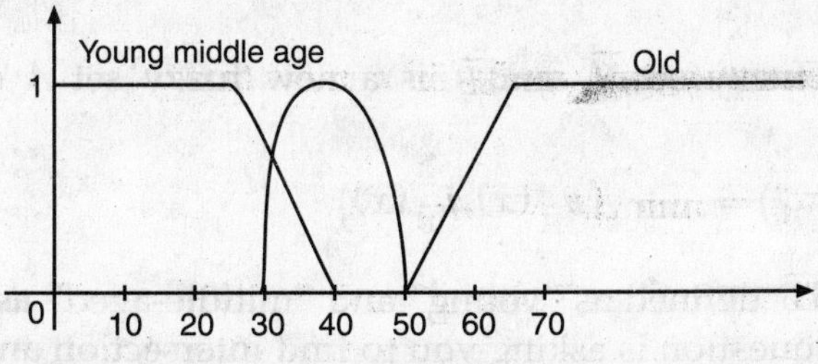

Fig. 3.15. Example of fuzzy sets expressing "young", "middle-aged", and "old"

Please note here, that 35 and even 39 and 40 falls in young and middle age as well, which was never possible with crisp sets.

3.3.2. Basic Fuzzy Set Operation

Give X to be the universe of discourse and $\tilde{A}$ and $\tilde{B}$ is a new fuzzy set $A \cup B$ also on X with a membership function defined as

$$\mu_{\tilde{A} \cup \tilde{B}}(x) = \max\left(\mu_{\tilde{A}}(x), \mu_{\tilde{B}}(x)\right)$$

Example: Let A be the fuzzy set of young people and B the fuzzy set of middle-aged people as illustrated in Fig. 3.16.

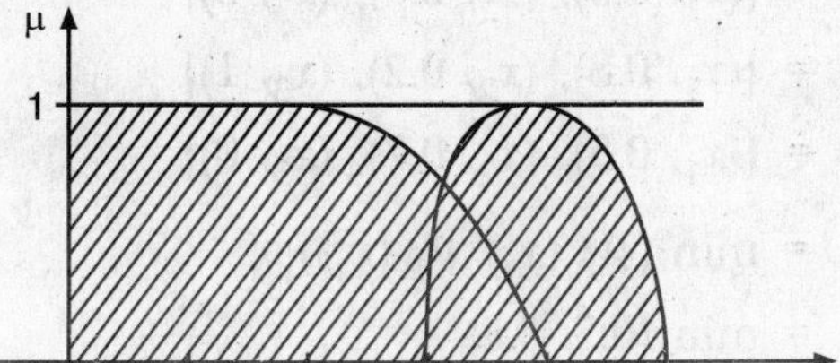

Fig. 3.16. $A \cup B$, the fuzzy set of "young or middle-aged"

In its discrete form, for x_1, x_2, x_3

If $\tilde{A} = \{(x_1, 0.5), (x_2, 0.7), (x_3, 0)\}$ and $\tilde{B} = \{(x_1, 0.8), (x_2, 0.2), (x_3, 1)\}$

$$\tilde{A} \cup \tilde{B} = \{(x_1, 0.8), (x_2, 0.7), (x_3, 1)\}$$

Since,
$$\mu_{\tilde{A}\cup\tilde{B}}(x_1) = \max\left(\mu_{\tilde{A}}(x_1), \mu_{\tilde{B}}(x_1)\right)$$
$$= \max(0.5, 0.8)$$
$$= 0.8$$

$$\mu_{\tilde{A}\cup\tilde{B}}(x_2) = \max\left(\mu_{\tilde{A}}(x_2)\right)$$
$$= \max(0.2, 0.7)$$
$$= 0.7$$

$$\mu_{\tilde{A}\cup\tilde{B}}(x_2) = \max\left(\mu_{\tilde{A}}(x_3), \mu_{\tilde{B}}(x_3)\right)$$
$$= \max(0, 1)$$
$$= 1$$

Intersection

The intersection of fuzzy sets $\tilde{A}$ and $\tilde{B}$ is a new fuzzy set $A \cap B$ with membership function defined as

$$\mu_{\tilde{A}\cap\tilde{B}}(x_1) = \min\left(\mu_{\tilde{A}}(x), \mu_{\tilde{B}}(x)\right)$$

Example: For $\tilde{A}$ and $\tilde{B}$ defined as "young" and "middle-aged" as illustrated in previous examples. Actually, the question is asking you to find intersection and this is a how it is has to be represented.

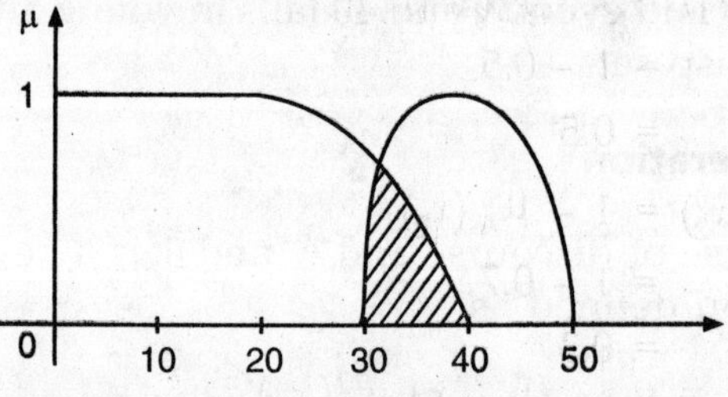

Fig. 3.17. Young and Middle aged group

In it's discrete form, for x_1, x_2, x_3

If
$$\tilde{A} = \{(x_1, 0.5), (x_2, 0.7), (x_3, 0)\}$$
$$\tilde{B} = \{(x_1, 0.8), (x_2, 0.2), (x_3, 1)\}$$
$$\tilde{A} \cap \tilde{B} = \{(x_1, 0.5), (x_2, 0.7), (x_3, 0)\}$$

Since,
$$\mu_{\tilde{A}\cap\tilde{B}}(x_1) = \min(\mu_{\tilde{A}}(x_1)\ \mu_{\tilde{B}}(x_1))$$
$$= \min(0.5, 0.8)$$
$$= 0.5$$

$$\mu_{\tilde{A}\cap\tilde{B}}(x_2) = \min(\mu_{\tilde{A}}(x_2)\ \mu_{\tilde{B}}(x_2))$$
$$= \min(0.7, 0.2)$$
$$= 0.2$$

$$\mu_{\tilde{A}\cap\tilde{B}}(x_3) = \min\ (\mu_{\tilde{A}}(x_3)\ \mu_{\tilde{B}}(x_3))$$
$$= \min\ (0, 1)$$
$$= 0$$

Complement

The complement of a fuzzy set $\tilde{A}$ is a new fuzz set $\tilde{A}$ with a membership function

$$\mu_{\tilde{A}}(x) = 1 - \mu_{\tilde{A}\cap\tilde{B}}(x)$$

Example: For the fuzzy set $\tilde{A}$ defined as "young" the complement "not young" is given b $\tilde{A}^c$. In its discrete form, for x_1, x_2 and x_3; it is drawn as given in Fig. 3.18.

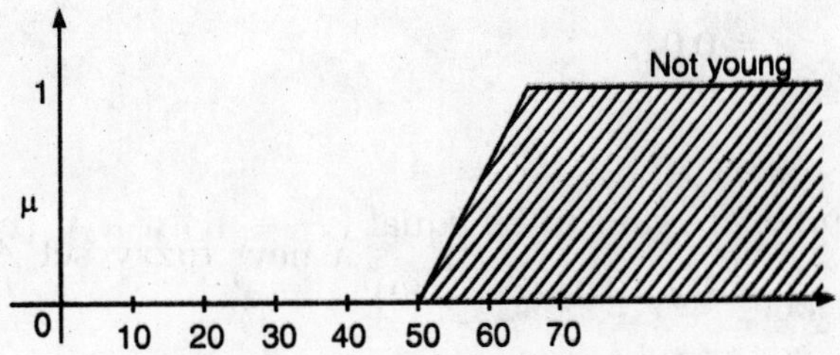

Fig. 3.18. Set of 'Not young'

$$\tilde{A} = \{(x_1, 0.5), (x_2, 0.7), (x_3, 0)\}$$
$$\tilde{A}^c = \{(x_1, 0.5), (x_2, 0.3), (x_3, 1)\}$$
$$\mu_{\tilde{A}c}(x_1) = 1 - \mu_{\tilde{A}}(x_1)$$
$$= 1 - 0.5$$
$$= 0.5$$
$$\mu_{\tilde{A}c}(x_2) = 1 - \mu_{\tilde{A}}(x_2)$$
$$= 1 - 0.7$$
$$= 0.3$$
$$\mu_{\tilde{A}c}(x_3) = 1 - \mu_{\tilde{A}}(x_3)$$
$$= 1 - 0$$
$$= 1$$

Other operations are,

Product of two fuzzy sets

The product of two fuzzy sets $\tilde{A}$ and $\tilde{B}$ is a new fuzzy sets $\tilde{A}\cdot\tilde{B}$ whose membership function is defined as

$$\mu_{\tilde{A}.\tilde{B}}(x) = \mu_{\tilde{A}}(x)\ \mu_{\tilde{B}}(x)$$

Example: $\tilde{A} = \{(x_1, 0.2), (x_2, 0.8), (x_3, 4)\}$

$$\tilde{B} = \{(x_1, 0.4), (x_2, 0), (x_3, 1)\}$$

$$\tilde{A} \cdot \tilde{B} = \{(x_1, 0.8), (x_2, 0), (x_3, 0.4)\}$$

Since, $\mu_{\tilde{A}.\tilde{B}}(x_1) = \mu_{\tilde{A}}(x_1) \,.\, \mu_{\tilde{B}}(x_2)$

$= (0.2) \,.\, (0.4)$

$= 0.08$

$\mu_{\tilde{A}.\tilde{B}}(x_2) = \mu_{\tilde{A}}(x_2) \,.\, \mu_{\tilde{B}}(x_2)$

$= (0.8) \,.\, (0)$

$= 0$

$\mu_{\tilde{A}.\tilde{B}}(x_3) = \mu_{\tilde{A}}(x_3) \,.\, \mu_{\tilde{B}}(x_3)$

$= (0.4) \,.\, (0.1)$

$= 0.04$

Equality

Two fuzzy sets $\tilde{A}$ and $\tilde{B}$ are said to be equal ($\tilde{A} = \tilde{B}$) if $\mu\tilde{A}\ (x) = \mu\tilde{B}(x)$

Example: $\tilde{A} = \{(x_1, 0.2)\ (x_2, 0.8)\}$

$\tilde{B} = \{(x_1, 0.6)\ (x_2, 0.8)\}$

$\tilde{C} = \{(x_1, 0.2)\ (x_2, 0.8)\}$

$\tilde{A} \neq \tilde{B}$

Since $\mu\tilde{A}\ (x_1) = \mu\tilde{B}(x_1)$ although

$\mu\tilde{A}\ (x_2) = \mu\tilde{B}(x_2)$

But $\tilde{A} \neq \tilde{C}$

Since $\mu\tilde{A}\ (x_1) = \mu\tilde{C}(x_1) = 0.2$

and $\mu\tilde{A}\ (x_2) = \mu\tilde{B}(x_2) = 0.8$

Product of a fuzzy set with a crisp number

Multiplying a fuzzy set $\tilde{A}$ by a crisp number a results in a new fuzzy set product $a.\ \tilde{A}$ with the membership function

$$\mu_{a\,.\,\tilde{A}}\ (x) = a \,.\, \mu_{\tilde{A}}\ (x)$$

Example: $\tilde{A} = \{(x_1, 0.4), (x_2, 0.6), (x_3, 0.8)\}$

For $a = 0.3$

$a\ .\ \tilde{A} = \{(x_1, 0.12), (x_2, 0.18), (x_3, 0.24)\}$

Since, $\mu_{a\,.\,\tilde{A}}\ (x_1) = a \,.\, \mu\tilde{A}\ (x_1)$

$= 0.3 \,.\, 0.4$

$= 0.12$

$\mu_{a\,.\,\tilde{A}}\ (x_2) = a \,.\, \mu\tilde{A}\ (x_2)$

$$= 0.3 \cdot 0.6$$
$$= 0.18$$

$$\mu_{a \cdot \tilde{A}}(x_3) = a \cdot \mu \tilde{A}(x_3)$$
$$= 0.3 \cdot 0.8$$
$$= 0.24$$

Power of a fuzzy set

The a power of a fuzzy set $\tilde{A}$ is a new fuzzy set A^α whose membership function is given by

$$\mu_{Aa}(x) = (\mu_{\tilde{A}}(x))^\alpha$$

Raising a fuzzy set to its second power is called concentration (CON) and taking the square root is called Delation (DIL).

Example: $\tilde{A} = \{(x_1, 0.4), (x_2, 0.2), (x_3, 0.7)\}$

For $\alpha = 2$

$$\mu_{\tilde{A}^2}(x) = (\mu_{\tilde{A}^2}(x))^2$$

Hence, $(\tilde{A})^2 = \{(x_1, 0.16), (x_2, 0.04), (x_3, 0.49)\}$

Since, $\mu_{\tilde{A}^2}(x_1) = (\mu_{\tilde{A}}(x_1))^2 = (0.4)^2 = 0.16$

$$\mu_{\tilde{A}^2}(x_2) = (\mu_{\tilde{A}}(x_2))^2 = (0.2)^2 = 0.04$$
$$\mu_{\tilde{A}^3}(x_3) = (\mu_{\tilde{A}}(x_3))^2 = (0.7)^2 = 0.49$$

Difference

The difference of two fuzzy sets $\tilde{A}$ and $\tilde{B}$ is a new fuzzy set $\tilde{A} - \tilde{B}$ defined as

$$\tilde{A} - \tilde{B} = (\tilde{A} \cap \tilde{B}^c)$$

Example: $\tilde{A} = \{(x_1, 0.2), (x_2, 0.5), (x_3, 0.6)\}$;

$$\tilde{B} = \{(x_1, 0.1), (x_2, 0.4), (x_3, 0.5)\}$$
$$\tilde{B}^c = \{(x_1, 0.9), (x_2, 0.6), (x_3, 0.5)\}$$
$$\tilde{A} - \tilde{B} = \tilde{A} \cap \tilde{B}^c$$
$$= \{(x_1, 0.2), (x_2, 0.5), (x_3, 0.5)\}$$

Disjunctive Sum

The disjunctive sum of the fuzzy sets $\tilde{A}$ and $\tilde{B}$ is a new fuzzy set $\tilde{A} \oplus \tilde{B}$ defined as

$$\tilde{A} \oplus \tilde{B} = (\tilde{A} \cap \tilde{B}) \cup (\tilde{A} \cap \tilde{B}^c)$$

Example: $\tilde{A} = \{(x_1, 0.4), (x_2, 0.8) (x_3, 0.6)\}$

$$\tilde{B} = \{(x_1, 0.2), (x_2, 0.6) (x_3, 0.9)\}$$

Now, $\tilde{A}^c = \{(x_1, 0.6), (x_2, 0.2) (x_3, 0.4)\}$

$$\tilde{B}^c = \{(x_1, 0.8), (x_2, 0.4) (x_3, 0.1)\}$$

$$\tilde{A}^c \cap \tilde{B}^c = \{(x_1, 0.2), (x_2, 0.2)\ (x_3, 0.4)\}$$
$$\tilde{A} \cap \tilde{B}^c = \{(x_1, 0.4), (x_2, 0.4)\ (x_3, 0.1)\}$$
$$\tilde{A} \oplus \tilde{B} = \{(x_1, 0.4), (x_2, 0.4)\ (x_3, 0.4)\}$$

3.4. PROPERTIES OF FUZZY SETS

Fuzzy sets follow some of the properties satisfied by crisp sets. In fact, crisp sets can be thought of as special instances of fuzzy sets. Any fuzzy set $\tilde{A}$ is a subset of the refreshment X. Also, the membership of any element belonging to the null set ϕ is 0 and the membership of any element belonging to the reference set is 1.

The properties satisfied by fuzzy sets are:

Commutativity: $\tilde{A} \cup \tilde{B} = \tilde{B} \cup \tilde{A}$

$\tilde{A} \cap \tilde{B} = \tilde{B} \cap \tilde{A}$

Associativity: $\tilde{A} \cup (\tilde{B} \cup \tilde{C}) = (\tilde{A} \cup \tilde{B}) \cup \tilde{C}$

$\tilde{A} \cap (\tilde{B} \cap \tilde{C}) = (\tilde{A} \cap \tilde{B}) \cap \tilde{C}$

Distributivity: $\tilde{A} \cup (\tilde{B} \cap \tilde{C}) = (\tilde{A} \cup \tilde{B}) \cap (\tilde{A} \cup \tilde{C})$

$\tilde{A} \cap (\tilde{B} \cup \tilde{C}) = (\tilde{A} \cap \tilde{B}) \cup (\tilde{A} \cap \tilde{C})$

Idempotence: $\tilde{A} \cup \tilde{A} = \tilde{A}$

$\tilde{A} \cap \tilde{A} = \tilde{A}$

Identity: $\tilde{A} \cup \phi = \tilde{A}$

$\tilde{A} \cup X = \tilde{A}$

$\tilde{A} \cap \phi = \phi$

$\tilde{A} \cup X = X$

Transitivity: If $\tilde{A} \subseteq \tilde{B} \subseteq \tilde{C}$, then $\tilde{A} \subseteq \tilde{C}$

Involution: $(\tilde{A})^c = \tilde{A}$

De Morgan's laws: $(\tilde{A} \cap)^c = (\tilde{A}^c \cup \tilde{B}^c)$

$(\tilde{A} \cup)^c = (\tilde{A}^c \cap \tilde{B}^c)$

Since fuzzy sets can overlap, the laws of excluded middle do not hold good.

Thus, $\tilde{A} \cup \tilde{A}^c \neq X$

$\tilde{A} \cap \tilde{A}^c \neq \phi$

EXAMPLE 3.5.

The task is to recognize English alphabetical characters (*F, E, x, Y, I, T*) in an image processing system. Define two fuzzy sets $\tilde{I}$ and $\tilde{F}$ to represent the identification of characters *I* and *F*.

$$\tilde{I} = \{(F, 0.4), (E, 0.3), (X, 0.1), (Y, 0.1), (I, 0.9), (T, 0.8)\}$$
$$\tilde{F} = \{(F, 0.99), (E, 0.8), (X, 0.1), (Y, 0.2), (I, 0.5), (T, 0.5)\}$$

Find the following:

(*a*) (*i*) $\tilde{I} \cup \tilde{F} = \{(F, 0.99), (E, 0.8), (X, 0.1), (Y, 0.2), (I, 0.9), (T, 0.8)\}$

(*ii*) $\tilde{I} - \tilde{F} = (\tilde{I} \cap \tilde{F}^c)$

$= \{(F, 0.01), (E, 0.2), (X, 0.1), (Y, 0.1), (I, 0.5), (T, 0.5)\}$

(*iii*) $\tilde{F} \cup \tilde{F}^c = \{(F, 0.09), (E, 0.8), (X, 0.9), (Y, 0.8), (I, 0.5), (T, 0.5)\}$

(*b*) De Morgan's law:

$$(\tilde{I} \cup \tilde{F})^c = \tilde{I}^c \cap \tilde{F}^c$$

$$\tilde{I} \cup \tilde{F} = \{(F, 0.09), (E, 0.8), (X, 0.1), (Y, 0.2), (I, 0.9), (T, 0.8)\}$$

$$(\tilde{I} \cup \tilde{F})^c = \{(F, 0.01), (E, 0.2), (X, 0.9), (Y, 0.8), (I, 0.1), (T, 0.2)\}$$

$$\tilde{I}^c = \{(F, 0.06), (E, 0.7), (X, 0.9), (Y, 0.9), (I, 0.1), (T, 0.2)\}$$

$$\tilde{F}^c = \{(F, 0.01), (E, 0.2), (X, 0.9), (Y, 0.8), (I, 0.5), (T, 0.5)\}$$

and $\tilde{I}^c \cap \tilde{F}^c = \{(F, 0.01), (E, 0.2), (X, 0.9), (Y, 0.8), (I, 0.1), (T, 0.2)\}$

Hence, $(\tilde{I} \cup \tilde{F})^c = \tilde{I}^c \cap \tilde{F}^c$

EXAMPLE 3.6.

Consider the fuzzy sets $\tilde{A}$ and $\tilde{B}$ defined on the interval $X = [0, 5]$ of real numbers, by the membership grade functions.

$$\mu_{\tilde{A}}(x) = \frac{x}{x+1}, \quad \mu_{\tilde{B}}(x) = 2^{-x}$$

Determine the mathematical formulae and graphs of the membership grade functions of each of the following sets.

(*a*) A^c, B^c

(*b*) $A \cup B$

(*c*) $A \cap B$

(*d*) $(A \cup B)^c$

Solution: (*a*) $\mu_{\tilde{A}^c}(x) = 1 - \mu_{\tilde{A}}(x) = 1 - \dfrac{x}{x+1} = \dfrac{1}{x+1}$

$$\mu_{\tilde{B}^c}(x) = 1 - \mu_{\tilde{B}}(x) = 1 - 2^{-x} = \frac{2^{x-1}}{2^x}$$

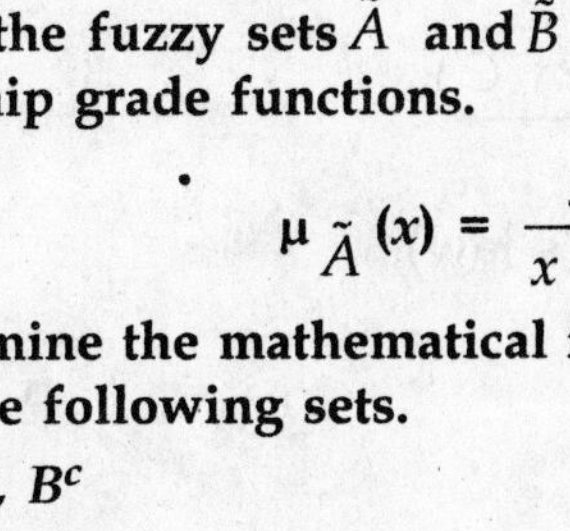

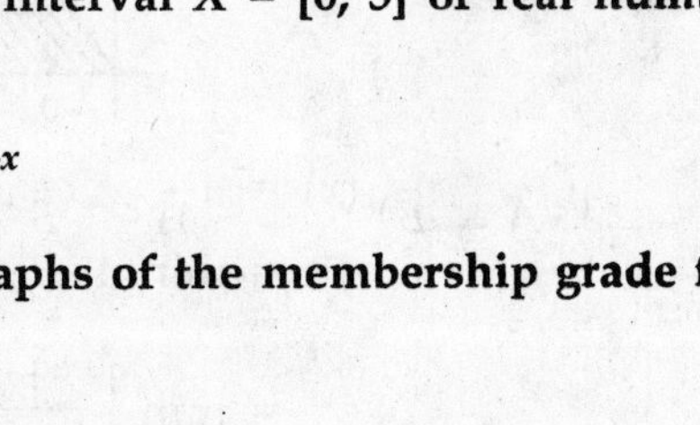

(*b*) $\mu_{\tilde{A} \cup \tilde{B}}(x) = \max(\mu_{\tilde{A}}(x), \mu_{\tilde{B}}(x))$

$$= \max\left(\frac{x}{x+1}, 2^{-x}\right)$$

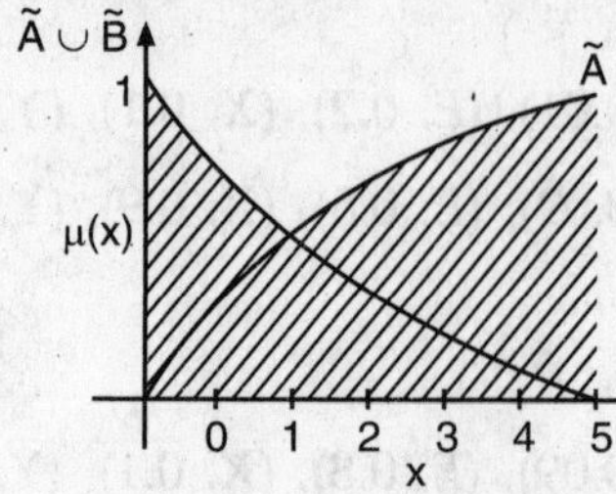

(c) $\mu_{\tilde{A} \cap \tilde{B}}(x) = \min(\mu_{\tilde{A}}(x), \mu_{\tilde{B}}(x))$

$$= \min\left(\frac{x}{x+1}, 2^{-x}\right)$$

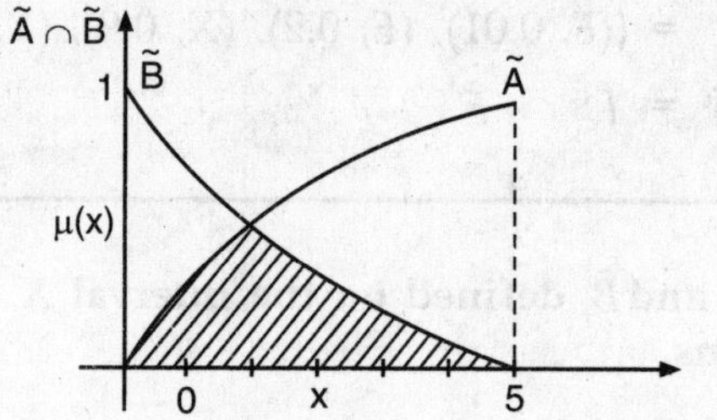

(d) $\mu_{(\tilde{A} \cup \tilde{B})^c}(x) = \mu_{\tilde{A}^c \cap \tilde{B}^c}(x)$ (De Morgan's law)

$$= \min(\mu_{\tilde{A}^c}(x), \mu_{\tilde{B}^c}(x))$$

$$= \min\left(\frac{1}{x+1}, \frac{2^x - 1}{2^x}\right)$$

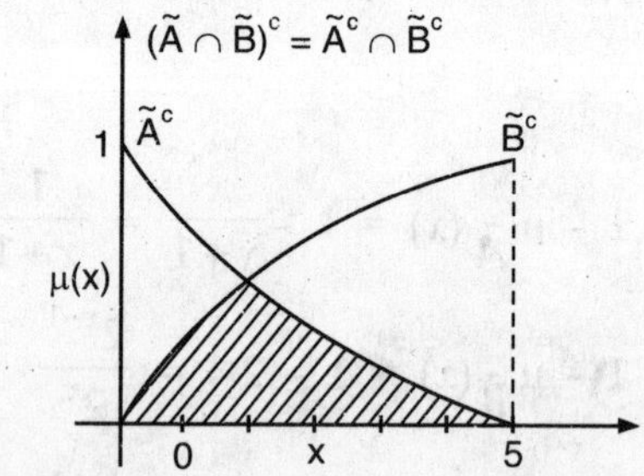

3.5. CRISP AND FUZZY RELATION

In this section, we review crisp relation as a prelude to fuzzy relations. The concept of relations between sets is built on the Cartesian product operator of sets.

3.5.1. Crisp Relation

Cartesian Product

The Cartesian product of two sets A and B denoted by $A \times B$ is the set of all ordered pairs such that the first element in the pair belongs to A and the second element belongs to B.

i.e., $A \times B = \{(a, b) / a \in A, b \in B\}$

If $A \neq B$ and A and B are non-empty then $A \times B \neq B \times A$.

The Cartesian product could be extended to n number of sets.

$$\mathop{X}_{i=1}^{n} A_i = \{(a_1, a_2, a_3, ..., a_n) / a_i \,\delta\, A_i \text{ for every } i = 1, 2, ..., n)$$

Observe that $$\left|\mathop{X}_{i=1}^{n} A_i\right| = \prod_{i=1}^{n} |A_i| \quad ... (6.46)$$

EXAMPLE 3.7.

Given $A_1 = \{a, b\}, \quad A_2 = \{1, 2\}, \quad A_3 = \{a\}$

$A_1 \times A_2 = \{(a, 1), (b, 1), (a, 2), (b, 2), |A_1 \times A_2| = 4$ and $(A_1) = |A_2| = 2$

Here, $|A_1 \times A_2| = |A_1| . |A_2|$

Also, $A_1 \times A_2 \times A_3 = \{(a, 1, a), (a, 2, a), (b, 1, a), (b, 1, a), (b, 2, a)\}$

$|A_1 \times A_2 \times A_3| = 4 = |A_1| . |A_2| . |A_3|$

Other Crisp Relations: An n-ary relation denoted as $R(X_1, X_2, ..., X_n)$ among crisp sets $X_1, X_2, ..., X_n$ is a subset of the Cartesian product $\mathop{X}_{i=1}^{n} X_i$ and is indicative of an association or relation among the tuple elements.

For $n = 2$, the relation $R(X_1, X_2)$ is termed as a binary relation; for $n = 3$, the relation is termed ternary; for $n = 4$, quarternary for $n = 5$, quinary and so on.

If the universe of discourse of sets are finite, the n-ary relation can be expressed as an n-dimensional relation matrix. Thus, for a binary relation $R(X, Y)$ where $X = \{x_1, x_2, ..., x_n\}$ and $Y = \{y_1, y_2, ..., y_n\}$, the relation matrix R is a two dimensional matrix where X represents the rows, Y represents the columns and $R(i, y) = 1$ if (xi, yj) ÎR and $R(i, j) = 0$ if (xi, yj) Ï R.

EXAMPLE 3.8.

Given $X = \{1, 2, 3, 4\}$,

$X \times X = \{(1, 1) (1, 2), (1, 3) (1, 4) (2, 1), (2, 2), (2, 3), (2, 4), (3, 1), (3, 2), (3, 3), (3, 4), (4, 1), (4, 2), (4, 3), (4, 4)\}$

Let the rotation R is defined as $R = \{(x, y) / y = x + 1, x, y \in X\}$

$R = \{(1, 2), (2, 3), (3, 4)\}$

The relation matrix R is given by

$$R = \begin{array}{c} \\ 1 \\ 2 \\ 3 \\ 4 \end{array}\!\!\begin{array}{c} \begin{matrix} 1 & 2 & 3 & 4 \end{matrix} \\ \begin{bmatrix} 0 & 1 & 0 & 0 \\ 0 & 0 & 1 & 0 \\ 0 & 0 & 0 & 1 \\ 0 & 0 & 0 & 0 \end{bmatrix} \end{array}$$

Operations on Relations

Given two relations R and S defined on $X \times Y$ and represented by relation matrices, the following operations are supported by R and S.

Union: $R \cup S$ $\quad R \cup S(x, y) = \max(R(x, y), S(x, y))$

Intersection: $R \cap S$ $\quad R \cap S(x, y) = \min(R(x, y), S(x, y))$

Complement: $\overline{R}$ $\quad \overline{R}(x, y) = 1 - R(x, y)$

Composition of Relations: $R \,.\, S$

Given R to be relation on X, Y and S to be a relation on Y, Z then $R \,.\, S$ is a composition of relation on X, Z defined as:

$R \,.\, S = \{(x, z)/(x, z) \in X \times Z, \exists\, y \in Y \text{ such that } (x, y) \in R \text{ and } (y, z) \in S\}$

A common form of the composition relation is the max-min composition.

Max-min Composition: Given the relation matrices of the relation R and S, the max-min composition is defined as

For $\quad T = R \,.\, S$

$$T(x, z) = \max_{y \in Y} (\min R(x, y), S(y, z)))$$

EXAMPLE 3.9.

Let R, S be defined on the sets $\{1, 3, 5\} \times \{1, 3, 5\{$

Let $\quad R: \{(x, y) \,/\, y = x + 2),\quad S: \{(x, y) \,/\, x < y\}$

$R = \{(1, 3), (3, 5)\},\ S = \{(1, 3), (1, 5), (3, 5)\}$

The relation matrices are:

$$R: \begin{array}{c|ccc} & 1 & 3 & 5 \\ \hline 1 & 0 & 1 & 0 \\ 3 & 0 & 0 & 1 \\ 5 & 0 & 0 & 0 \end{array}$$

$$S: \begin{array}{c|ccc} & 1 & 3 & 5 \\ \hline 1 & 0 & 1 & 1 \\ 3 & 0 & 0 & 1 \\ 5 & 0 & 0 & 0 \end{array}$$

Using max-min composition:

$$R_0\, S: \begin{array}{c|ccc} & 1 & 3 & 5 \\ \hline 1 & 0 & 0 & 1 \\ 3 & 0 & 0 & 0 \\ 5 & 0 & 0 & 0 \end{array}$$

Since $\quad R_0\, S(1, 1) = \max\{\min(0, 0), \min(1, 0) \min(0, 0)\}$

$= \max(0, 0, 0) = 0$

$R_0\, S(1, 3) = \max\{0, 0, 0\} = 0$

$R_0\, S(1, 5) = \max\{0, 1, 0\} = 1$

Similarly, $\quad R_0\, S(3, 1) = 0$

$R_0\, S(3, 3) = R_0\, S(3, 5) = R_0\, S(5, 1) = R_0\, S(5, 3) = R_0\, S(5, 5) = 0$

R_0 S from the relation matrix is {(1, 5)}.

Also S_0 R:

$$\begin{array}{c|ccc} & 1 & 3 & 5 \\ \hline 1 & 0 & 0 & 1 \\ 3 & 0 & 0 & 0 \\ 5 & 0 & 0 & 0 \end{array}$$

3.5.2. Fuzzy Relations

Fuzzy relation is a fuzzy set defined on the Cartesian product of crisp sets $X_1, X_2, ..., X_n$ where the n-tupples $(x_1, x_2, ..., x_n)$ may have varying degree of membership within the relation. The membership values indicate the strength of the relation between the tupples.

EXAMPLE 3.10.

Let R be fuzzy relation between two sets X_1 and X_2 where X_1 is the set of desires and X_2 is the set of symtoms.

X_1 = {Typhoid, viral fever, common cold}

X_2 = {running nose, high temperature, shivering}

The fuzzy relation R may be defined as

	Running nose	High temperature	Shivering
Typhoid	0.1	0.9	0.8
Viral fever	0.2	0.9	0.7
Common cold	0.9	0.4	0.6

Fuzzy Cartesian Product

Let $\tilde{A}$ be a fuzzy set defined on the universe X and $\tilde{B}$ be a fuzzy set defined on the universe Y, the Cartesian product between the fuzzy set $\tilde{A}$ and $\tilde{B}$ indicated as $\tilde{A} \times \tilde{B}$ and resulting in a fuzzy relation $\tilde{R}$ is given by

$$\tilde{R} = \tilde{A} \times \tilde{B} \subset X \times Y$$

where R has its membership function given by

$$\mu \tilde{R}(x, y) = \mu \tilde{A} \times \tilde{B} (x, y)$$
$$= \min (\mu_{\tilde{A}}(x), \mu \tilde{B}(y))$$

EXAMPLE 3.11.

Let $\tilde{A} = \{(x_1, 0.2), (x_2, 0.7), (x_3, 0.4)\}$ and

$\tilde{B} = \{(y_1, 0.5), (y_2, 0.6)\}$

be two fuzzy sets defined on the universe of discourse $X = \{x_1, x_2, x_3\}$ and $Y = \{y_1, y_2\}$ respectively. Then the fuzzy relation $\tilde{R}$ resulting out of the fuzzy Cartesian product $\tilde{A} \times \tilde{B}$ is given by

$$\tilde{R} = \tilde{A} \times \tilde{B} = \begin{array}{c} \\ x_1 \\ x_2 \\ x_3 \end{array}\begin{array}{c} \begin{array}{cc} y_1 & y_2 \end{array} \\ \begin{bmatrix} 0.2 & 0.2 \\ 0.5 & 0.6 \\ 0.4 & 0.4 \end{bmatrix} \end{array}$$

Since, $\tilde{R}(x_1, y_1) = \min(\mu_{\tilde{A}}(x_1), \mu_{\tilde{B}}(y_1)) = \min(0.2, 0.5) = 0.2$

$\tilde{R}(x_1, y_2) = \min(0.2, 0.6) = 0.2$

$\tilde{R}(x_2, y_1) = \min(0.7, 0.5) = 0.5$

$\tilde{R}(x_2, y_2) = \min(0.7, 0.6) = 0.6$

$\tilde{R}(x_3, y_1) = \min(0.4, 0.5) = 0.4$

$\tilde{R}(x_3, y_2) = \min(0.4, 0.6) = 0.4$

3.5.3. Operations on Fuzzy Relations

Let $\tilde{R}$ and $\tilde{S}$ be fuzzy relations on $X \times Y$.

Union: $\mu_{\tilde{R} \cup \tilde{S}}(x, y) = \max(\mu_{\tilde{R}}(x, y), \mu_{\tilde{S}}(x, y))$

Intersection: $\mu_{\tilde{R} \cap \tilde{S}}(x, y) = \min(\mu_{\tilde{R}}(x, y), \mu_{\tilde{S}}(x, y))$

Complement: $\mu_{\tilde{R}^c}(x, y) = 1 - \mu_{\tilde{R}}(x, y), \mu_{\tilde{S}}(x, y))$

Composition of Relations

The definition is similar to that of crisp relation. Suppose $\tilde{R}$ is a fuzzy relation defined on $X \times Y$ and $\tilde{S}$ is a fuzzy relation defined on $Y \times Z$, then $\tilde{R}_0\tilde{S}$ is a fuzzy relation on $Y \times Z$. The fuzzy max-min composition is defined as:

$$\mu_{\tilde{R}\,0\,\tilde{S}}(x, z) = \max_{y \in Y} (\min(\mu_{\tilde{R}}(x, y), \mu_{\tilde{S}}(y, z)))$$

Example: $X = \{x_1, x_2, x_3\}$ $Y = \{y_1, y_2\}$ $Z = \{z_1, z_2, z_3\}$

Let $\tilde{R}$ be a fuzzy relation

$$\begin{array}{c} \\ x_1 \\ x_2 \\ x_3 \end{array}\begin{array}{c} \begin{array}{cc} y_1 & y_2 \end{array} \\ \begin{bmatrix} 0.2 & 0.1 \\ 0.2 & 0.9 \\ 0.8 & 0.6 \end{bmatrix} \end{array}$$

Let $\tilde{S}$ be a fuzzy relation

$$\begin{array}{c} \\ y_1 \\ y_2 \end{array}\begin{array}{c} \begin{array}{ccc} z_1 & z_2 & z_3 \end{array} \\ \begin{bmatrix} 0.6 & 0.4 & 0.7 \\ 0.5 & 0.8 & 0.9 \end{bmatrix} \end{array}$$

Then $\tilde{R}_0\tilde{S}$ by max-min composition yields

$$\tilde{R}_0\tilde{S} = \begin{array}{c} \\ x_1 \\ x_2 \\ x_3 \end{array}\begin{array}{c} \begin{array}{ccc} z_1 & z_2 & z_3 \end{array} \\ \begin{bmatrix} 0.5 & 0.4 & 0.5 \\ 0.5 & 0.8 & 0.9 \\ 0.6 & 0.6 & 0.7 \end{bmatrix} \end{array}$$

$$\begin{aligned} \mu_{\tilde{R}\,0\,\tilde{S}}(x_1, z_1) &= \max(\min(0.5, 0.6), \min(0.1, 0.5)) \\ &= \max(0.5, 0.1) \\ &= 0.5 \end{aligned}$$

$$\begin{aligned} \mu_{\tilde{R}\,0\,\tilde{S}}(x_1, z_2) &= \max(\min(0.5, 0.4), \min(0.1, 0.8)) \\ &= \max(0.4, 0.1) \\ &= 0.4 \end{aligned}$$

Similarly,

$$\begin{aligned} \mu_{\tilde{R}\,0\,\tilde{S}}(x_1, z_3) &= \max(0.5, 0.1) = 0.5 \\ \mu_{\tilde{R}\,0\,\tilde{S}}(x_2, z_1) &= \max(0.2, 0.5) = 0.5 \\ \mu_{\tilde{R}\,0\,\tilde{S}}(x_2, z_2) &= \max(0.2, 0.8) = 0.8 \\ \mu_{\tilde{R}\,0\,\tilde{S}}(x_2, z_3) &= \max(0.2, 0.9) = 0.9 \\ \mu_{\tilde{R}\,0\,\tilde{S}}(x_3, z_1) &= \max(0.6, 0.5) = 0.6 \\ \mu_{\tilde{R}\,0\,\tilde{S}}(x_3, z_2) &= \max(0.4, 0.6) = 0.6 \\ \mu_{\tilde{R}\,0\,\tilde{S}}(x_3, z_3) &= \max(0.7, 0.6) = 0.7 \end{aligned}$$

EXAMPLE 3.12.

Consider a set of $P = \{P_1, P_2, P_3, P_4\}$ of four varieties of paddy plants, set $D = \{D_1, D_2, D_3, D_4\}$ of the various diseases affecting the plants and $S = \{S_1, S_2, S_3, S_4\}$ be the common symptoms of the diseases.

Let

$$\tilde{R} = \begin{array}{c} \\ P_1 \\ P_2 \\ P_3 \\ P_4 \end{array}\begin{array}{c} \begin{array}{cccc} D_1 & D_2 & D_3 & D_4 \end{array} \\ \begin{bmatrix} 0.6 & 0.6 & 0.9 & 0.8 \\ 0.1 & 0.2 & 0.9 & 0.8 \\ 0.9 & 0.3 & 0.4 & 0.8 \\ 0.9 & 0.8 & 0.1 & 0.2 \end{bmatrix} \end{array}$$

$$\tilde{S} = \begin{array}{c} \\ D_1 \\ D_2 \\ D_3 \\ D_4 \end{array}\begin{array}{c} \begin{array}{cccc} S_1 & S_2 & S_3 & S_4 \end{array} \\ \begin{bmatrix} 0.1 & 0.2 & 0.7 & 0.9 \\ 1 & 1 & 0.4 & 0.6 \\ 0 & 0 & 0.5 & 0.9 \\ 0.9 & 1 & 0.8 & 0.2 \end{bmatrix} \end{array}$$

Obtain the association of the plants with the different symptoms of the diseases using max-min composition.

Solution: To obtain the association of the plants with the symptoms, R_0S which is a relation on the sets P and S is to be computed.

Using max-min composition,

$$R_0 S = \begin{array}{c} \\ P_1 \\ P_2 \\ P_3 \\ P_4 \end{array} \begin{array}{c} \begin{array}{cccc} S_1 & S_2 & S_3 & S_4 \end{array} \\ \begin{bmatrix} 0.8 & 0.8 & 0.8 & 0.9 \\ 0.8 & 0.8 & 0.8 & 0.9 \\ 0.8 & 0.8 & 0.8 & 0.9 \\ 0.8 & 0.8 & 0.7 & 0.9 \end{bmatrix} \end{array}$$

3.6. FUZZY TO CRISP CONVERSION

Crisp to Fuzzy conversion is said to be Fuzzy fication which is required at output whereas Fuzzy to crisp conversion in similar to analog to digital conversion which is required in input, details of this article are covered in next chapter (article 4.5.).

SOLVED EXAMPLES

EXAMPLE 3.13.

Let $$A = \{(x_1, 0.2), (x_2, 0.7), (x_3, 0.4)\}$$
$$B = \{(y_1, 0.5), (y_2, 0.6)\}$$

are two fuzzy sets defined on the universe of discourse respectively. Determine the relation $A \times B$.

$$X = \{x_1, x_2, x_3\}$$
$$Y = \{y_1, y_2\}$$
$$A \times B = \min(x_1, y_1) = 0.2$$
$$\min(x_1, y_2) = 0.2$$
$$\min(x_2, y_1) = 0.5$$
$$\min(x_2, y_2) = 0.6$$
$$\min(x_3, y_1) = 0.4$$
$$\min(x_3, y_2) = 0.4$$

EXAMPLE 3.14.

Let $$A = \{(3, 0.5), (5, 1), (7, 0.6)\}$$
$$B = \{(3, 1), (5, 0.6)\}$$

Find $A \times B$, A^2, $A + B$, A/B, $A \cdot B$.

$$\min(0.5, 1) = 0.5$$
$$\min(0.5, 0.6) = 0.5$$
$$\min(1, 1) = 1$$
$$\min(1, 0.6) = 0.6$$

$$\min(0.6, 1) = 0.6$$
$$\min(0.6, 0.6) = 0.6$$

EXAMPLE 3.15.

Consider a set $P = \{P_1, P_2, P_3, P_4\}$ of four varieties of plants, $D = \{d_1, d_2, d_3, d_4\}$ is a set of various discretes affecting the plants and $S = \{S_1, S_2, S_3, S_4\}$ be the common symptoms of the diseases.

Let $\tilde{R}$ be a relation on $P \times D$ and $\tilde{S}$ be a relation of $D \times S$

$$\tilde{R} = \begin{array}{c} \\ P_1 \\ P_2 \\ P_3 \\ P_4 \end{array}\begin{array}{c} \begin{array}{cccc} d_1 & d_2 & d_3 & d_4 \end{array} \\ \begin{bmatrix} 0.6 & 0.6 & 0.9 & 0.8 \\ 0.1 & 0.2 & 0.9 & 0.8 \\ 0.9 & 0.3 & 0.4 & 0.8 \\ 0.9 & 0.8 & 0.1 & 0.2 \end{bmatrix} \end{array}$$

$$\tilde{S} = \begin{array}{c} \\ d_1 \\ d_2 \\ d_3 \\ d_4 \end{array}\begin{array}{c} \begin{array}{cccc} S_1 & S_2 & S_3 & S_4 \end{array} \\ \begin{bmatrix} 0.1 & 0.2 & 0.7 & 0.9 \\ 0.1 & 1 & 0.4 & 0.6 \\ 0 & 0 & 0.5 & 0.9 \\ 0.9 & 1 & 0.8 & 0.2 \end{bmatrix} \end{array}$$

Obtain the max-min composition.

$$\mu_{\tilde{R}_0\tilde{S}}(p, s) = \max_{d \in D} = (\min(\mu_{\tilde{R}}(p, d), \mu_{\tilde{S}}(d, s))$$

$$\mu_{\tilde{R}_0\tilde{S}}(p_1, s_1) = \max[\min(0.6, 0.1), \min(0.6, 1), \min(0.9, 0), \min(0.8, 0.9)]$$
$$= \max(0.1, 0.6, 0, 0.8) = 0.8$$

$$\mu_{\tilde{R}_0\tilde{S}}(p_2, s_2) = \max[\min(0.6, 0.2), \min(0.6, 1), \min(0.9, 0), \min(0.8, 1)]$$
$$= \max(0.2, 0.6, 0, 0.8) = 0.8$$

$$\mu_{\tilde{R}_0\tilde{S}}(p_1, s_3) = \max[\min(0.6, 0.7), \min(0.6, 0.4), \min(0.9, 0.5), \min(0.8, 0.8)]$$
$$= \max(0.6, 0.4, 0.5, 0.8) = 0.8$$

$$\mu_{\tilde{R}_0\tilde{S}}(p_1, s_4) = \max[\min(0.6, 0.9), \min(0.6, 0.6), \min(0.9, 0.9), \min(0.8, 0.2)]$$
$$= \max(0.6, 0.6, 0.9, 0.2) = 0.9$$

$$\mu_{\tilde{R}_0\tilde{S}}(p_2, s_1) = \max[\min(0.1, 0.1), \min(0.2, 0.1), \min(0.9, 0), \min(0.8, 0.9)]$$
$$= \max(0.1, 0.2, 0, 0.8) = 0.8$$

$$\mu_{\tilde{R}_0\tilde{S}}(p_2, s_2) = \max[\min(0.1, 0.2), \min(0.2, 0.1), \min(0.9, 0), \min(0.8, 1)]$$
$$= \max(0.1, 0.2, 0, 0.8) = 0.8$$

$$\mu_{\tilde{R}_0\tilde{S}}(p_2, s_3) = \max[\min(0.1, 0.7), \min(0.2, 0.4), \min(0.9, 0.5), \min(0.8, 0.8)]$$
$$= \max(0.1, 0.2, 0.5, 0.8) = 0.8$$

$$\mu_{\tilde{R}_0\tilde{S}}(p_2, s_4) = \max[\min(0.1, 0.9), \min(0.2, 0.6), \min(0.9, 0.9), \min(0.8, 0.2)]$$
$$= \max(0.1, 0.2, 0.9, 0.2) = 0.9$$

$$\mu_{\tilde{R}_0\tilde{S}}(p_3, s_1) = \max[\min(0.9, 0.1), \min(0.3, 0.1), \min(0.4, 0), \min(0.8, 0.9)]$$
$$= \max(0.1, 0.3, 0, 0.8) = 0.8$$

$$\mu_{\tilde{R}_0\tilde{S}}(p_3, s_2) = \max[\min(0.9, 0.2), \min(0.3, 1), \min(0.4, 0), \min(0.8, 1)]$$
$$= \max(0.2, 0.3, 0, 0.8) = 0.8$$

$$\mu_{\tilde{R}_0\tilde{S}}(p_3, s_3) = \max[\min(0.9, 0.7), \min(0.3, 0.4), \min(0.4, 0.5), \min(0.8, 0.8)]$$
$$= \max(0.7, 0.3, 0.4, 0.8) = 0.8$$

$$\mu_{\tilde{R}_0\tilde{S}}(p_3, s_4) = \max[\min(0.9, 0.9), \min(0.3, 0.6), \min(0.4, 0.9), \min(0.8, 0.2)]$$
$$= \max(0.9, 0.3, 0.4, 0.2) = 0.9$$

$$\mu_{\tilde{R}_0\tilde{S}}(p_4, s_1) = \max[\min(0.9, 0.1), \min(0.8, 1), \min(0.1, 0), \min(0.2, 0.9)]$$
$$= \max(0.1, 0.8, 0, 0.2) = 0.8$$

$$\mu_{\tilde{R}_0\tilde{S}}(p_4, s_2) = \max[\min(0.9, 0.2), \min(0.8, 1), \min(0.1, 0), \min(0.2, 1)]$$
$$= \max(0.2, 0.8, 0, 0.2) = 0.8$$

$$\mu_{\tilde{R}_0\tilde{S}}(p_4, s_3) = \max[\min(0.9, 0.7), \min(0.8, 0.4), \min(0.1, 0.5), \min(0.2, 0.8)]$$
$$= \max(0.7, 0.4, 0.1, 0.2) = 0.7$$

$$\mu_{\tilde{R}_0\tilde{S}}(p_4, s_4) = \max[\min(0.9, 0.9), \min(0.8, 0.6), \min(0.1, 0.9), \min(0.2, 0.2)]$$
$$= \max(0.9, 0.6, 0.1, 0.2) = 0.9$$

Then, R_0S by max-min composition yields.

$$R_0S = \begin{array}{c} \\ P_1 \\ P_2 \\ P_3 \\ P_4 \end{array}\begin{array}{c} \begin{array}{cccc} S_1 & S_2 & S_3 & S_4 \end{array} \\ \begin{bmatrix} 0.8 & 0.8 & 0.8 & 0.9 \\ 0.8 & 0.8 & 0.8 & 0.9 \\ 0.8 & 0.8 & 0.8 & 0.9 \\ 0.8 & 0.8 & 0.7 & 0.9 \end{bmatrix} \end{array}$$

EXAMPLE 3.16.

Let $X = \{A, B, C, D\}, \quad Y = \{1, 2, 3, 4\}$

$$\tilde{A} = \{(a, 0), (b, 0.8), (c, 0.6), (d, 1)\}$$

$$\tilde{B} = \{(1, 0.2), (2, 1), (3, 0.8), (4, 0)\}$$

$$\tilde{C} = \{(1, 0), (2, 0.4), (3, 1), (4, 0.8)\}$$

Determine the implication relation, when

$$\tilde{A} = \{(a, 1), (b, 0.2), (c, 0.4), (d, 0)\}$$

(*i*) if X is $\tilde{A}$ then y is $\tilde{B}$

(*ii*) if X is $\tilde{A}$ then y is $\tilde{B}$ else y is $\tilde{C}$.

(*i*)
$$\tilde{A} \times \tilde{B} = \begin{array}{c} \\ a \\ b \\ c \\ d \end{array}\begin{array}{c} \begin{array}{cccc} 1 & 2 & 3 & 4 \end{array} \\ \begin{bmatrix} 0 & 0 & 0 & 0 \\ 0.2 & 0.8 & 0.8 & 0 \\ 0.2 & 0.6 & 0.6 & 0 \\ 0.2 & 1 & 0.8 & 0 \end{bmatrix} \end{array} \qquad Y = [1, 1, 1, 1]$$

$$\tilde{A} \times Y = \begin{array}{c} \\ a \\ b \\ c \\ d \end{array}\begin{array}{c} \begin{array}{cccc} 1 & 2 & 3 & 4 \end{array} \\ \begin{bmatrix} 1 & 1 & 1 & 1 \\ 0.2 & 0.2 & 0.2 & 0.2 \\ 0.4 & 0.4 & 0.4 & 0.4 \\ 0 & 0 & 0 & 0 \end{bmatrix} \end{array}$$

$$= \begin{array}{c} \\ a \\ b \\ c \\ d \end{array}\begin{array}{c} \begin{array}{cccc} 1 & 2 & 3 & 4 \end{array} \\ \begin{bmatrix} 1 & 1 & 1 & 1 \\ 0.2 & 0.8 & 0.8 & 0.2 \\ 0.4 & 0.6 & 0.6 & 0.4 \\ 0.2 & 1 & 0.8 & 0 \end{bmatrix} \end{array}$$

(*ii*)
$$\tilde{A} \times \tilde{B} = \begin{bmatrix} 0 & 0 & 0 & 0 \\ 0.2 & 0.8 & 0.8 & 0 \\ 0.2 & 0.6 & 0.6 & 0 \\ 0.2 & 1 & 0.8 & 0 \end{bmatrix}$$

$$\tilde{A} = \{(a, 1), (b, 0.2), (c, 0.4), (d, 0)\}$$

$$\tilde{C} = \{(1, 0), (2, 0.4), (3, 1), (4, 0.8)\}$$

$$\begin{array}{c} \\ a \\ b \\ c \\ d \end{array}\begin{array}{c} \begin{array}{cccc} 1 & 2 & 3 & 4 \end{array} \\ \begin{bmatrix} 0 & 0.4 & 1 & 0.8 \\ 0 & 0.2 & 0.2 & 0.2 \\ 0 & 0.4 & 0.4 & 0.4 \\ 0 & 0 & 0 & 0 \end{bmatrix} \end{array} = \tilde{A} \times \tilde{C}$$

$$(\tilde{A} \times \tilde{B}) \cup (\tilde{A} \times \tilde{C}) = \begin{bmatrix} 0 & 0.4 & 1 & 0.8 \\ 0.2 & 0.8 & 0.8 & 0.2 \\ 0.2 & 0.6 & 0.6 & 0.4 \\ 0.2 & 1 & 0.8 & 0 \end{bmatrix}$$

EXAMPLE 3.17.

Apply the fuzzy modes ponens rule to deduce relation is quite slow.

(i) If the temperature is high then the relation is slow.

(ii) the temperature is very high.

$\tilde{H}$ (High)

$V\tilde{H}$ (Very high)

$\tilde{S}$ (Slow)

$Q\tilde{S}$ (Quite slow)

Indicate the associated fuzzy set as follows:

The set of temperatures,

For x = {30, 40, 50, 60, 70, 80, 90, 100}

The set of relations per minute,

Y = {10, 20, 30, 40, 50, 60}

$\tilde{H}$ = {(70, 1), (80, 1), (90, 0.3)}

$V\tilde{H}$ = {(90, 0.9), (100, 1)}

$\tilde{H}$ = {(70, 0), (80, 0), (90, 0.7)}

$Q\tilde{S}$ = {(10, 1), (20, 0.8)}

$\tilde{S}$ = {(30, 0.8), (90, 1), (50, 06)}

$\tilde{H} \times \tilde{S}$ = {(70, 1)}

	10	20	30	40	50	60
30	0	0	0	0	0	0
40	0	0	0	0	0	0
50	0	0	0	0	0	0
60	0	0	0	0	0	0
70	0	0	0.8	1	0.6	0
80	0	0	0.8	1	0.6	0
90	0	0	0.3	0.3	0.3	0
100	0	0	0	0	0	0

$\tilde{H} \times Y$ =

10	20	30	40	50	60
30	1	1	1	1	1	1
40	1	1	1	1	1	1
50	1	1	1	1	1	1
60	1	1	1	1	1	1
70	0	0	0	0	0	0
80	0	0	0	0	0	0
90	0.7	0.7	0.7	0.7	0.7	0.7
100	1	1	1	1	1	1

$\tilde{R}(x, y)$ =

10	20	30	40	50	60
30	1	1	1	1	1	1

40	1	1	1	1	1	1
50	1	1	1	1	1	1
60	1	1	1	1	1	1
70	0	0	0.8	1	0.6	0
80	0	0	0.8	1	0.6	0
90	0.7	0.7	0.7	0.7	0.7	0.7
100	1	1	1	1	1	1

To deduce relation is quite slow, we make composition,

$[0\ 0\ 0\ 0\ 0\ 0\ 0.9\ 1] \times [\tilde{R}(x, y)]$

$$[0\ 0\ 0\ 0\ 0\ 0\ 0.9\ 1] \begin{bmatrix} 1 & 1 & 1 & 1 & 1 & 1 \\ 1 & 1 & 1 & 1 & 1 & 1 \\ 1 & 1 & 1 & 1 & 1 & 1 \\ 1 & 1 & 1 & 1 & 1 & 1 \\ 0 & 0 & 0.8 & 1 & 0.6 & 0 \\ 0 & 0 & 0.8 & 1 & 0.6 & 0 \\ 0.7 & 0.7 & 0.7 & 0.7 & 0.7 & 0.7 \\ 1 & 1 & 1 & 1 & 1 & 1 \end{bmatrix}$$

$[1\ \ 1\ \ 1\ \ 1\ \ 1\ \ 1]$

SHORT QUESTIONS WITH ANSWERS

Q.1. Given A_1 = {a, b} and A 2 {1, 2}.Find the Cartesian product $A_1 \times A_2$.

Ans. $A_1 \times A_2 = \{a,1\}, \{a, 2\}, \{b, 1\}, \{b,2\}$

Q.2. Consider fuzzy set $A = \{(x_1, 0.4), (x_2, 0.2), (x_3, 0.7)\}$ for $\alpha = 2$ find the power of fuzzy set.

Ans. We know that, $\mu_A{}^{\alpha(x)} = (\mu_A(x))^{\alpha}$

$\mu_A{}^{2(x)} = \mu_A(x)^2$

Hence $(A)^2 = \{(x_1, 0.16), (x_2, 0.04), (x_3, 0.49)\}$

$\mu_A{}^{2(x_1)} = (\mu_A(x_1))^2 = (0.4)^2 = 0.16$

$\mu_A{}^{2(x_2)} = (\mu_A(x_2))^2 = (0.2)^2 = 0.04$

$\mu_A{}^{2(x_3)} = (\mu_A(x_3))^2 = (0.7)^2 = 0.49$

Q.3. Given two fuzzy sets

A = {1/2 + 0.5/3 + 0.3/4 + 0.2/5}

B = {0.5/2 + 0.7/3 + 0.2/4 + 0.4/5}

Find: (*i*) *AUB* (*ii*) *A′*, *B′*

Ans. (*i*) *AUB*= {1/2 + 0.7/3 + 0.3/4 + 0.4/5}

(*ii*) *A′* = {0/2 + 0.5/3 + 0.7/4 + 0.8/5}

B′ = {0.5/2 + 0.3/3 + 0.8/4 + 0.6/5}

Q.4. What are the different shapes of membership functions have?

Ans. Traingular

Trapazoidal

Gaussian

Sigmoid

Polynomial

Q.5. What are the different features of membership functions?

Ans. Core

Support

Boundary

LONG QUESTIONS WITH ANSWER

Q.1. What are the different operations on Fuzzy sets?

Ans. 1. Union: The union of two fuzzy sets A and B is a new fuzzy set also on X with a membership function defined as.

$$\mu_{A \cup B}(x) = \max(\mu_A(x), \mu_B(x))$$

or

$$\mu_{A \cup B}(x) = \mu_A(x) \vee \mu_B(x)$$

2. Intersection: The intersection of two fuzzy sets A and B is a new fuzzy $A \cup B$ with membership function defined as

$$\mu_{A \cap B}(x) = \mu_A(x) \wedge \mu_B(x)$$

or

$$\mu_{A \cap B}(x) = \min(\mu_A(x), \mu_B(x))$$

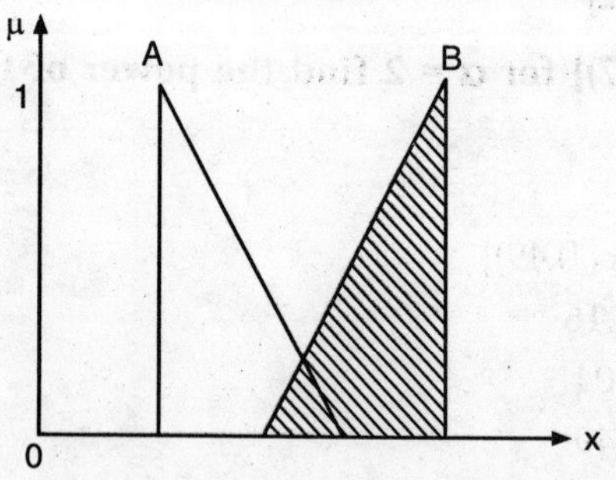

Fig. 3.19. Union of fuzzy sets

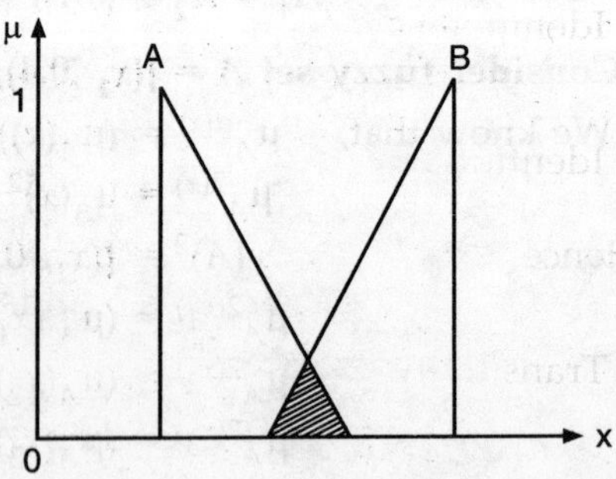

Fig. 3.20. Intersection of fuzzy sets

3. Complement: The complement of a fuzzy set A is a new fuzzy set is a new fuzzy set A with A membership function

$$\mu_{\bar{A}}(x) = 1 - \mu_A(x)$$

4. Product of two fuzzy sets: The product of two fuzzy sets A and B is a new fuzzy $A.B$ whose membership function is defined as

$$\mu_{A \cap B}(x) = \mu_A(x), \mu_B(x)$$

5. Equality: Two fuzzy sets *A* and *B* are said to be equal $(A = B)$ $\mu_A(x) = \mu_B(x)$.

6. Product of a fuzzy set with a crusp number: Multiplying a fuzzy set *A* by a crisp number result in a new fuzzy set product a.A with the membership function $\mu_{a.A}(x) = a.\ \mu_A(x)$.

7. Power of a fuzzy set: The α power of fuzzy set is a new fuzzy set *A* whose membership function is given by

$$\mu_A^{\alpha}(x) = (\mu_A(x))^{\alpha}$$

Raising a fuzzy set to its second power is called concent (CON) taking the square rool is called Dilation (DIL).

8. Difference: The difference of two fuzzy sets *N* and *B* is a new fuzzy set *A* – *B* defined as

$$\underset{\sim}{A} - \underset{\sim}{B} = (\underset{\sim}{A} \cap \underset{\sim}{B})$$

9. Disjunctive SUM: The disjunctive sum of two fuzzy sets $\underset{\sim}{A}$ and $\bar{\underset{\sim}{B}}$ is a fuzzy set $\underset{\sim}{A} \oplus \underset{\sim}{B}$ defined a

$$\underset{\sim}{A} \oplus \underset{\sim}{B} = (\bar{A} \cap B) \cup (A \cap \bar{B}).$$

Q.2. What are the different properties of fuzzy sets?

Ans. 1. Commutativity

$$\underset{\sim}{A} \cup \underset{\sim}{B} = \underset{\sim}{B} \cup \underset{\sim}{A}$$

$$\underset{\sim}{A} \cap \underset{\sim}{B} = \underset{\sim}{B} \cap \underset{\sim}{A}$$

2. Associativity

$$\underset{\sim}{A} \cup (\underset{\sim}{B} \cup \underset{\sim}{C}) = (\underset{\sim}{A} \cup \underset{\sim}{B}) \cup \underset{\sim}{C}$$

$$\underset{\sim}{A} \cap (\underset{\sim}{B} \cap \underset{\sim}{C}) = (\underset{\sim}{A} \cap \underset{\sim}{B}) \cap (\underset{\sim}{A} \cup \underset{\sim}{C})$$

3. Distruibutivity

$$\underset{\sim}{A} \cup (\underset{\sim}{B} \cap \underset{\sim}{C}) = (\underset{\sim}{A} \cup \underset{\sim}{B}) \cap (\underset{\sim}{A} \cup \underset{\sim}{C})$$

$$\underset{\sim}{A} \cap (\underset{\sim}{B} \cup \underset{\sim}{C}) = (\underset{\sim}{A} \cap \underset{\sim}{B}) \cup (\underset{\sim}{A} \cap \underset{\sim}{C})$$

4. Idempotency

$$\underset{\sim}{A} \cup \underset{\sim}{A} = \underset{\sim}{A} \text{ and } \underset{\sim}{A} \cap \underset{\sim}{A} = \underset{\sim}{A}$$

5. Identitity

$$\underset{\sim}{A} \cup \phi = \underset{\sim}{A} \text{ and } \underset{\sim}{A} \cap X = \underset{\sim}{A}$$

$$\underset{\sim}{A} \cap \phi = \phi \text{ and } \underset{\sim}{A} \cap X = X$$

6. Transitivity

$$\text{If } \underset{\sim}{A} \subseteq \underset{\sim}{B} \subseteq \underset{\sim}{C} \text{ then } \underset{\sim}{A} \subseteq \underset{\sim}{C}$$

7. Involution

$$\bar{\bar{\underset{\sim}{A}}} = \underset{\sim}{A}$$

8. Demorgan's Law

$$\overline{(\underset{\sim}{A} \cap \underset{\sim}{B})} = (\bar{\underset{\sim}{A}} \cup \bar{\underset{\sim}{B}})$$

$$\overline{(\underset{\sim}{A} \cup \underset{\sim}{B})} = (\bar{\underset{\sim}{A}} \cap \bar{\underset{\sim}{B}})$$

9. Cardinality: The cardinality of a set is the total number of elements in the set.

$$\text{card}(\infty) = \sum_{x_i} \mu_n(x_i)$$

The cardinality of a fuzzy set is used in defining other properties such as normalisation factor.

10. **Height (Normal vs Subnormal):** The height of a fuzzy set is the highest membership value of its membership function.

$$\text{Height}(A) = \max_{x_i} \mu_A(x_i)$$

A fuzzy set with height 1 is called a Normal fuzzy set. In contrast a fuzzy set whose height is less than 1 is called subnormal fuzzy set.

Q.3. Consider a set $P = \{P_1, P_2, P_3, P_4\}$ of four varieties of plants, this is set of various diseases affecting plants. $D = \{D_1, D_2, D_3, D_4\}$ and $S = \{S_1, S_2, S_3, S_4\}$ be the common symptoms of diseases. Let $\tilde{R}$ be a relation on PXD and $\tilde{S}$ be a relation. $\tilde{D} \times \tilde{S}$.

$$\tilde{R} = \begin{array}{c} \\ P_1 \\ P_2 \\ P_3 \\ P_4 \end{array} \begin{array}{c} \begin{array}{cccc} D_1 & D_2 & D_3 & D_4 \end{array} \\ \begin{bmatrix} 0.6 & 0.6 & .9 & .8 \\ .1 & .2 & .9 & .8 \\ .9 & .3 & .4 & .8 \\ .9 & .8 & .1 & .2 \end{bmatrix} \end{array} \qquad \tilde{S} = \begin{array}{c} \\ D_1 \\ D_2 \\ D_3 \\ D_4 \end{array} \begin{array}{c} \begin{array}{cccc} S_1 & S_2 & S_3 & S_4 \end{array} \\ \begin{bmatrix} .1 & .2 & .7 & .9 \\ 1 & 1 & .4 & .6 \\ 0 & 0 & .5 & .9 \\ .9 & 1 & .8 & .2 \end{bmatrix} \end{array}$$

Obtain the max-min composition.

Ans.

$$\tilde{R} \circ \tilde{S} = \begin{array}{c} \\ P_1 \\ P_2 \\ P_3 \\ P_4 \end{array} \begin{array}{c} \begin{array}{cccc} S_1 & S_2 & S_3 & S_4 \end{array} \\ \begin{bmatrix} .8 & .8 & .8 & .9 \\ .8 & .8 & .8 & .9 \\ .8 & .8 & .8 & .9 \\ .8 & .8 & .7 & .9 \end{bmatrix} \end{array}$$

Since, min (0.6, 0.1) = 0.1, min (0.9, 0) = 0

min (0.6, 1) = 0.6, min (0.8, 0.9) = 0.8

min (0.1, 0.6, 0, 0.8) = 0.5

max (0.2, 0.6, 0, 0.8) = 0.8

max (0.6, 0.4, 0.5, 0.8) = 0.8

max (0.6, 0.6, 0.5, 0.8) = 0.9

max (.1, .2, 0, .8) = .8

max (.1, .3, 0, .8) = .8

max (.2, .3, 0, .8) = .8

max (.7, .3, .4, .8) = .8

max (.7, .3, .4, .8) = .8

max (.9, .6, .1, .2) = 0.9

max (.1, .8, 0, .2) = 0.8

max (.2, .8, 0, .2) = 0.8

max (.7, .4, .1, .2) = 0.1
max (.9, .6, .1, ..2) = .9

Q.4. Differentiate between crisp sets and fuzzy sets.

Ans.

Crisp sets	Fuzzy sets
1. Probability of occuring will be other 0 or 1 and not other than that	1. Probability of occoring will be between 0 and 1 *i.e.,* (0-1).
2. A crisp set is defined by crisp boundaries, *i.e.,* there is no uncertainity in location of boundaries of set.	2. A fuzzy set then is a set containing elements that have varying degrees of membership.
3. For example: IS WATER COLOURLESS? The answer will be either yes or no *i.e.,* 1 and 0 and hence this is an example of crisp set.	3. For example: IS RAM HONEST? There can be so many possibility *i.e.,* extremely honest, very honest extremely dishonest, honest at etc. Hence, this is fuzzy set.
4. It does not contain membership function	4. It contains membership function.

Q.5. Verify the crisp relation:

$$\mathbf{R\ (i, j) = 1 \quad if \quad (x_i, y_i)\ \varepsilon\ R}$$

$$\mathbf{R\ (i, j) = 0 \quad if \quad (x_i, y_i) \not\varepsilon R}$$

Ans. For binary relation $R\ (X, Y)$ where $X = \{x_1, x_2, \ldots x_n\}$ and $Y = \{y_1, y_2, \ldots y_n\}$. The relation matrix R is a two-dimeonal matrix where X represents the rows, Y represents the column and

$$R\ (i, j) = 1 \text{ if } (x_i, y_i)\ \varepsilon\ R \text{ and}$$

$$R\ (i, j) = 0 \text{ if } (x_i, y_i) \not\varepsilon R \text{ and}$$

Example, $X = \{1, 2, 3, 4\}$

$X \times X = \{(1, 1), (1, 2), (1, 3), (1, 4), (2, 1), (2, 2), (2, 3), (2, 4), (3, 1), (3, 2), (3, 3), (3, 4), (4, 1), (4, 2), (4, 3), (4, 4)\}.$

$R = \{(x, y) \mid y = x + 1, x, y\ \varepsilon X\}$

$R = \{(1, 2), (2, 3), (3, 4)\}$

$$R = \begin{array}{c} \\ 1 \\ 2 \\ 3 \\ 4 \end{array}\begin{array}{c} \begin{array}{cccc} 1 & 2 & 3 & 4 \end{array} \\ \begin{bmatrix} 0 & 1 & 0 & 0 \\ 0 & 0 & 1 & 0 \\ 0 & 0 & 0 & 1 \\ 0 & 0 & 0 & 0 \end{bmatrix} \end{array} \quad \textbf{Ans.}$$

Since R is given by (1, 2), (2, 3), (3, 4), therefore, in relational matrix R_{12}, R_{23}, R_{34} accquires truth value and remaining got truth values 0.

Q.6. What are the different operations on fuzzy relations?

Ans. 1. UNION: Let $\underset{\sim}{R}$ and $\underset{\sim}{S}$ be fuzzy relations of $X \times Y$.

$$\mu\, \underset{\sim}{R} \cup \underset{\sim}{S}\ (x, y) = \max\ (\mu_R\ (x, y), \mu_S\ (x, y))$$

2. **INTERSECTION:**

$$\mu_{\underset{\sim}{R} \cap \underset{\sim}{S}}(x, y) = \min(\mu_R(x, y), \mu_S(x, y))$$

3. **COMPLEMENT:**

$$\mu_{\overline{\underset{\sim}{R}}}(x, y) = 1 - \mu_R(x, y)$$

4. COMPOSITION OF RELATIONS: Suppose $\underset{\sim}{R}$ is a fuzzy relation defined on $X \times Y$ and $\underset{\sim}{S}$ is fuzzy relation defined on $Y \times Z$, then $\underset{\sim}{R} \cdot \underset{\sim}{S}$ is a fuzzy.

Q.7. Define membership in detail function.

Ans. An membership function is a curve that defines how the (fuzzy) values of a fuzzy variable in a certain region are mapped to a membership value between 0 and 1. An membership function can have different shapes. The simplest and most commonly uses membership function is the triangular type which can be symmetric or asymmetrical. Different types of membership functions are:

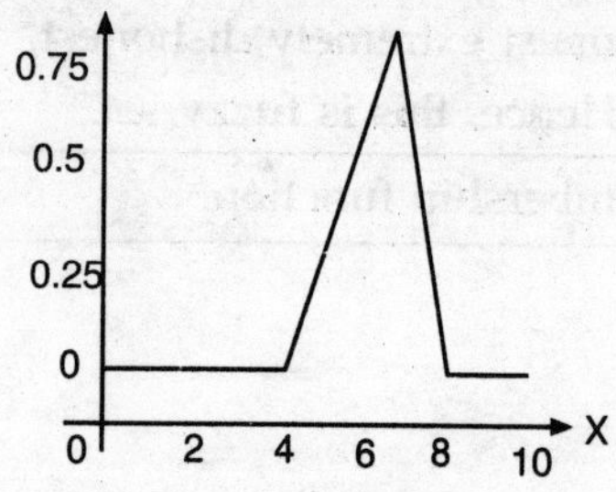

(a) Triangular

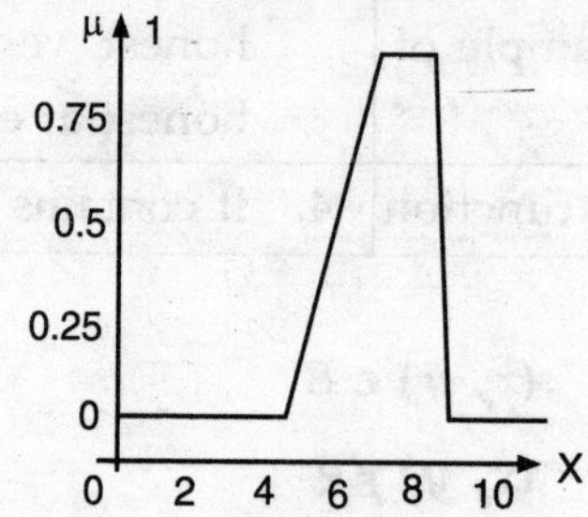

(b) Trapezoid

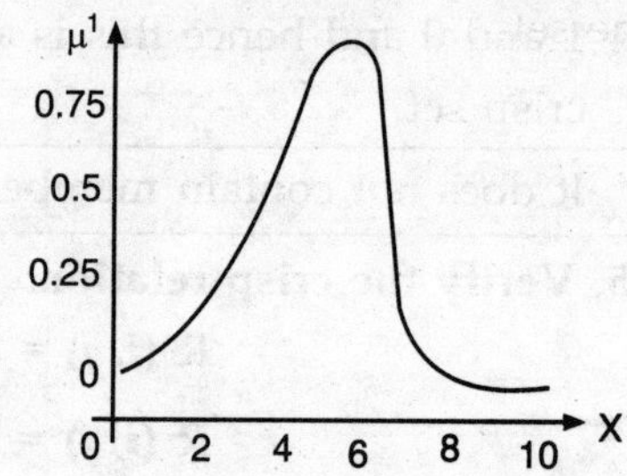

(c) Gaussian

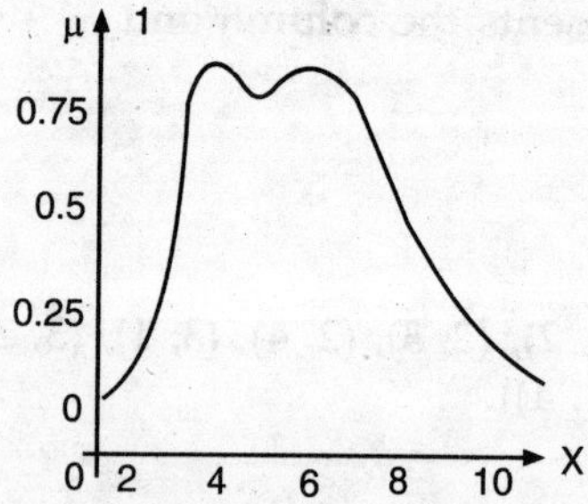

(d) Two sided gaussion

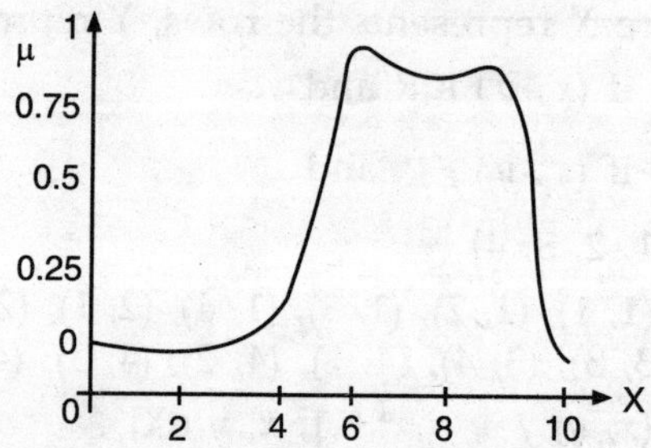

(e) Generalized bell

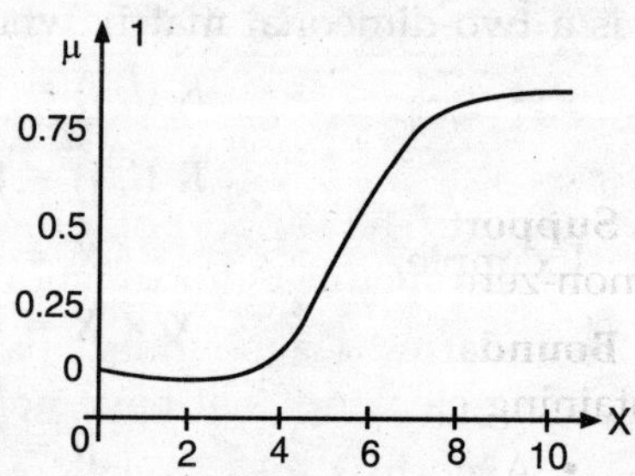

(f) Sigmoid right

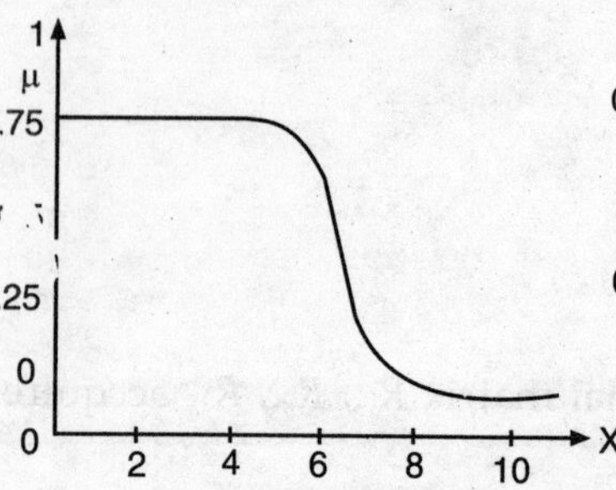

(g) Sigmoid left

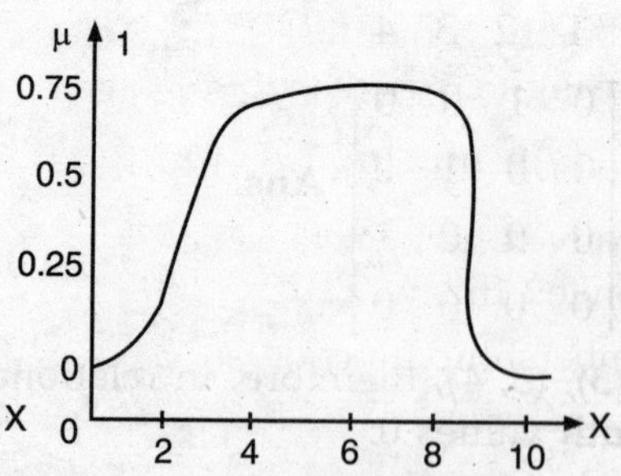

(h) Difference sigmoid

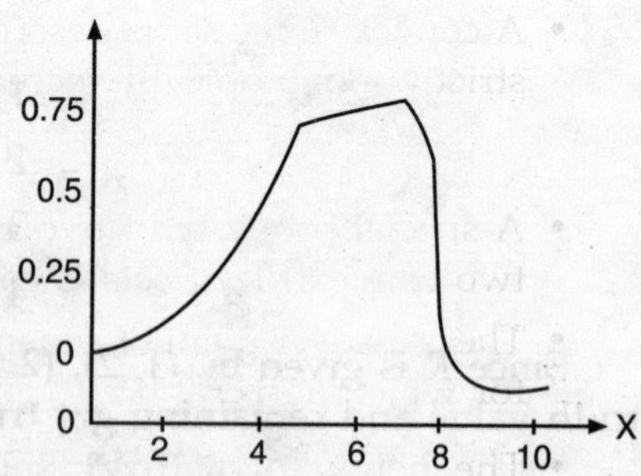

(i) Product-sigmold

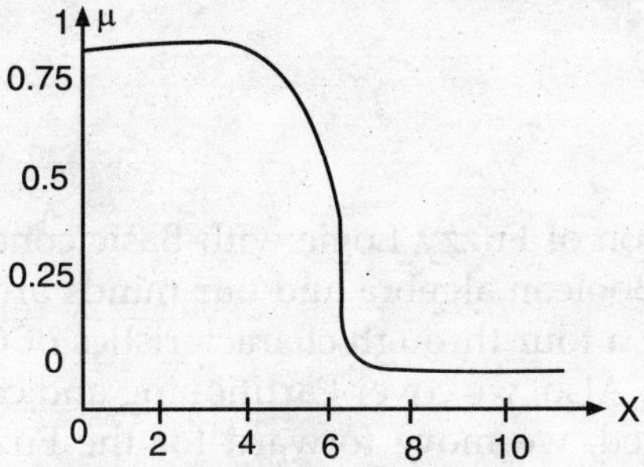

(j) Polynomial -z

Fig. 3.21.

The membership function essentially embodies all fuzziness for a particularly fuzzy set, its description is the essence of fuzzy proper or operation.

Features of Membership Function: The core of a membership function for some fuzzy set A is defined as that region of the universe that is characterized by complete and full membership in the set A.

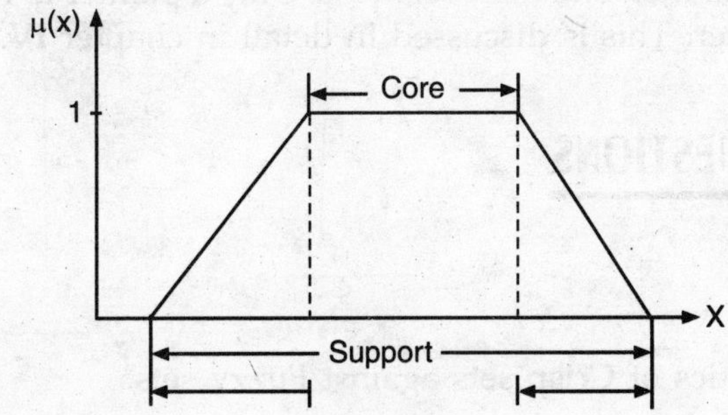

Fig. 3.22.

$$\mu_A(x) = 1$$

Support: The support of a membership function is that region of universe that characterized by non-zero membership function in set A, *i.e.*, $\mu_A(x) = 1$

Boundaries: The boundary of a membership function is defined as the region of the universe containing elements that have non-zero membership but not complete membership.

- A normal fuzzy set is one whose membership function at least one element x in the universe whose membership value is unity.
- A convex fuzzy set is described by a membership function those membership values are strictly monotonically increasing or strictly monotonically decreasing.

$$\mu_A(y) \geq \min[\mu_A(x), \mu_A(z)]$$

- A special property of two convex fuzzy sets, say A or B is that the intersection of these two convex fuzzy sets is convex fuzzy .
- The crossover point of a membershhip function are defined as the elements in the universe for which $\mu_A(x) = 0.5$.
- The height of a fuzzy set $\underset{\sim}{A}$ is the maximum value of the membership function.

$$\max\{\mu_{\underset{\sim}{A}}(x)\}$$

If the height is less than unity, fuzzy set is used to be subnormal.

SUMMARY

We start with the introduction of Fuzzy Logic with Basic concept. Since our school days we learn crisp sets, venn diagram, Boolean algebra and our minds are trained for 'ones' and 'zeros'. We start with the same and take a tour through characteristics of Crisp Sets, operations on Crisp Sets and properties of Crisp Sets. Also, we cover Partitioning and covering, Rules of Inclusion and Exclusion. Once all this is revised, we move forward for the Fuzzy set theory and operations. Membership functions discussed are linear, triangular, trapezoidal and Gaussian. Thereafter, Basic Fuzzy Operations as Intersection, Complement, Product of two fuzzy sets, Product of a fuzzy set with a crisp number, Power of Fuzzy set, difference and disjunctive sum discussed.

We have also discussed the Properties of fuzzy sets as Commutative Property, Associative Property, Distributivity, Impotence, Identity, Transitivity, Involution and finally the famous De Morgan's law. Again for analogy and prelude to fuzzy relations we discuss both, *i.e.*, Fuzzy and Crisp relations, as union, intersection, complement, composition of relations, max min composition all on Crisp relations first and then on Fuzzy relations. Only a pointer to Fuzzy to Crisp conversion is given at last in the chapter. This is discussed in detail in chapter IV.

REVIEW QUESTIONS

1. What is fuzzy logic?
2. Discuss the characteristics of Crisp sets against Fuzzy sets.
3. Discuss a crisp membership function against Fuzzy membership function.
4. Discuss various operations as Union, Intersection, Complement and Difference in Crisp sets. Chalk them out again in Fuzzy world.
5. Discuss the Properties of Crisp sets.
6. How partitioning is different from covering?
7. Show with example how will you find the product of two fuzzy sets? Let the two fuzzy sets be $\{(x_1, 0.3), (x_2, 0.5), (x_3, 0.7)\}$ and $\{(x_1, 0.2), (x_2, 0.4), (x_3, 0.6)\}$
8. Find the product of a fuzzy set with crisp number. Say the fuzzy set is {(x.3,.4,.5} and crisp number is 0.2.
9. Find the square of the fuzzy set $\{(x_1, 0.3), (x_2, 0.4), (x_3, 0.5)\}$
10. Exploit various properties of Fuzzy sets to derive De Morgan's law for fuzzy sets.

NUMERICAL PROBLEMS

1. In a class of 120 students numbered 1 to 120, all even numbered students opt for Physics, whose numbers are divisible by 5 opt for Chemistry and those whose numbers are divisible by 7 opt for Math. How many opt for none of the three subjects?
2. Of the 200 candidates who were interviewed for a position at a call center, 100 had a two-wheeler, 70 had a credit card and 140 had a mobile phone. 40 of them had both, a two-

wheeler and a credit card, 30 had both, a credit card and a mobile phone and 60 had both, a two wheeler and mobile phone and 10 had all three. How many candidates had none of the three?

3. In a class of 40 students, 12 enrolled for both English and German. 22 enrolled for German. If the students of the class enrolled for at least one of the two subjects, then how many students enrolled for only English and not German?

4. In a class 40% of the students enrolled for Math and 70% enrolled for Economics. If 15% of the students enrolled for both Math and Economics, what % of the students of the class did not enroll for either of the two subjects?

5. In a room of 50 people whose dresses have either red or white color, 30 are wearing red dress, 16 are wearing a combination of red and white. How many are wearing dresses that have only white color?

6. Suppose A, B, C are sets. Prove that

$$A \times (B \cup C) = (A \times B) \cup (C \times A)$$

7. There are 150 students at High School. 66 students play baseball, 45 play basketball, and 42 play soccer. 27 students play exactly two sports, and three students play all three of the sports. How many of the 150 students play none of the three sports?

8. Prove that

$$\begin{aligned} R \cup S\ (x, y) &= \max\ (R(x, y), S(x, y)) \\ R \cap S\ (x, y) &= \min\ (R(x, y), S(x, y)) \\ R^C(x, y) &= 1 - R(x, y) \\ R.S.(x, y) &= \max\ (\min(R(x, y), S(y, z)))\ y \in Y \end{aligned}$$

9. Say fuzzy sets $\tilde{A}$ and $\tilde{B}$ are defined in an interval X = [0.6] of real number by the membership grade function defined by

$$\mu_{\tilde{A}}(x) = 1 - \frac{1}{1+2x} \quad \text{and} \quad \mu_{\tilde{B}}(x) = 3^{-x}$$

Determine mathematical formulae and plots of membership grade function for :

(a) A^c, B^c

(b) $A \cup B$

(c) $A \cap B$

(d) $(A \cup B)^C$

(e) $(A \cap B)^C$

10. Let R be fuzzy relation between two sets X_1 and X_2 where X_1 is the set of fabrics and X_2 be the amount of dirt so

$$X_1 = \{\text{Cotton, silk, nylon}\}$$
$$X_2 = \{\text{less dirty, very dirty}\}$$

Define R interms of amount of detergent used.

□□□

4 CHAPTER FOUR

FUZZY LOGIC-II (FUZZY MEMBERSHIP, RULES)

CHAPTER OUTLINE

- 4.1 Membership Function
- 4.2. Inference in Fuzzy Logic
- 4.3. Fuzzy IF- THEN Rules
- 4.4. Fuzzy Implications and Fuzzy Algorithm
- 4.5. Fuzzification and Defuzzification
- 4.6. Fuzzy Controller
- 4.7 Analysis and Design For T-S Fuzzy Model

4.1. MEMBERSHIP FUNCTION

Practically all real values measured by different procedures (in particular expert methods) are approximated, with rare exception. Except an inaccuracy of measurement there are some characteristics of measured values such as indetermination, multiplicity and incorrectness [Nariniani]. All mentioned above characteristics we shall combine by the term "fuzziness" Fuzzy can be: values of parameters, value of functions, ratio between objects etc.

Different kinds of uncertainty are to some extent characteristics of practically any situation of making decision in which the expert information is used. The nature of uncertainty is essentially different. Firstly, it's necessary to point out the objective uncertainty which is peculiar to all real constants and is connected with our world's "organization" itself [Nariniani, 1994]. Secondly, the subjective uncertainty is peculiar to human's nature on the whole and to his abilities to evaluate the information in particular. The reasons of the origin of the second type of uncertainty are: expert's lack of knowledge about the characteristics of the objects; his unsatisfied confidence degree in correctness of his marks; knowledge contradiction; in distinction of the information spresentation and the semantic uncertainty corrected with different are indirect meanings of the natural language, indefiniteness of the keywords and definitions [Nariniani, 1994; Orlovski, 1981]. Beside, indefiniteness appears in the process of marks integration coming from different experts.

The result of the research [Larichev, 2002; Borisov, 1989] show that the main difficulty is caused by the necessity to appraise numerical values of the objects (variants, criteria) or to give numerical evaluation on the ratio scale between them. But the person fully much more confident [Orlovski, 1981; Larichev, 2002] if the has an opportunity to give unclear marks in the form of the intervals or with the help of rough sets pointing out the degree of belonging of the objects to these sets. In particular, more often in a form of package the additive criterion is used and its objectivation is connected with defining of weight coefficients of the objects [Voloshyn, 2003]. This approach seemed to be even more justified because in prevailing number of the cases it is enough to have the approximated characteristic of the data set and the expert info does not demand high accuracy.

So, the application of fuzzy sets and the function of belonging allow formalizing factors of uncertainty and unclearness with might happen in expert evaluation situation by some means. The function of belonging was interpreted in various works differently: as "subjective probability" [Orlovski, 1981], expert's confidence degree in object's belonging to the concept described by fuzzy set [Borisov, 1989], the opportunity of its interpretation by this concept [Melechov, 1990] and so on. For all this was traditionally considered reflection. But "unnatural" distinctiveness, and simple concept that according to its destination, is called to reflect distinction and inexactness of subjective marks. According to the author's opinion it is incorrect to demand precise meanings of the function of belonging from the expert. It is more logical to demand rough marks of the function of belonging because the situation of subjective evaluation is connected with the principle inexactness ("objective", "situational", "semantic" and so on [Nariniani, 1994]) in the definition of the value of function of belonging.

Definition

The membership function $\mu_A(x)$ describes the membership of the elements x of the base set X in the fuzzy set A, whereby for $\mu_A(x)$ a large class of functions can be taken. Reasonable functions are often piecewise linear functions, such as triangular or trapezoidal functions.

The grade of membership $\mu_A(x_0)$ of a membership function $\mu_A(x)$ describes for the special element $I = I_0$, to which grade it belongs to the fuzzy set A. This value is in the unit interval [0, 1]. Of course, x_o can simultaneously belong to another fuzzy set B, such that $\mu_B(x_o)$ characterises the grade of membership of x_o to B. This case is shown in Fig. 4.1.

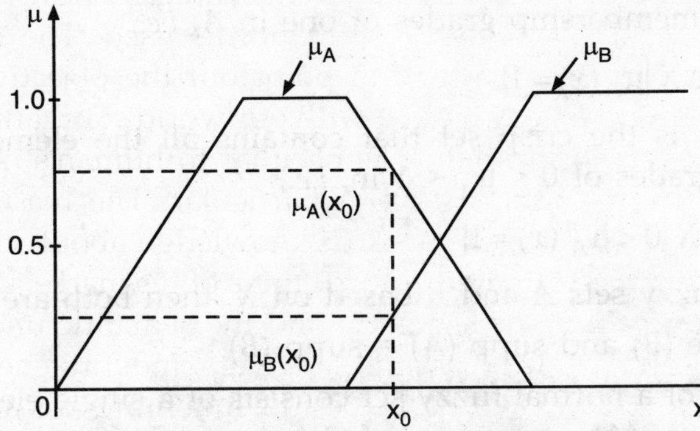

Fig. 4.1. Membership function

Membership grades of x_0 in the sets A and B: $\mu_A(x_o) = 0.75$ and $\mu_B(x_o) = 0.25$.

In the following, a set of important properties and characteristics of fuzzy sets will be described.

- Having two fuzzy sets and based on , then both are equal if their membership functions are equal, *i.e.,*

$$A = B \Leftrightarrow \mu_A(x) = \mu_B(x), x \varepsilon X$$

- The universal set is defined as
- The height of a fuzzy set A is the largest membership grade obtained by any element in that set, *i.e.,*

$$\text{hgt }(A) = \sup_{x \varepsilon X} \mu_A(I)$$

- A fuzzy set A is called normal when hgt $(A) = 1$, and it is subnormal when hgt $(A) < 1$.
- The support of a fuzzy set A is the crisp set that contains all the elements of that have nonzero membership grades in A, *i.e.,*
- An illustration is shown in Fig. 4.2.

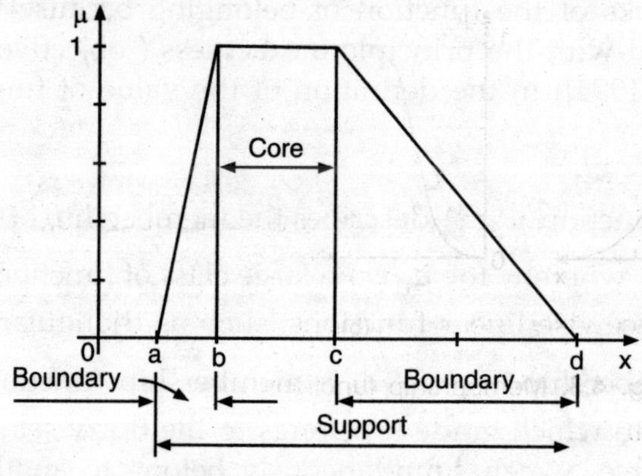

Fig. 4.2. Some characteristicss of a membership function

- The core of a normal fuzzy set A is the crisp set that contains all the elements of X that have the membership grades of one in A, *i.e.,*

  core $(A) = \{x \varepsilon X | \mu_A(x) = 1\}$

- The boundary is the crisp set that contains all the elements of X that have the membership grades of $0 < \mu_A < 1$ in, *i.e.,*

  $\text{bnd}(A) = \{x \varepsilon X | 0 < \mu_A(x) = 1\}$

- Having two fuzzy sets A and B based on X, then both are similar if

  core (A) = core (B) and supp (A) = supp (B).

- If the support of a normal fuzzy set consists of a single element x_o of X, which has the property supp(A) = core $(A) = (x_o)$.
- this set is called a singleton.

The type of representation of the membership function depends on the base set. If this set consists of many values, or is the base set a continuum, then a *parametric* representation is appropriate. For that functions are used that can be adapted by changing the parameters. Piecewise linear membership functions are preferred, because of their simplicity and efficiency with respect to computability. Mostly these are trapezoidal or triangular functions, which are defined by four and three parameters, respectively. Figure 4. shows a trapezoidal function formally described by

$$\mu\ (x, a, b, c, d) = \begin{cases} 0, & x < a, x, > d \\ \dfrac{x-a}{b-a}, & a \leq x \leq b \\ 1, & b < x < c \\ \dfrac{d-x}{d-c} & c \leq x \leq d \end{cases}$$

which migrates for the case $b = c$ into a triangular membership function. For some applications the modelling requires continuously differentiable curves and therefore smooth transitions, which the trapezoids do not have. Here, for example, three of these functions are mentioned, which are shown in Fig. 4.3.

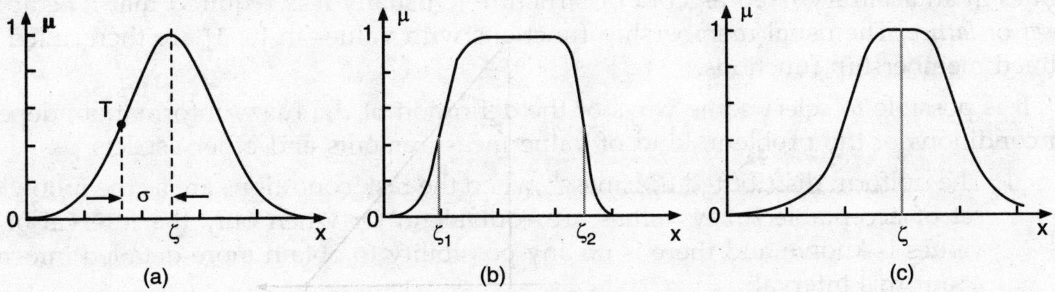

Fig. 4.3. Membership functions with smooth transitions

These are:

- the normalised Gaussian function

$$\mu(x\,, \zeta, \sigma) = e^{-\frac{(x-\zeta)^2}{2\sigma^2}}$$

- the difference of two sigmoidal functions

$$\mu(x, a_1, \zeta_1, a_2, \zeta_2) = [1 + e^{-\alpha_1(x-\zeta_1)}]^{-1} - [1 + e^{-\alpha_2(x-\zeta_2)}]^{-1}$$

- the generalised bell function

$$\mu(x, \alpha, \beta, \zeta) = \left[1 + \left|\frac{x-\zeta}{\alpha}\right|^{2\beta}\right]^{-1}$$

Definition

For any set X, a membership function on X is any function from X to the real unit interval [0,1].

Membership functions on X represent fuzzy subsets of X. The membership function which represents a fuzzy set $\tilde{A}$ is usually denoted by μA. For an element x of X, the value $\mu A(x)$ is called thememembership degree of x in the fuzzy set The membership degree $\mu A(x)$ quantifies the grade of membership of the element x to the fuzzy set $\tilde{A}$. The value 0 means that x is not a member of the fuzzy set; the value 1 means that x is fully a member of the fuzzy set. The values between 0 and 1 characterize fuzzy members, which belong to the fuzzy set only partially.

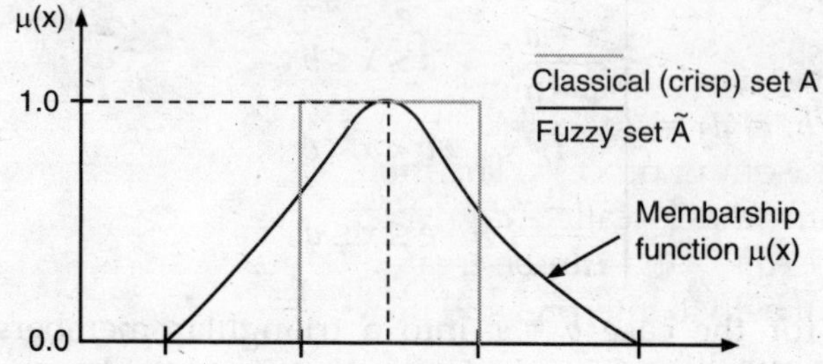

Fig. 4.4. Membership function of a fuzzy set

Sometimes, [1] a more general definition is used, where membership functions take values in an arbitrary fixed algebra or structure L; usually it is required that L be at least a *poset* or *lattice*. The usual membership functions with values in [0, 1] are then called [0, 1]-valued membership functions.

It is possible to select some ways of the definition of the fuzzy information, depending on conditions of the problem, kind of value measurements and other aspects:

1. The uniform distributed fuzziness", when the task conditions are as the following: all set of acceptable fuzzy values are equivalent. Or when only the interval of fuzzy values is known and there is no any possibility to obtain more detailed information about this interval.
2. Point fuzziness, when the outcomes of separate measurements of the fuzzy value are obtained by the way of point estimations which are not conterminous among themselves.
3. Three points fuzziness, when the fuzzy value boundaries and the its most probable or most advisable value are known. Such way of the definition of fuzzy value can be described by the triangular membership function.
4. Nonuniform distributed fuzziness", when the fuzzy value of the membership function is described by some no triangular function.

The development of mathematical apparatus for operating with unclear functions of belonging is very actual because it would allow solving the tasks of decisions taking on the basis of expert information with the help of the methods more adequate to the nature of expert mark which is characterized by different types of inexactness and uncertainty.

4.2. INFERENCE IN FUZZY LOGIC

The core section of a fuzzy system is that part, which combines the facts obtained from the fuzzification with the rule base and conducts the fuzzy reasoning process. This is called a *fuzzy inference machine*. Here rule and composition operations are applied from sections

In the following, for simplicity it is assumed that there is only one input $x_1 = x$ and the rule base is described with max/min operators from as follows:

$$\mu_R\ (x, u) = \max_r \{\min \{\mu p_r(x), \mu_{B_r}(u)\}\}$$

The operations can be reordered such that only the relevant operands are on the right-hand side. Then

$$\mu_B\ (u) = \max_r \{\min \{\min \{\underbrace{\max_{x \varepsilon X} \{\min \{\mu_A(x), \mu_{P_r}(x)\}\}}_{H_r}, \mu_{B_r}(\mu)\}\}$$

$$= \max_r \{\min \{H_r, \mu_{B_r}(u)\}\}$$

is obtained for the reasoning process. The inner term *Hr*, which combines the fact with the premise, is a constant and is called *degree of relevance* of the rule *r*. It characterises the relevance of the fired rule and can be treated as a de-normalised universal fuzzy set. The following example will help to interpret and provide understanding of this reasoning operation by applying graphical means.

For example, a simple fuzzy system is given, which models the brake behaviour of a car driver depending on the car speed. The inference machine should determine the brake force for a given car speed. The speed is specified by the two linguistic terms ``low'' and ``medium'', and the brake force by ``moderate'' and ``strong''. The rule base includes the two rules

1. If the car speed is low THEN the brake force is moderate
2. If the car speed is medium THEN the brake force is strong

For the linguistic terms in the premises ``speed is low'' and ``speed is medium'' or ``force is moderate'' and ``force is strong'' are defined in Fig. 4.5.

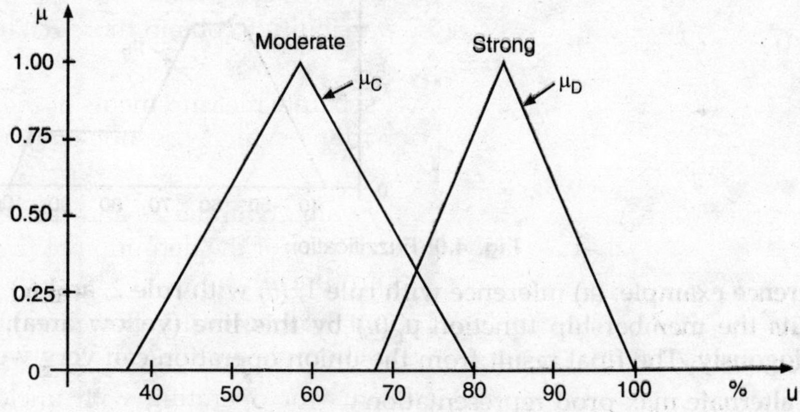

Fig. 4.5. Fuzzy set of a car brake force

For these two rules the membership functions for the premises are $\mu p_1 = \mu_A$, $\mu p_2 = \mu_B$, and for the conclusions μ_{B_1}, μ_C, $\mu_{B_2} = \mu_D$. The car speed is $x_o = 70$ km/h, which is a crisp value and therefore can be represented in fuzzy notation by the singleton $\mu_A(x) = \{70\ \text{km/h}\}$. Inserting these values into Eq. 4. one obtains

$$\mu_B(u) = \max \{\min \{H_1, \mu_C(u)\}, \min \{H_2, \mu_D(u)\}\}$$

with

$$H_1 = \max_{x \varepsilon X} \{\min \{\mu_A(x), \mu_A(x)\}\}$$

and $$H_1 = \max_{x \varepsilon X} \{\min \{\mu_A(x), \mu_B(x)\}\}$$

In rule 1, there is only a relation between μ_A and μ_C. The degree of relevance H_1 for rule 1 is a horizontal line of height 0.75, which is the value from the fuzzification as shown in following figure. The reasoning operation min $\{H_1, \mu_C(u)\}$.

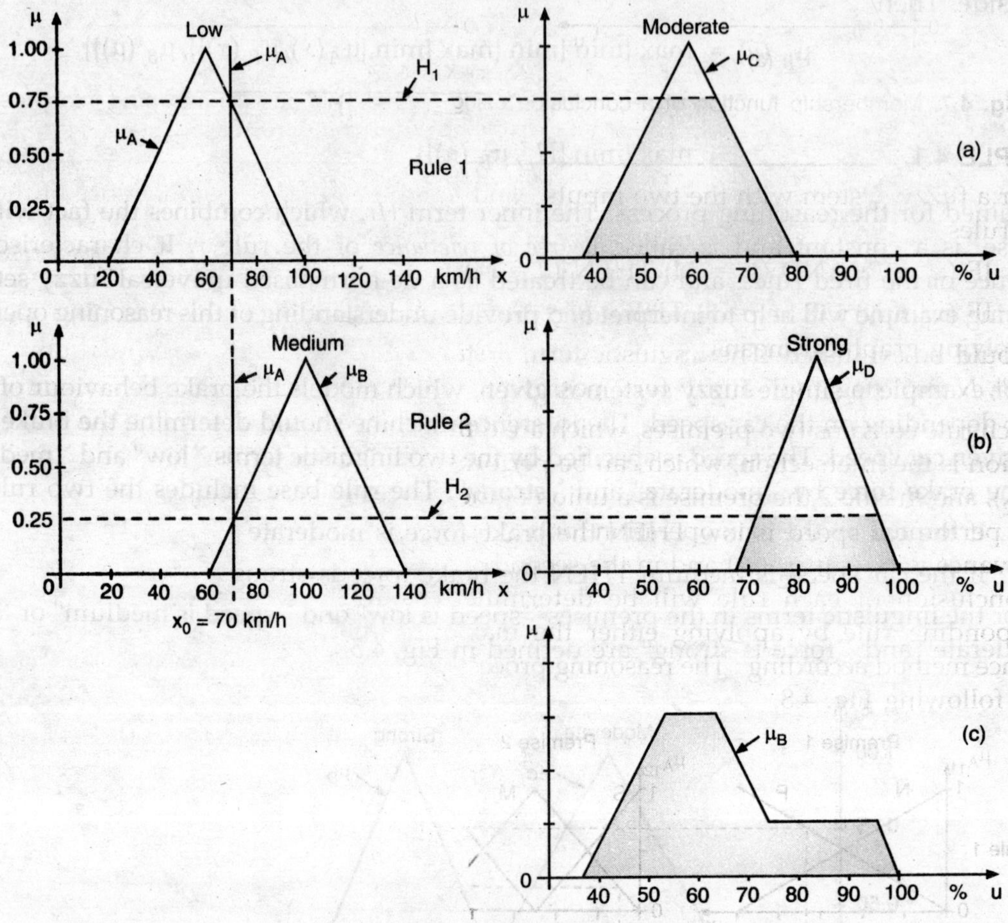

Fig. 4.6. Fuzzification

Fuzzy inference example: (*a*) inference with rule 1, (*b*) with rule 2, and (*c*) final fuzzy set with rule 1 cuts the membership function $\mu_C(u)$ by this line (yellow area). The rule 2 is evaluated analogously. The final result from the union operation can very well be inferred.

Taking an alternate max-prod representation

$$\mu_B(u) = \max_r \underbrace{\max_{x \varepsilon X} \{\min \{\mu_A{}'(x), \mu p_r(x)\}\}}_{H_r} \mu_{B_r}(u)$$

$$= \max_r \{H_r \, \mu_{B_r}(u)\}$$

is obtained for the max-prod inference. The difference between the max/min and max-prod inference is that in the first case the membership function of the conclusion is cut and in the second case scaled. Following figure illustrates this.

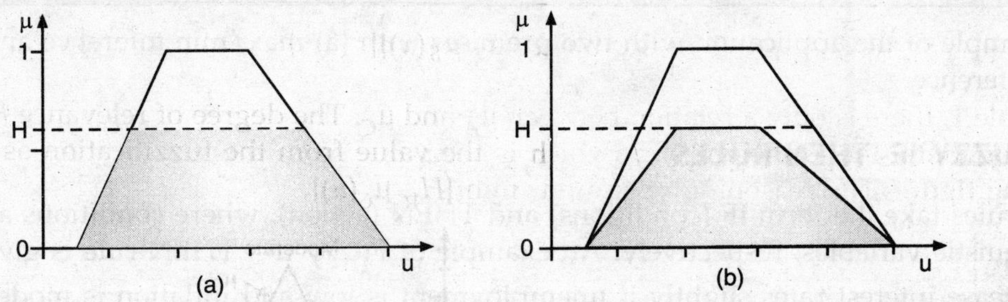

Fig. 4.7. Membership function of a conclusion using (a) max/min and (b) max-prod inference

EXAMPLE 4.1.

For a fuzzy system with the two inputs and and the output the inference of the two fuzzy rules

1. IF (x_1 = P) AND (x_2 = M) THEN (u = M)
2. IF (x_1 = N) OR (x_2 = S) THEN (u = S)

should be evaluated. The linguistic term S stands for ``small'', $M$ for ``medium'', L for ``large'', $N$ for ``negative'' and P for ``positive''.

Each rule contains two premises, which are differently connected. In rule 1 the connective operation is the intersection, which can be performed by the min operation for $\mu_{A_{11}}(x_1)$ and $\mu_{A_{12}}(x_2)$, and in rule 2 the premise is a union of the two premises (x_1 = N) and (x_2 = S), which can be performed by the max operation for $\mu_{A_{21}}(x_1)$, and $\mu_{A_{22}}(x_2)$. In the first case the degree of relevance is $H_1 = \mu_{A_{12}}(x_2)$ and in the second case $\mu_{A_{22}}(x_2)$. The membership functions of the conclusion of each rule will be determined using the degree of relevance of the corresponding rule by applying either the max/min inference method or the max-prod inference method according . The reasoning process using both inference methods is visualised in the following Fig. 4.8.

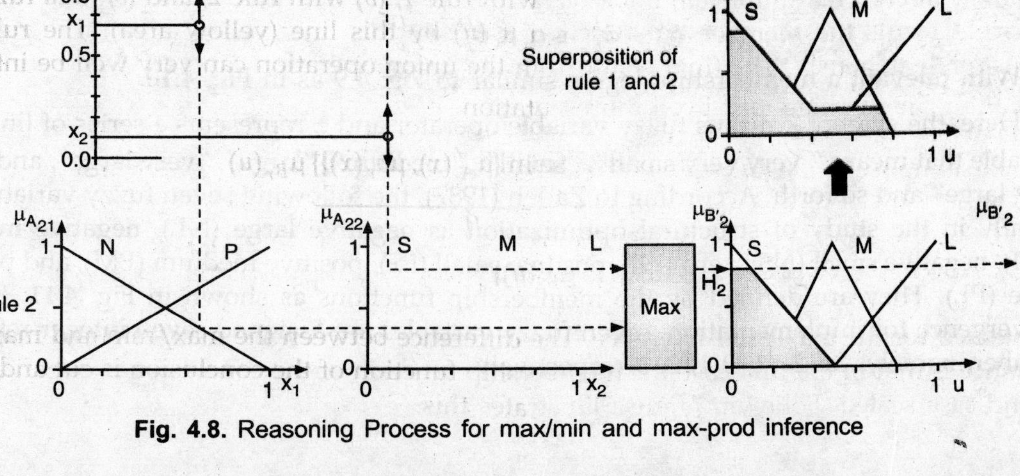

Fig. 4.8. Reasoning Process for max/min and max-prod inference

Example of the application with two premises with (a) max/min inference and (b) max-prod inference

4.3. FUZZY IF- THEN RULES

Fuzzy rules take the form IF (conditions) and THEN (action), where conditions and actions are linguistic variables, respectively. An example of Fuzzy IF - THEN rule is given below.

Increase interest rates slightly if unemployment is low and inflation is moderate.

Increase interest rates sharply if unemployment is low and inflation is moderate but rising sharply.

Decrease interest rates slightly if unemployment is low but increasing and inflation rate is low and stable.

The primary format of IF- THEN rules is given in Fig. 4.9.

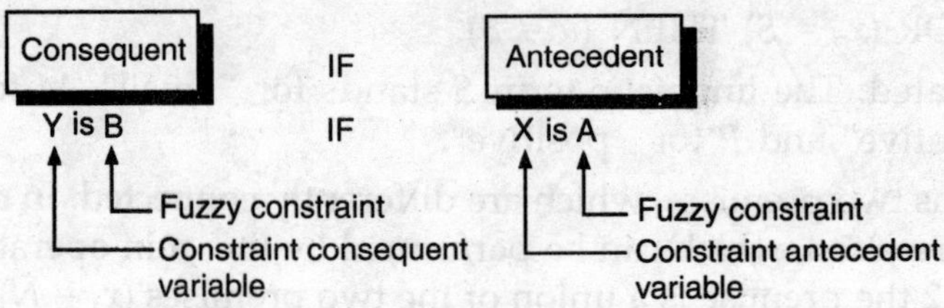

Fig. 4.9. Format of IF-THEN Rule

EXAMPLE 4.2.

Volume is small if pressure is high.

Usually in civil engineering, most of these specifications in codes and the functional requirements set by the users must be given in natural language to describe the expert's knowledge of design modification.

In usual structural optimization, the stress constraint written as

$$\sigma_i^L \le \sigma_i \le \sigma_i^U$$

Is a member stress constraint, where σ_i is the stress in member 'i' and σ_i^L and σ_i^U are the lower and upper bounds of the allowable stress. But in case of fuzzy optimization, the Eq. (15.11) is replaced as

$$= \tilde{\sigma}_i^L - \tilde{\sigma}_i^{Le} \lesssim \sigma_i \le \tilde{\sigma}_i^U + \tilde{\sigma}_i^{Ue}$$

With relevant a membership degree similar to Fig. 4.9 as in Fig. 4.10.

Here, the symbol $\lesssim$ means fuzzy variable operator and ± represents a series of linguistic variable that means " very very small", "small" , "medium", "large", "very large", and "very very large" and so forth. According to Zadeh (1987), the following seven fuzzy variables are usually in the study of structural optimization as negative large (NL), negative medium (NM), negative small (NS), zero (ZE), positive small (PS), positive medium (PM), and positive large (PL). They are defined by the membership functions as shown in Fig. 4.11. For the convergence for implementation, seven fuzzy variables are assigned seven integer reference number, namely -3, -2, -1, 0 1, 2, 3 respectively.

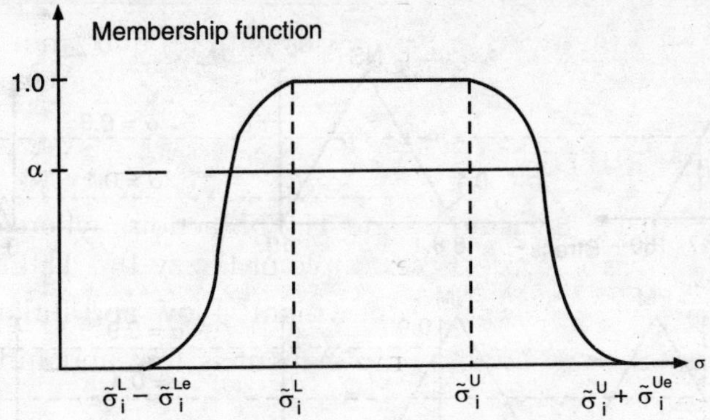

Fig. 4.10. Fuzzy constraint for stress

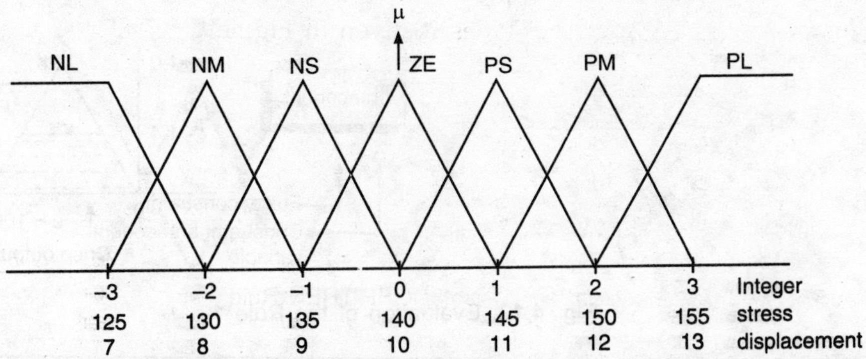

Fig. 4.11. Membership function for fuzzy variable

If the allowable stress is 140 MPa and the tolerance for each unit is 5Mpa, *X* axis is also given in terms of stress values. Similarly, one can define seven fuzzy membership functions for displacements as well as for any other variable.

As explained before, heuristic fuzzy rules can be written as

Rule 1: IF the maximum of violated member stress constraints is PS and all the displacement constraints are inactive

THEN the change of the corresponding member cross sectional area is PS

Rule 2: IF all constraints are inactive and the minimum of member stress constraints in NL . THEN the change of the corresponding member cross-sectional area in NS.

As an input, the constraint c_{ij} is usually classified as

1. active for ZE.
2. inactive for NL, NM, AND NS, AND
3. violated for PS, PM, and PL.

On the other hand for the output, the modification of the member cross-sectional areas has the possibilities, NL, NM, NS, ZE, PS, PM and PL.

Fuzzy controller inputs are usually crisp numbers. Fuzzy inputs may also be considered in the case of uncertain or noisy measurements and crisp numbers may be defuzzified, Fig. 15.10 shows the application to Rule 1. The degree of fulfillment (DIF) of Rule 1 is 0.4. The total fuzzyoutput μ_{out} is the union of the two outputs shown in Fig. 4.12.

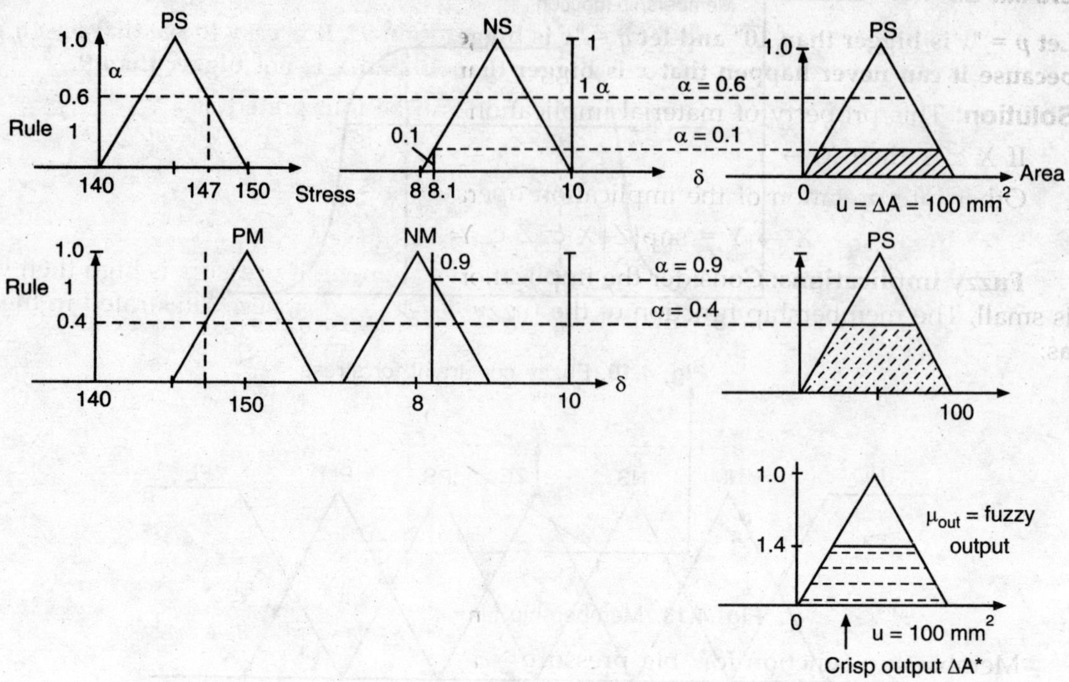

Fig. 4.12. Evaluation of the Rule 1

4.4. FUZZY IMPLICATIONS AND FUZZY ALGORITHM

The material implication

Let p = "x is in A" and q = "y is in B" are crisp propositions, where A and B are crisp sets for the moment.

The implication $p \rightarrow q$ is interpreted as $\neg(p \wedge \neg q)$. "p entails q" means that it can never happen that p is true and q is not true. It is easy to see that $p \rightarrow q = \neg p \vee q$

The full interpretation of the material implicationp $\rightarrow q$ is that the degree of truth of $p \rightarrow q$ quantifies to what extend q is at least as true as p, *i.e.*,

$$p \rightarrow q \text{ is true} \Rightarrow \tau(p) \leq \tau(q)$$

$$1p \rightarrow q = 1 \text{ if } \tau(p) \leq \tau(q)$$

$$0 \text{ otherwise}$$

p	q	$p \rightarrow q$
1	1	1
0	1	1
0	0	1
1	0	0

The truth table for the material implication.

EXAMPLE 4.3.

Let p = "x is bigger than 10" and let q = "x is bigger than 9". It is easy to see that $p \to q$ is true, because it can never happen that x is bigger than 10 and x is not bigger than 9.

Solution: This property of material implication can be interpreted as:

If $X \subset Y$ then $X \to Y$

Other interpretation of the implication operator is

$$X \to Y = \sup\{Z \mid X \subset Z \subset Y\}.$$

Fuzzy implications: Consider the implication statement: if pressure is high then volume is small, The membership function of the fuzzy set A, big pressure, illustrated in the figure as:

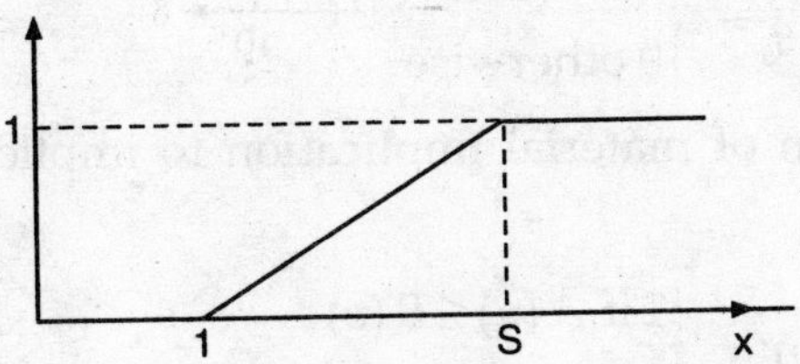

Fig. 4.13. Membership function for "big pressure"

Membership function for "big pressure" scan be interpreted as

- 1 is in the fuzzy set big pressure with grade of membership 0
- 2 is in the fuzzy set big pressure with grade of membership 0.25
- 4 is in the fuzzy set big pressure with grade of membership 0.75
- x is in the fuzzy set big pressure with grade of membership 1, for all $x \leq = 5$

$$A(u) = \begin{cases} 1 \text{if } u \geq 5 \\ 1-(5-u)/ \text{if } 1 \leq u \leq 5 \\ 0 \text{ otherwise} \end{cases}$$

The membership function of the fuzzy set B, small volume, can be interpreted as

- 5 is in the fuzzy set small volume with grade of membership 0
- 4 is in the fuzzy set small volume with grade of membership 0.25
- 2 is in the fuzzy set small volume with grade of membership 0.75
- x is in the fuzzy set small volume with grade of membership 1, for all x = 1

$$B(v) = \begin{cases} 1 \text{if } u \leq 5 \\ 1-(v-1)/4 \text{ if } 1 \leq v \leq 5 \\ 0 \text{ otherwise} \end{cases}$$

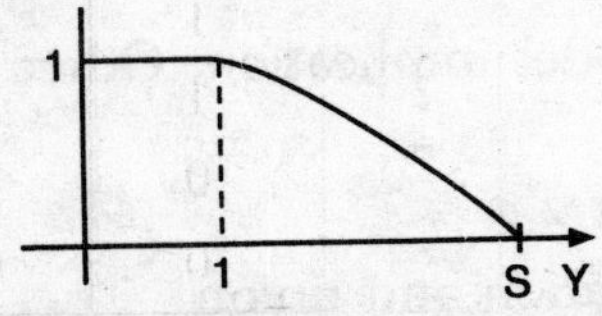

Fig. 4.14. Membership function for "small volume"

Membership function for "small volume".

If p is a proposition of the form x is A where A is a fuzzy set, for example, big pressure and q is a proposition of the form y is B for example, small volume then we define the fuzzy implication $A \to B$ as a fuzzy relation.

It is clear that $(A \to B)(u, v)$ should be defined pointwise and likewise, *i.e.* $(A \to B)(u, v)$ depends only on $A(u)$ and $B(v)$.

That is $(A \to B)(u, v) = I(A(u), B(v)) = A(u) \to B(v)$ In our interpretation $A(u)$ is considered as the truth value of the proposition: "u is big pressure", and $B(v)$ is considered as the truth value of the proposition: "v is small volume". That is u is big pressure $\to$ v is small volume $\equiv A(u) \to B(v)$. Remembering the full interpretation of the material implication

$$p \to q = \begin{cases} 1 \text{ if } \tau(p) \le \tau(q) \\ 0 \text{ otherwise} \end{cases}$$

One possible extension of material implication to implications with intermediate truth values can be

$$A(u) \to B(v) = \begin{cases} 1 \text{ if } A(u) \le B(v) \\ 0 \text{ otherwise} \end{cases}$$

"4 is big pressure" $\to$ "1 is small volume"

$$A(4) \to B(1) = 0.75 \to 1 = 1$$

However, it is easy to see that this fuzzy implication operator (called Standard Strict) sometimes is not appropriate for real-life applications. Namely, let

$A(u) = 0.8$ and $B(v) = 0.8$. Then we have

$$A(u) \to B(v) = 0.8 \; ? \; 0.8 = 1$$

Suppose there is a small error of measurement in $B(v)$, and instead of 0.8 we have 0.7999. Then

$$A(u) \to B(v) = 0.8 \to 0.7999 = 0$$

This example shows that small changes in the input can cause a big deviation in the output, *i.e.* our system is very sensitive to rounding errors of digital computation and small errors of measurement.

A smoother extension of material implication operator can be derived from the equation

$$X \to Y = \sup\{Z \mid X \cap Z \subset Y\}$$

That is

$$A(u) \to B(v) = \sup\{z \mid \min\{A(u), z\} \le B(v)\}$$

so, $$A(u) \to B(v) = \begin{cases} 1 \text{ if } A(u) \le B(v) \\ B(v) \text{ otherwise} \end{cases}$$

This operator is called G¨odel implication. Other possibility is to extend the original definition,

$$p \to q1 = p \vee q$$

Using the definition of negation and union

$A(u) \to B(v) = \max\{1 - A(u), B(v)\}$. This operator is called *Kleene-Dienes implication*. In many practical applications they use *Mamdani's implication* operator to model causal relationship between fuzzy variables.

This operator simply takes the minimum of truth values of fuzzy predicates

$$A(u) \to B(v) = \min\{A(u), B(v)\}$$

It is easy to see this is not a correct extension of material implications, because $0 \to 0$ yields zero.

However, in knowledge-based systems, we are usually not interested in rules, where the antecedent part is false. Following are the main Fuzzy Implications:

Larsen $x \to y = xy$

Lukasiewicz $x \to y = \min\{1, 1 - x + y\}$

Mamdani $x \to y = \min\{x, y\}$

Standard Strict $x \odot y = 1$ if $x = y$
0 otherwise

G¨odel $x \to y = 1$ if $x = y$
y otherwise

Gaines $x \to y = 1$ if $x = y$
y/x otherwise

Kleene-Dienes $x \to y = \max\{1 - x, y\}$

Kleene-Dienes-Luk. $x \to y = 1 - x + xy$

Modifiers

Definition 1

Let A be a fuzzy set in X. Then we can define the fuzzy sets "very A" and "more or less A". By (very A)$(x) = A(x)^2$ (more or less A)$(x) = A(x)^{1/2}$.

The use of fuzzy sets provides a basis for a systematic way for the manipulation of vague and imprecise concepts.

In particular, we can employ fuzzy sets to represent linguistic variables.

A linguistic variable can be regarded either as a variable whose value is a fuzzy number or as a variable whose values are defined in linguistic terms.

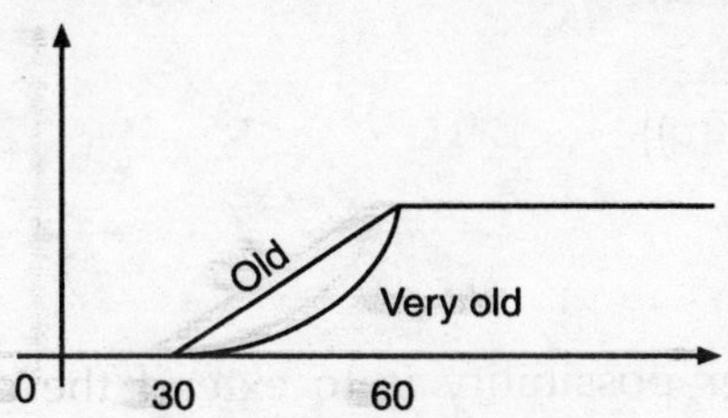

Fig. 4.15. Fuzzy set

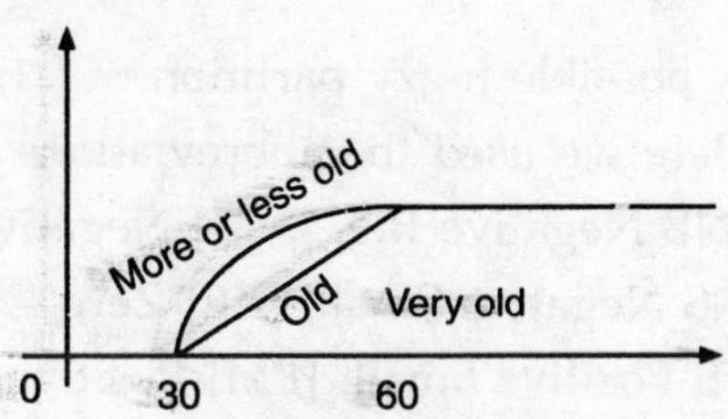

Fig. 4.16. Fuzzy set

Definition 2

Linguistic Variables:

A linguistic variable is characterized by a quintuple $(x, T(x), U, G, M)$ in which

- x is the name of variable;
- $T(x)$ is the term set of x, that is, the set of names of linguistic values of x with each value being a fuzzy number defined on U;
- G is a syntactic rule for generating the names of values of x;
- and M is a semantic rule for associating with each value its meaning. For example, if speed is interpreted as a linguistic variable, then its term set T (speed) could be

T = {slow, moderate, fast, very slow, more or less fast, . . . } where each term in T (speed) is characterized by a fuzzy set in a universe of discourse $U = [0, 100]$.

We might interpret "slow" as "a speed below about 40 mph," "moderate" as "a speed close to 55 mph," and "fast" as "a speed above about 70mph". These terms can be characterized as fuzzy sets whose membership functions are shown in the figure be-

Values of linguistic variable speed.

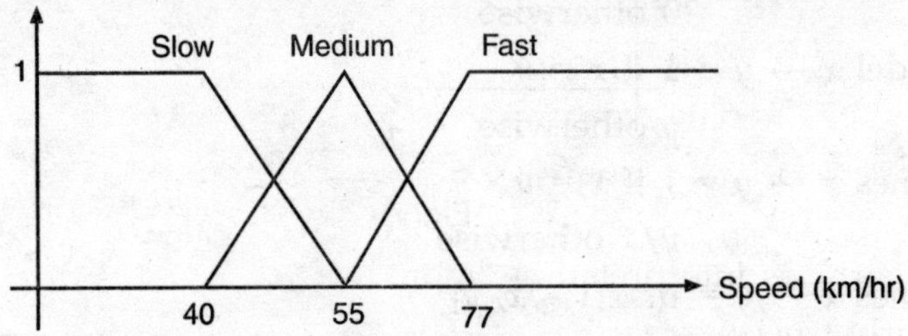

Fig. 4.17. Lignuistic variables

In many practical applications we normalize the domain of inputs and use the following type of fuzzy

Partition

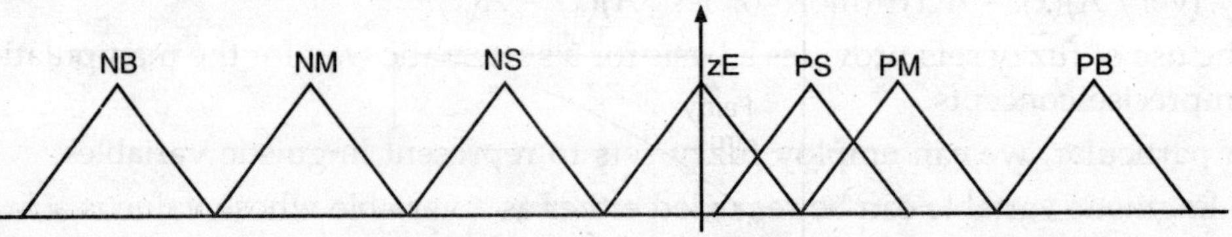

Fig. 4.18. Partition

A possible fuzzy partition of [-1, 1].

Here we used the abbreviations

`NB Negative Big, [NM] Negative Medium

NS Negative Small, [ZE] Zero,

PS Positive Small, [PM] Positive Medium,

PB Positive Big.

The linguistic variable Truth.

Truth = {Absolutely false, Very false, False, Fairly true, True, Very true, Absolutely true}

One may define the membership function of linguistic terms of truth as True(u) = u for each u ε [0, 1]. False(u) = 1 – u

for each u ε [0, 1].

Absolutely false(u) =1 if u = 0

0 otherwise

Absolutely true(u) = 1 if u = 1

0 otherwise

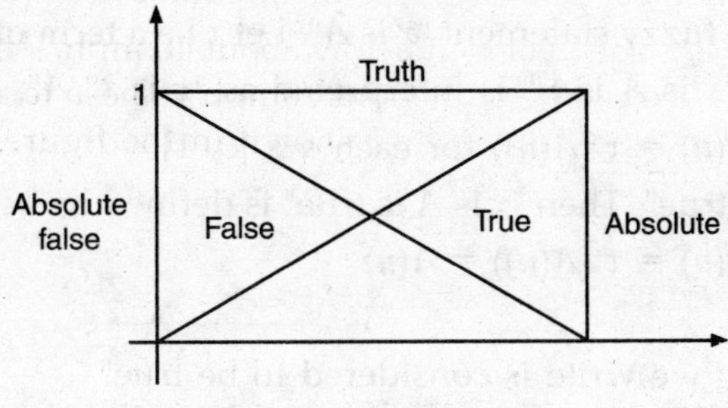

Fig. 4.19.

The world "Fairly" is interpreted as "more or less".

Fairly true(u) = vu

for each u ε [0, 1].

Very true(u) = u^2

for each u ε [0, 1].

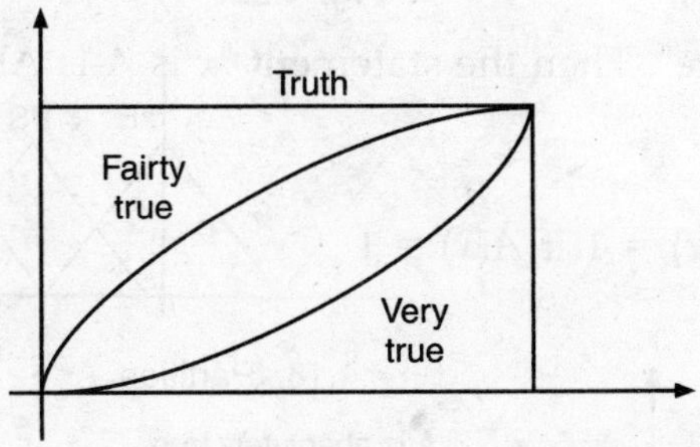

Fig. 4.20.

The world "Fairly" is interpreted as "more or less".

Fairly false (u) = $\sqrt{(1-u)}$

for each u ε [0, 1].

Very false(u) = (1 – u)2

for each u ε [0, 1].

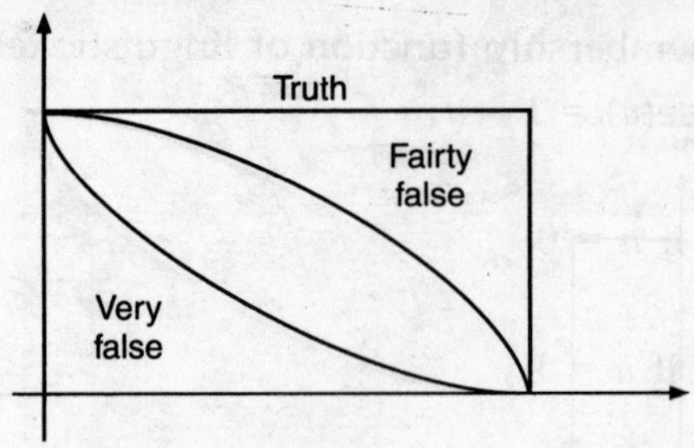

Fig. 4.21.

Suppose we have the fuzzy statement "x is A". Let τ be a term of linguistic variable Truth. Then the statement "x is A is t " is interpreted as "x is τ° A". Where

$$(t \circ A)(u) = \tau\ (A(u)) \text{ for each } u\ \varepsilon\ [0, 1].$$

For example, let τ = "true". Then "x is A is true" is defined by "x is τ° A" = "x is A" because

$$(\tau^\circ A)(u) = \tau\ (A(u)) = A(u)$$

for each u ε [0, 1].

It is why "everything we write is considered to be true".

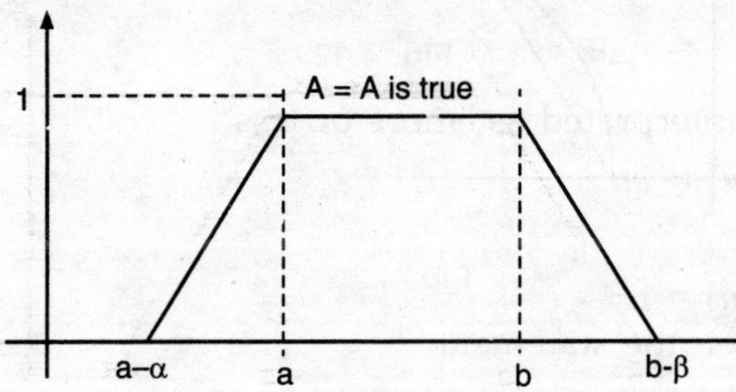

Fig. 4.22.

Let τ = "absolutely true". Then the statement "x is A is Absolutely true" is defined by "x is τ° A",

A is Absolutely true

Where $(\tau^\circ A)(x) = 1$ if $A(x) = 1$

0 otherwise

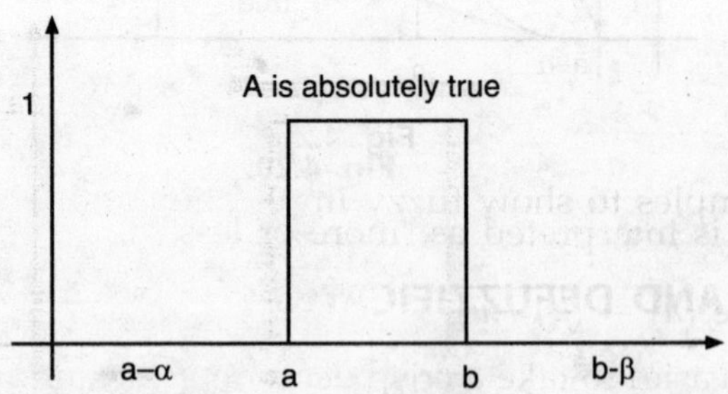

Fig. 4.23.

Let τ = "absolutely false". Then the statement "x is A is Absolutely false" is defined by "x is τ ° A",

Where $(\tau \circ A)(x) = 1$ if $A(x) = 0$

0 otherwise

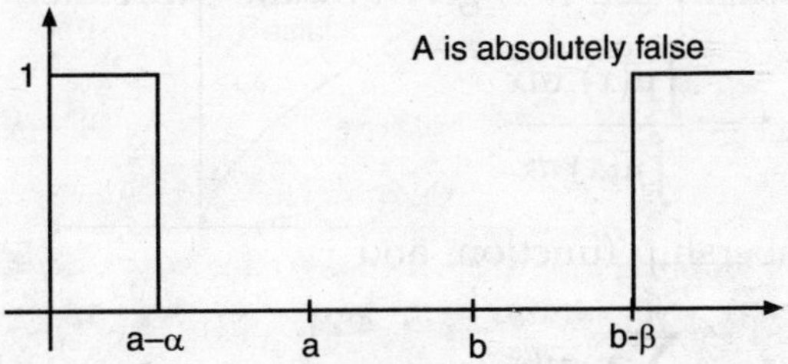

Fig. 4.24.

Let τ = "Fairly true". Then the statement "*x* is *A* is Fairly true" is defined by "*x* is *t* is τ° *A*", where

$$(\tau \circ A)(x) = A(x)$$

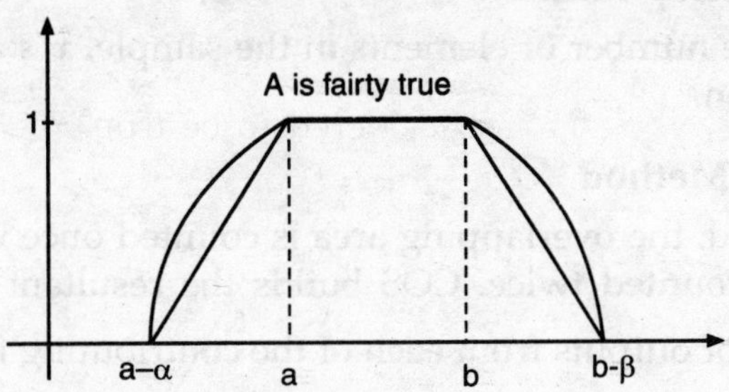

Fig. 4.25.

Let τ = "Very true". Then the statement "*x* is *A* is Fairly true" is defined by "*x* is τ° *A*", where $(\tau \circ A)(x) = (A(x))^2$

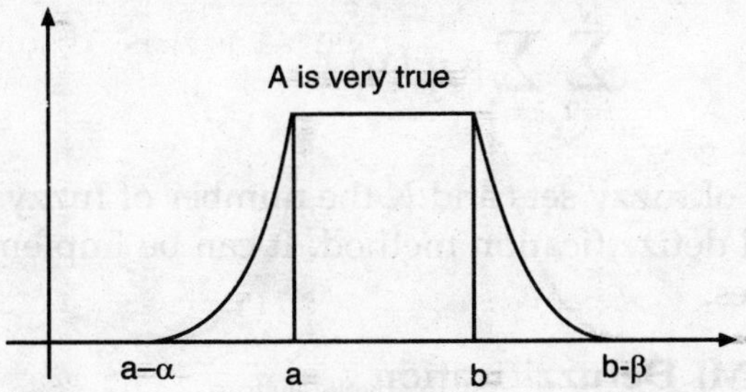

Fig. 4.26.

This is the best examples to show fuzzy Implication and Algorithms.

4.5. FUZZIFICATION AND DEFUZZIFICATION

For a fuzzy output, it is easier to take a crisp decision if the output is represented as a single scalar quantity. This conversion of a fuzzy set to single crisp value is called *defuzzification* and is the reverse process of fuzzification. Following are the three methods namely centroid method, centre of sums, and mean of maxima.

Centroid Method

Also known as the *centre of gravity* or the *centre of area* method, it obtains the centre of area (x^*) occupied by the fuzzy set. It is given by the expression

$$x^* = \frac{\int \mu(x)\, x dx}{\int \mu(x)\, dx}$$

for a continuous membership function, and

$$x^* = \frac{\sum_{i=1}^{n} x_i . \mu(x_i)}{\sum_{i=1}^{n} \mu(x_i)}$$

for a discrete membership function.

Here, n represents the number of elements in the sample, x_i's are the elements and $\mu(x_i)$ is its membership function.

Centre of Sums (COS) Method

In the centroid method, the overlapping area is counted once whereas in centre of *sums*, the overlapping area is counted twice. COS builds the resultant membership function by taking the algebraic sum of outputs from each of the contributing fuzzy sets $\tilde{A}_1$, $\tilde{A}_2$, ... , etc. The defuzzified value x^* is given by

$$x^* = \frac{\sum_{i=1}^{N} x_i . \sum_{k=1}^{n} \mu_{\tilde{A}_k}(x_i)}{\sum_{i=1}^{N} \sum_{k=1}^{n} \mu_{\tilde{A}_k}(x_i)}$$

Here n is the number of fuzzy sets and N the number of fuzzy variables. COS is actually the most commonly used defizzification method. It can be implemented easily and leads to rather fast inference cycles.

Mean of Maxima (MOM) Defuzzification

Sometimes to defuzzifying the output, crisp value is taken with the highest degree of membership. In case with more than one element having the maximum value, the mean value of the maxima is taken. The equation of the defuzzified value z^* is given by

$$x^* = \frac{\sum_{x_i \, \varepsilon \, M}^{N} x_i}{|M|}$$

where $M = \{x_i \mid \mu(x_i) \text{ is equal to the height of fuzzy set}\}$

$|M|$ is the cardinality of the set M. In the continuous case, M could be defined as

$$M = \{x \; \varepsilon \; [-c, c] \mid \mu(x) \text{ is equal to the height of the fuzzy set}\}.$$

In such a case, the *mean of maxima* is the arithmetic avarage of mean values of all intervals contained in M including zero length intervals.

The height of a fuzzy set A, *i.e.*, $h(A)$ is the largest membership grade obtained by any element in that set.

Example: $\tilde{A}_1$, $\tilde{A}_2$ and $\tilde{A}_3$ are three fuzzy sets as shown in Fig. 4.27 (*a*), (*b*) and (*c*). Figure 4.27(*d*) illustrates the aggregate of the fuzzy sets.

The defuzzification using (*i*) centroid method, (*ii*) centre of sums method, and (*iii*) mean of maxima method is illustrated as follows.

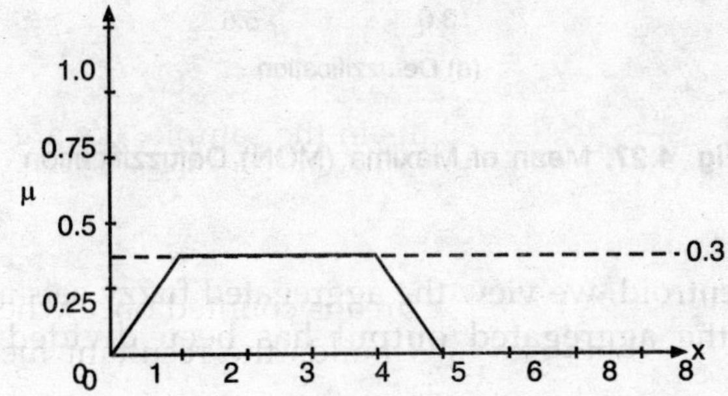

(a) $\tilde{A}_1$ Fuzzy set

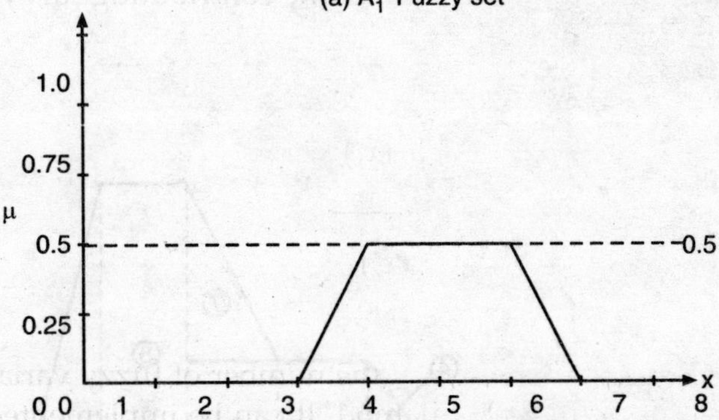

(b) $\tilde{A}_2$ Fuzzy set

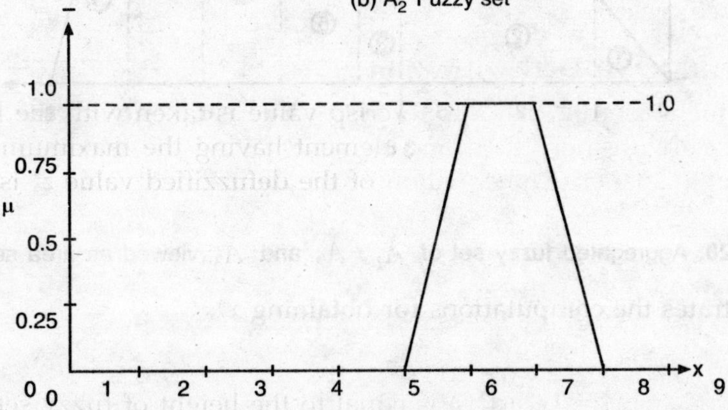

(c) $\tilde{A}_3$ Fuzzy set

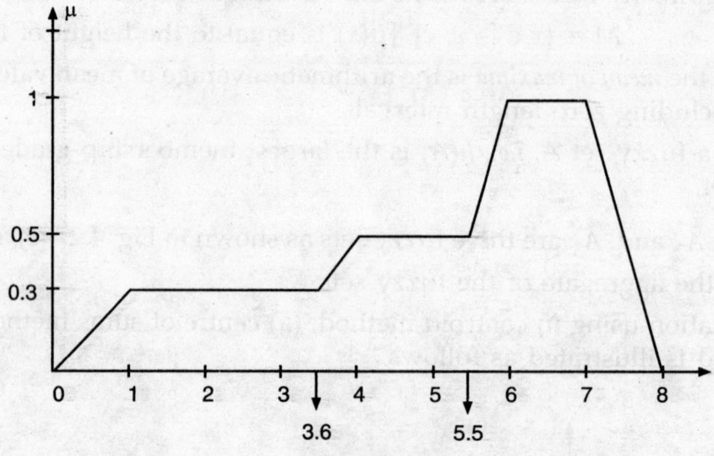

(d) Defuzzification

Fig. 4.27. Mean of Maxima (MON) Defuzzification

Centroid Method

To compute x^*, the centroid, we view the aggregated fuzzy sets as shown in Figs 4 and 4. Note that in Fig 4. the aggregated output has been divided into areas for better understanding.

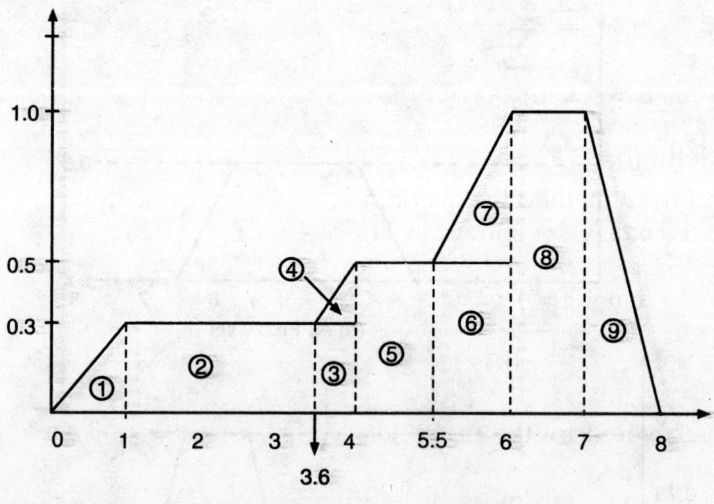

Fig. 4.28. Aggregated fuzzy set of $\tilde{A}_1$, $\tilde{A}_2$ and $\tilde{A}_3$ viewed as area segments

Table 4. illustrates the computations for obtaining x^*.

Table 4.1.

Area segment no.	Area (A)	$\bar{x}$	$A\bar{x}$
1.	$\frac{1}{2} \times 0.3 \times 1 = 0.15$	0.67	0.1005
2.	$2.6 \times 0.3 = 0.78$	2.3	1.794
3.	$0.3 \times 0.4 = 0.12$	3.8	0.456
4.	$\frac{1}{2} \times 0.4 \times 0.2 = 0.04$	3.8667	0.1546
5.	$1.5 \times 0.5 = 0.75$	4.75	3.5625
6.	$0.5 \times 0.5 = 0.25$	5.75	1.4375
7.	$\frac{1}{2} \times 0.5 \times 0.5 = 0.125$	5.833	0.729
8.	$1 \times 1 = 1$	6.5	6.5
9.	$\frac{1}{2} \times 1 \times 1 = 0.5$	7.33	3.665

In table 4. Area (A) shows the area of the segments of the aggregated fuzzy st and $\bar{x}$ shows the corresponding centroid. Now,

$$x^* = \frac{\sum A\bar{x}}{\sum A}$$

$$x^* = 18.353/3.715$$

Cenre of Sums Method

Here, unlike, centroid method the overlapping area is counted not once but twice. Making use of the aggregated fuzzy set shown in Fig. 4. the centre, of sums, x^* is given by

$$x^* = \frac{\frac{1}{2} \times 0.3 \times (3+5) \times 2.5 + \frac{1}{2} \times 0.5 \times (4+2) \times 5 + \frac{1}{2} \times 1 \times (3+1) \times 6.5}{\frac{1}{2} \times 0.3 \times (3+5) + \frac{1}{2} \times 0.5 \times \frac{1}{2} \times 1 \times (3+1)} = 5$$

Here, the areas covered by the fuzzy sets $\tilde{A}_1$, $\tilde{A}_2$, $\tilde{A}_3$ (Refer Figs. ??) are given by

$\frac{1}{2} \times 0.3 \times (3+5)$, $\frac{1}{2} \times 0.5 \times (4+2)$, and $\frac{1}{2} \times 1 \times (3+1)$ respectively.

Mean of Maxima Method

Since the aggregated fuzzy set shown in Fig. 4. is a continuous set, x^* the mean of maxima is computed as $x^* = 6.5$.

Here, $M = \{X \ \varepsilon \ [6, 7] \mid \mu(x) = 1\}$ and the height of the aggregated fuzzy set is 1.

Figure 4 shows the defuzzified outputs using the above three methods.

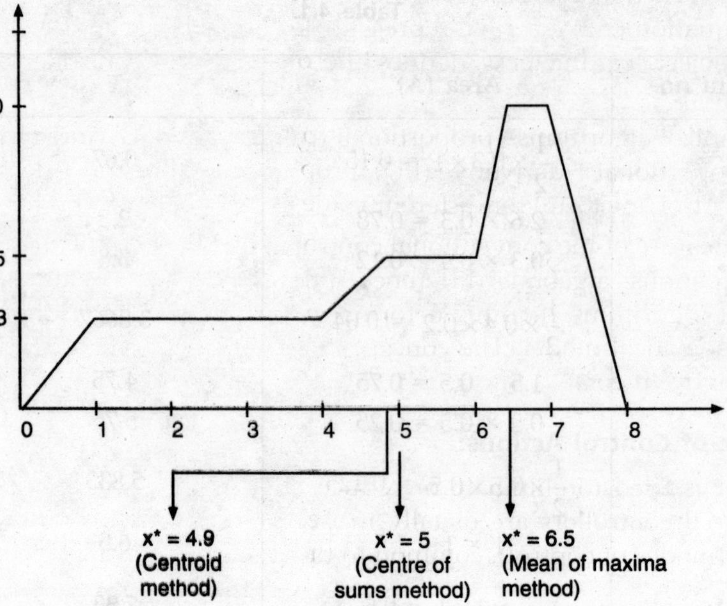

Fig. 4.29. Defuzzified outputs of the aggregate of $\tilde{A}_1$, $\tilde{A}_2$ and $\tilde{A}_3$

4.6. FUZZY CONTROLLER

4.6.1. Conventional Control System

The elementary basic diagram of a feed back control system is presented in Fig. (4.30). It consists of an error detector (or comparator) unit which compares the actual measured values of the variable coming through the feedback circuit with the set point (r) and produces an error signal, if any to guarantee a desired response of the output y. The process of keeping the output y close to the set point (reference input) r, despite the presence of disturbances fluctuations of the system parameters, and noise measurement is called regulation. The law governing corrective action of the controller is called the control algorithm. The output of controller is the control law is

$$u(k) = f\,(e(k),\, e\,(k-1),\, \ldots\, e\,(k-v),\, u\,(k-1),\, u\,(k-2),\, \ldots\, u\,(k-v))$$

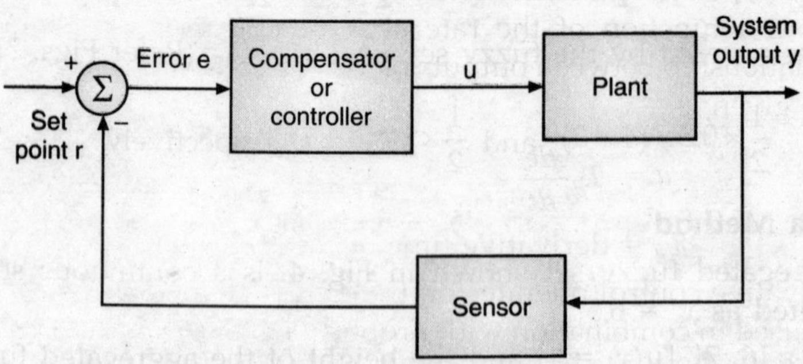

Fig. 4.30. Generalized conventional control system

providing a control action that describes the relationship between the input and output of the controller. In equation (2.2) $e = r - y$ represents the error between the desired set point and output of the object, parameter v defines the order of the controller and f is in general nonlinear function.

Different control algorithms– proportional (p) derivative (D), integral (I), proportional + integral (PI), proportional + derivative (PD), proportional+ integral + derivation (PID) can be derived from control law (3.16) for different values of parameter v and for different functions.

A common feature of the conventional control is that the control algorithm is analytically described by equations– algebraic, difference, differential, and so on. In general, the synthesis of such control algorithms requires a formalized analytical description of the controlled system by mathematical model. The concept of analyticity is one of the main paradigms of conventional control theory.

4.6.1.1. Modes of Control Actions:

A controller is used to eliminate or reduce the error by generating a correction signal Modern industrial controllers are usually made to produce one or a combination of basic control actions (modes of control), common to the industrial automatic controllers, described below.

***(i)* Proportional (P) Control Action:** A proportional (P) controller continuously adjusts the manipulated variable so that input to the process is approximately in balanced with the process demand. In proportional control, the output of the controller is proportional to the error. The relationship between output of the controller m and the error signal is given as–

$$u = k_p\, e$$

Where, u = output

k_p = a constant, called proportional sensitivity

***(ii)* Integral (*I*) or Reset Control Action:** In integral (I) control mode; the value of the controller output v is changed at a rate proportional to error signal *e*. The mathematical expression of integral– only controller is given by

$$u = \frac{1}{T_i}\int_0^t e\, dt$$

Where, T_i = integral time setting of the controller.

***(iii)*Derivative (D) or Rate Control Action:** In derivative control action, the output signal of the controller is a function of the rate at which the error is changing. Mathematical expression (the relationship between outputs of the controller to the input error) of derivative-only controller is given as

$$u = T_d \frac{de}{dt}$$

Where, T_d = derivative time of the controller.

Since the derivative control operates on the rate of change of the error and not on error itself, it is always used in combination with proportional or proportion– plus –integral control action.

***(iv)* Proportional–plus–Integral– plus derivative (PID) control Action:** In proportional plus–integral- plus-derivative (PID) control, also called three mode controller, the mathematical expression is given as:

$$u = k_p \left(e + \frac{1}{T_i} \int_0^t e \, dt + T_d \frac{de}{dt} \right)$$

***(v)* Proportional-plus-Integral (PI) Control Action:** In proportional–plus–integral (PI) Control, the output is proportional to a linear combination of the input error and time integral of the input error. Mathematical expression of PI controller is given as.

$$u = k_p \left(e + \frac{1}{T_i} \int_0^t e \, dt \right)$$

Where the integral time T_i adjusts the integral control action,

***(vi)* Proportional-plus-Derivative (PD) Control Action:** In proportional-plus–derivative (PD) control the derivative action causes the controller output to vary as the rate of change of the actuating error signal varies. Mathematical expression of PD controller given as.

$$u = k_p \left(e + T_d \frac{de}{dt} \right)$$

Where T_d is the derivative time, the time interval by which the rate action advances the offset of proportional action.

4.6.2. Fuzzy ControL

The seminal work by L. Zadeh on fuzzy algorithms [2] introduced the idea of formulating the control algorithm by logical rules. Since then, fuzzy logic control has gradually been recognized as the most significant and fruitful application for fuzzy logic and fuzzy set theory. In the past few years, advances in microprocessors and hardware technologies have created an even more diversified application domain for fuzzy controllers. Indeed, for complex and /or ill-defined system that are not easily subjected to conventional automatic control methods, fuzzy logic control provide a feasible alternative since they can capture the approximate, qualitative aspects of human reasoning and decision-marking processes.

So the main paradigm of fuzzy is that the control algorithm is knowledge-based algorithm, described by the methods of fuzzy logic. The fuzzy logic control system is a kind of expert knowledge-based system that contains the control algorithm in a simple rule base. The knowledge encoded in the rule-base is derived from human experience and intuition, and from theoretical and practical understanding of the dynamics of the controlled object, what makes the fuzzy control special and conceptually different from conventional control is the lack of an analytical description.

So regarding a fuzzy system as either an approximate reasoner or a functional approximator, it can be utilized in various ways in control systems. The most popular and the simplest usage is to use it as a direct controller as shown in Fig. 4.31. A basic requirement for implementing such a fuzzy controller is the availability of a control that provides the necessary knowledge for the present control problem.

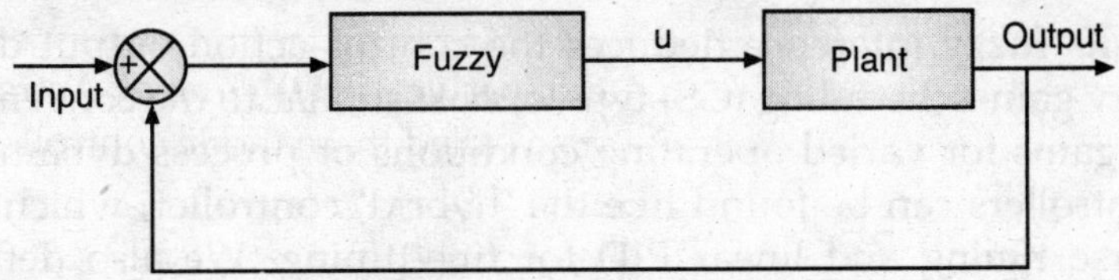

Fig. 4.31. Direct fuzzy controller

However, there are cases where either no expert exists or it is very difficult and expensive to derive control rules due, for example, to the complexity and the uncertainty of the controlled process. Under these circumstances a suitable rule-base must be constructed automatically by directly operating the process being controlled. So next in this chapter we will discuss the direct fuzzy controller which operates on some rule base.

4.6.3. Fuzzy Logic Controller: Basic Concept

By a fuzzy logic controller (FLC) we mean a control law that is described by knowledge based system consisting of if-then rules with vague predicates and a fuzzy logic inference mechanism.

The rule-based is the main part of the FLC. It is formed by a family of logical rules that describes the relationship between input and output u of the controller. In general we can represent the FLC in a form similar to the conventional control law.

$$u(k) = F\ (e(k),\ e\ (k-1), \ldots .e\ (k-v),\ u(k-1),\ u(k-2),\ \ldots u(k-v))$$

Where the function *F*, the control law, is described by the rule base. However, this doesn't mean that the FLC is a kind of transfer function or difference equation. The knowledge-based nature of the FLC dictates a limited usage of the past values of error e and control u because it is rather unreasonable to expect meaningful linguistic statement for $e(k-3)$, $e(k-4)$, $u(k-3)$, $u(k-4)$ and so on. In addition different combinations of values of e (.) and u (.) that have physical meaning for example, error change $\Delta e\ (k) = e(k) - e(k-1)$, error sum $\Sigma e(k) = \Sigma e(k)$, Control change $Du(k) = u(k) - u(k-1)$, and So, on can be considered in the FLC

A typical FLC describes the relationship between the change of the control $Du(k) = u(k) - u(k-1)$ on one hand, and the error $e(k)$ and its change $De(k) = e\ (k-1) - e\ (k-1)$ on the other hand. Such as control law can be formalized as

$\Delta\ u(k) = F(e\ (k), \Delta e(k)$ and is a manifestation of the general FLC of expression (2.8) with $v = 1$.

4.6.4. Classification of Fuzzy Controllers

In view of PID control principles, we suggest a classification of the existing fuzzy controllers as shows in Fig. 4.32. If a fuzzy controller is designed (or implied equivalently) to generate the control action within the proportional/ integral/ derivative (P/I/D) concept (s) like a conventional PID controller, we call it a fuzzy PID-like (or fuzzy PID) controller. The first fuzzy controller developed by Mamdani [16] is a fuzzy PI controller. The type of fuzzy non-PID controllers can be found in model- based fuzzy controllers, such as the MIMO models using the T-S-K consequent representation. We term direct action type for the Mandani's controller, since its fuzzy inference deduces the control–action output directly to drive the process. The fuzzy gain–scheduling (GS) type is most similar to the conventional GS controller in changing the gains for varied operating conditions or process dynamics. The other class of fuzzy PID controllers can be found like the 'hybrid' controller, which employs the fuzzy PID for the coarse tuning and linear PID for fine-tuning. We also define the Mamdani's controller to be composed–action (CA) type since its fuzzy inference output is a composed force of proportional and integral actions.

MIMO model is using the T-S-K cnosequent representation. We term direct action type for the Mandani's controller, since its fuzzy inference deduces the control-action output directly to drive the process. The fuzzy gain-scheduling (GS) type is most similar to the conventional GS controller in changing the gains for varied operating conditions or process dynamics. The other class of fuzzy PID controllers can be found like the 'hybrid' controller, which employs the fyzzy PID for the coarse tuning and linear PID for fine-tuning. We also define the Mamdani's controller to be composed-action (CA) type since its fuzzy inference output is a composed force of proportional and integral actions.

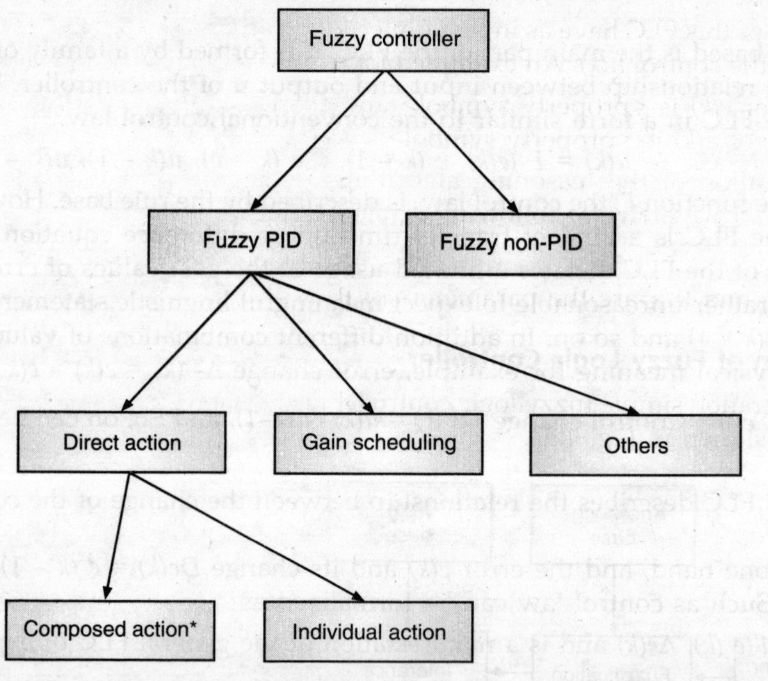

Fig. 4.32. Classification of fuzzy controllers inview of control actions (The asterisk indicates the commonly used type of controllers compared with its counterparts)

Another type of classification of fuzzy controllers can be made based on the dimensionality (defined as the total number of input variables) of the inference engines. In this dissertation work we adopt the Mamdani's two–input fuzzy inference structure. They used e and De as the input variables to the fuzzy inference.

4.6.5. Fuzzy PI Controller

In this section, we briefly describe the design of the fuzzy PI controller: its mathematical derivation and technical aspects. The fuzzy PI controller is actually a specially designed conventional digital controller.

The PI– like FLC describes with the aid of fuzzy if-then rules the relationship between the change of the control Du_{PI} *(k)*, and the error *e(k)* and its change $De(k) = e(k) - e(k-1)$ on other hand.

$$\Delta u_{pI}\ (k) = f\ (e(k),\ \Delta e\ (k)$$

i.e., $$\Delta u_{pI}\ (k) = k_{cen}\ \Delta e\ (k) + k_{en}\ \ e(k)$$

Where, $$u_{PI}\ (k) = u_{PI}\ (k-1) + k_u\ \Delta_{uPI}\ (k)$$

Where k_{cen}, k_{en} and k_u are the parameters of the fuzzy PI (FPIC) controllers. Thus for PI–like FLC consists the rules of the form.

If *e(k)* is < property symbol > and Δe is < property symbol > then Δu_{PI} *(k)* < property symbol >

4.6.6. Fuzzy PD Controller

If the input variable of the FLC is not the change of the control action *Du(k),* but the control value *u(k),* itself, we obtain an alternative types of FLC, which realizes a control law.

$$u\ (k) = f\ (e\ (k),\ \Delta e\ (k))$$

The rules of this FLC have as inputs (antecedents) the error *e(k)* and its changes $\Delta e(k)$, and as on output the control *u(k).* An example of a rule base of such an FLC is in following form.

If the error *e(k)* is <property symbol> and change in error $\Delta e(k)$ is <property symbol> then the control *u(k)* is <property symbol>

By application of the reasoning algorithm considering *u(k)* as an output of the FLC, instead of $\Delta u(k)$ we arrive at following implementing law for PD–like FLC.

$$u(k = k_{en}\ \ e\ (k) + k_{cen}\ \ \Delta e(k)$$

Where k_{en} and k_{cen} are the parameters of the fuzzy PD controller

4.6.7. Design of Fuzzy Logic Controller

First generation simple fuzzy logic controller can generally be depicted by a block diagram such as that shown in Fig. 4.33.

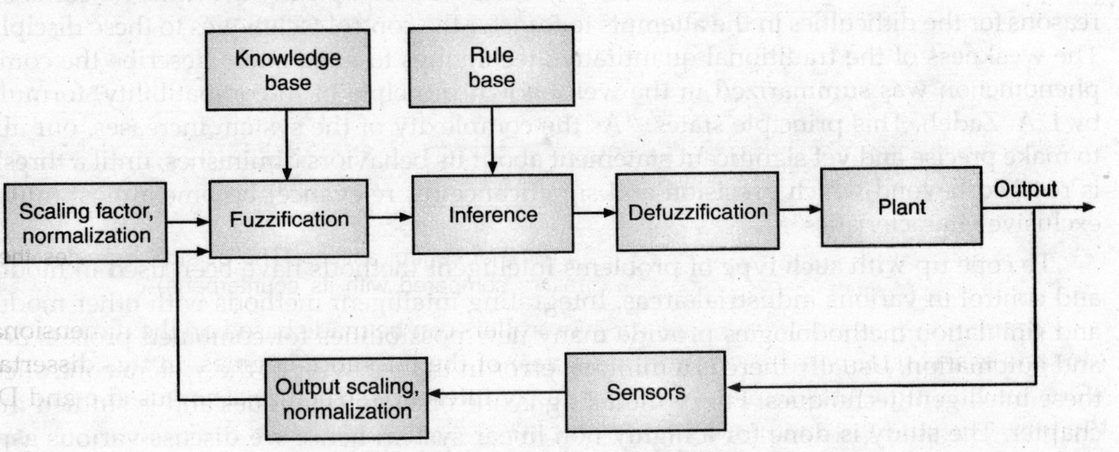

Fig. 4.33. A simple fuzzy ontrol system block diagram

The knowledge base module in Fig. 4.33. contains knowledge about all the input and output fuzzy partitions. It will include the term set and the corresponding membership functions defining the input variables to the fuzzy rule–base system and the output variables, or control actions, to the plant under control.

The steps in designing a simple fuzzy logic system are as follows.

1. Identify the variables (inputs, states, and outputs) of the plant.
2. Partition the universe of discourse or the interval spanned by each variable into a number of fuzzy subsets, assigning each linguistic label (subsets include all the elements in the universe)
3. Assign or determine a membership function for each fuzzy subset.
4. Assigning the fuzzy relationships between the inputs or states fuzzy subsets on the one hand and outputs fuzzy subsets on the other hand, thus forming the rule–base.
5. Choose appropriate scaling factors for the input and output variables in order to normalize the variable to [0, 1] or the [–1, 1] interval.
6. Fuzzify the inputs to the controller.
7. Use fuzzy approximate reasoning to infer the output contributed from each rule.
8. Aggregate the fuzzy output recommended by each rule.
9. Apply defuzzification to form a crisp output.

The above nine steps are fixed for the simple fuzzy logic controller but in an adaptive fuzzy controller, they are adaptively modified based on some adaptation law in order to optimize

4.7. ANALYSIS AND DESIGN FOR T-S FUZZY MODEL

4.7.1. Introduction

The complexity of this real world is most dramatically reflected in soft areas such as social, economic, ecological and biological system. Multidimensionality, hierarchal structures, mutual interactions, internal feedback mechanisms and unpredictable dynamics are only a part of the characteristics of such complex systems. This complexity accounts for some of the reasons for the difficulties in the attempts to transfer the control techniques to these disciplines. The weakness of the traditional quantitative techniques to adequately describe the complex phenomenon was summarized in the well known principle of in compatibility, formulated by L.A. Zadeh. This principle states, "As the complexity of the system increases, our ability to make precise and yet significant statement about its behaviors diminishes, until a threshold is reached beyond which precision and significance (or relevance) become almost mutually exclusive characteristics".

To cope up with such type of problems intelligent methods have been used in modeling and control in various industrial areas. Integrating intelligent methods with other modeling and simulation methodologies provide many new possibilities for combined process design and automation. Usually there is a minimal error in the behavior of the system controlled by these intelligent techniques. Fuzzy modeling is one of such techniques and is studied in this chapter. The study is done for a highly non linear system hence we discuss various aspects about fuzzy modeling and design a fuzzy logic controller (FLC) a highly complex non linear system i.e. inverted pendulum. Fuzzy logic control systems consist of a plant and a fuzzy logic controller as shown in Fig. 4.34. The output of the FLC is made by defuzzification method. So, the output of the FLC is a function of the degrees of membership of the fuzzy rules, and these degrees of membership are function of input variables that are the system states. Therefore, the control system become highly non-linear and the analysis of system stability for this system are very difficult. It is proved that if each fuzzy logic control systems consist of a plant and each fuzzy logic rule is stable in the sense of Lyapunov under common Lyapunov function, the overall system is also stable in the sense of Lyapunov. Tanaka and Sugeno proposed a stability design approach, which first modeled the plant by a Takagi–Sugeno (TS) fuzzy model. This fuzzy model represents the plant as a weighted sum of a set of linear state equations. An FLC is designed based on this fuzzy plant model. Then, Lyapunov's direct method can be applied to each fuzzy subsystem that is formed by each rule of the fuzzy plant model and the FLC. The stability of the whole system can be ensured if a required positive-definite matrix exists. Other stability design approaches related to this fuzzy-model-based approach can also be found in various text and research papers.

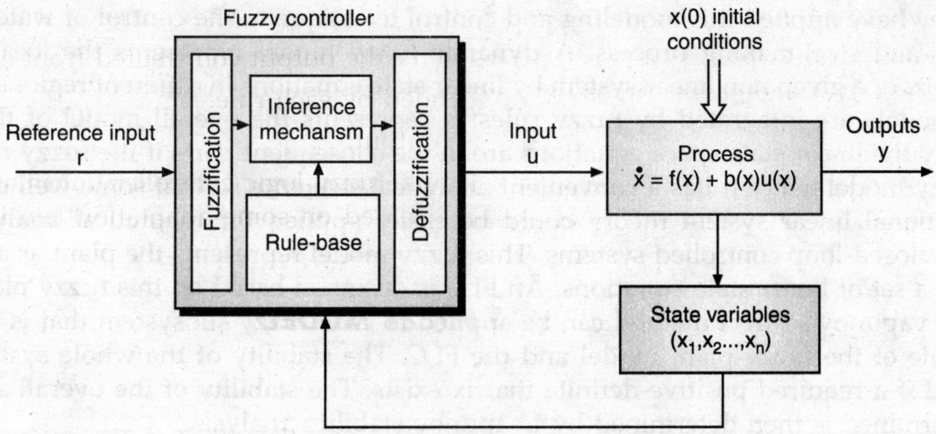

Fig. 4.34. Fuzzy logic control system

4.7.2. Fuzzy Modeling

Fuzzy modeling is an extension of expert system technique to uncertain and vague systems. The advantages in modeling and translate directly into more advance methodology, better understanding of fundamental concepts of fuzzy sets, and design practices of user friendly models.

Briefly speaking, fuzzy models of system comprise two main properties:

- They operate at the level of linguistic terms or fuzzy sets and similarly all systems dependencies can be portrayed in the same linguistic format.
- They represent and process uncertainity.

Fuzzy system models basically are of two types, which differ fundamentally in their ability to represent different types of information. The first type includes linguistic models (LMs) that are base on collections of if then rules with vague predicates and use fuzzy reasoning. In these models, fuzzy quantities are associated with linguistic labels and they are essentially qualitative expressions of the system. Models of this type form the basis for qualitative that describes the system behavior by using a natural language. The fuzzy logic controller is a prototypical example of a linguistic model. In that model the rules give a linguistic expression of the control strategy that is based on common sense.

The second strategy of fuzzy models is based on the Takgai-Sugeno-Kang (TSK) method of reasoning that was proposed by Sugeno and his coworkers. These models are formed by logical rules that have a fuzzy antecedent's part and functional consequent; essentially they are the combination of fuzzy anf non fuzzy models. Fuzzy moels based on the TSK method of reasoning, integrate the ability of LMs for qualitative knowledge representation with an effective potential or expressing qualitative information as well. In addition, this type of fuzzy models permits a relatively easy application of powerful learning techniques for their identification from data.

Thus among various kinds of fuzzy control methods, fuzzy model based control is widely used because the design and analysis of the overall fuzzy system can be systematically performed using the well established classical linear system theory. For e.g. Takagi and Sugeno suggested representing a non-linear system by a combination of some local linear sub systems.

They have applied this modeling and control technique to the control of water cleaning process and steel-making process. A dynamic fuzzy model represents the local dynamic behaviors of a given non linear system by linear state equations in different regions and these local models are integrated by fuzzy rules to represents the overall model of the system. Actually the linear state space equations are in the consequent parts if the fuzzy rules of the TS fuzzy model hence it has a convenient analytical structure so that some well established conventional linear system theory could be easily applied for theoretical analysis of the overall closed-loop controlled systems. This fuzzy model represents the plant as a weighted sum of a set of linear state equations. An FLC is designed based on this fuzzy plant model. Then, Lyapunov's direct method can be applied to each fuzzy subsystem that is formed by each rule of the fuzzy plant model and the FLC. The stability of the whole system can be ensured if a required positive-definite matrix exists. The stability of the overall system can be determined is then determined by Lyapunov stability analysis.

4.7.3. Fuzzy logic Control System

Let X be a universe of discourse. Consider a single-input nth-order nonlinear system of the following form:

$$\dot{x} = f(x) + b(x)u$$

where, $x \; \varepsilon \; X$, $x = [x_1, x_2, \ldots x_n(x)]$ is the state vector

$$f(x) = [f_1(x), f_2(x), \ldots f_n(x)]^T$$

and $$b(x) = [b_1(x), b_2(x), \ldots b_n(x)]^T$$

are functions describing the dynamics of the plant.

u is the control input of which the value is determined by an FLC.

The ith IF–THEN rule in the fuzzy rule base of the FLC is of the following form:

Rule i: IF x_1 is X_{i1} AND x_2 is X_{i2} AND…AND x_n is X_{in}, THEN $u = u_i(x)$, $i = \overline{1, r}$ (4.2)

where, $X_{i1}, X_{i2}, \ldots X_{in}$ are fuzzy sets which describe the linguistic terms (LT) of input variables and u_i describe the linguistic terms of output variables. Each fuzzy rule generates an activation degree: $\alpha_i \; \varepsilon = i = \overline{1, r}$

$\alpha_i(x(t)) = \min(\mu_{i,1}(x_1(t)), \mu_{i,2}(x_2(t)) \ldots \mu_{i,n}(x_n(t))$ u_i can be a single value for a function of state vector, $x(t)$.It is assumed that for any $x \varepsilon X$ in the input universe of discourse X, there exists at least one $\alpha_i \varepsilon[0, 1]$, $i = \overline{1, r}$ among all rules that is not equal to zero. The control input u, which must be applied to process, is a function of α_i and u_i .By applying the weighted sum defuzzification method, the output of the FLC is given by: $u = \frac{\sum_{i=1}^{r} a_i u_i}{\sum_{i=1}^{r} a_i}$ where r is the total number of the ru!es.

Property 1: For any input $x_0 \varepsilon X$, exist two rules p, q such that $u_{\min}(x_0) = u_p(x_0) \le u_0(x_0) \le u_q(x_0) = u_{\max}(x_0)$

Proof: Let $x_0 \varepsilon X$ then among all rules, can be found two rules p and q such that

$$u_p(x_0) = u_{\min}(x_0),\ u_q(x_0) = u_{\max}(x_0) \text{ where,}$$

$$u_{\min}(x_0) = \min_{i=1,2\ldots r}(u_i(x_0)),\ u_{\max}(x_0) = \max_{i=1,2\ldots r}(u_i(x_0))$$

$$\therefore u_{\min}(x_0) = \frac{\sum_{i=1}^{r}\alpha_i(x_0)\,u_{\min}(x_0)}{\sum_{i=1}^{r}\alpha_i(x_0)} \le \frac{\sum_{i=1}^{r}\alpha_i(x_0)\,u_i(x_0)}{\sum_{i=1}^{r}\alpha_i(x_0)} \le \frac{\sum_{i=1}^{r}\alpha_i(x_0)\,u_{\max}(x_0)}{\sum_{i=1}^{r}\alpha_i(x_0)} = u_{\max}(x_0)$$

$$u_{\min}(x_0) \le u_{\max}(x_0) \Rightarrow u_{\min}(x) \le u(x) \le u_{\max}(x),\ \forall\, x\, \varepsilon X$$

In conclusion, we can say that condition given by (3) holds true for all rules.

4.7.4. Design of Stable Fuzzy Logic Controllers

The method for stability analysis proposed in this paper is based on the following theorem. Each subsystem consist from one fuzzy rule and the process described. It is proved that if each subsystem is stable in the sense of Lyapunov, under a common Lyapunov function, the overall system is also stable in sense of Lyapunov.

Theorem: If a quadratic and positive-definite matrix P exists and:

1. $V(x) = x^T Px \to \infty$ as $\|x\| \to \infty$, $V(0) = 0$
2. $\dot{V}(x) < 0, \forall x \neq 0, \dot{V}(0) = 0$

in respect with any rule, then the system composes by FLC of type Takagi-Sugeno and the process described by, is globally asymptotically stable in at the origin. Based on this theorem we can be found the algorithm for designed stable fuzzy logic controllers.

Definition: A fuzzy subsystem associated with fuzzy rule i is a system with a plant of form (4.1) controlled by only u_i, which is the output of fuzzy rule i of the form.

4.7.5. Algorithm for Design of Stable Fuzzy Logic Controllers

The idea of the proposed stability analysis algorithm is to break down the problem of analyzing the stability of the whole fuzzy logic control system into analyzing the stability of the fuzzy subsystems individually. The complexity of the analysis is drastically decreased as it is easier to check whether every fuzzy subsystem can give a negative-definite V & for a given Lyapunov function V.

The steps of the algorithm are:

Step 1: Determine the equation of the dynamics of the plant;

Step 2: Choose the membership functions of the input linguistics terms;

Step 3: Implement fuzzy rules database;

Step 4: Define the Lyapunov function and we find the derivative of this function;

Step 5: analyzing the stability of the fuzzy subsystems individually, and determine the control signal u_i such that each fuzzy subsystems to be stabilizing.

Next, from the theorem 2 results that the whole fuzzy logic control system is stable. The stability proof of fuzzy logic control systems can be carried out by first applying Lyapunov direct method to each rule. If every rule individually applying to the plant of (4.1) results in a stable sub-system in the sense of Lyapunov subject to a common Lyapunov function, the whole fuzzy logic control system is stable.

4.7.6. Illustrative Example

Consider a nonlinear inverted pendulum system Fig. 4.35. The problem is to balance the inverted pendulum. Here, a simulated inverted pendulum system is used. The system amounts to an inverted pendulum mounted on a cart. The diagram below depicts this system.

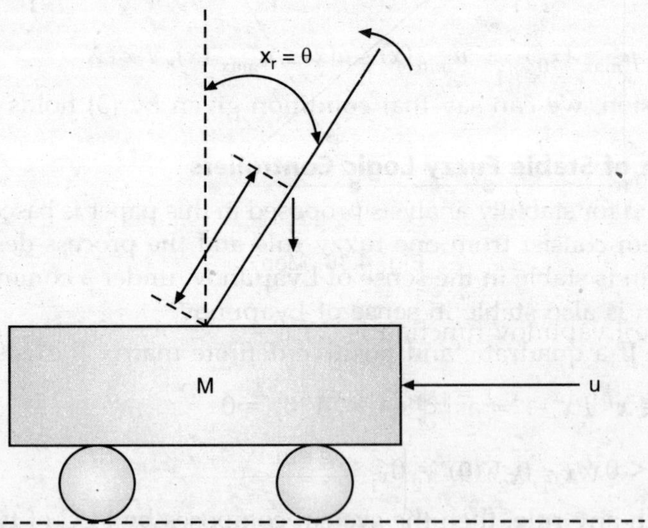

Fig. 4.35. A cart pendulum inverted problem

where, m – the mass of the pendulum;

M – the mass of the cart;

l – the length of the pendulum;

g – the acceleration due to gravity;

θ – the angle of the pendulum from the vertical;

u – is the force applied to the cart (in N).

Step 1: The angle of the pendulum from the vertical $x_1 = \theta$ and the angular velocity $x_2 = \dot{\theta}$ are chosen as the state variables (input linguistics terms).The system to be controlled by:

$$\begin{cases} \dot{x}_1 = x_2 \\ \dot{x}_2 = \dfrac{g \sin x_1 - amlx_2^2 \sin x_1 \cos x_1 - ua \cos x_1}{4l/3 - aml \cos^2 x_1} \end{cases}$$

Step 2: The membership functions of the input linguistics terms are shown in Fig 4.36.

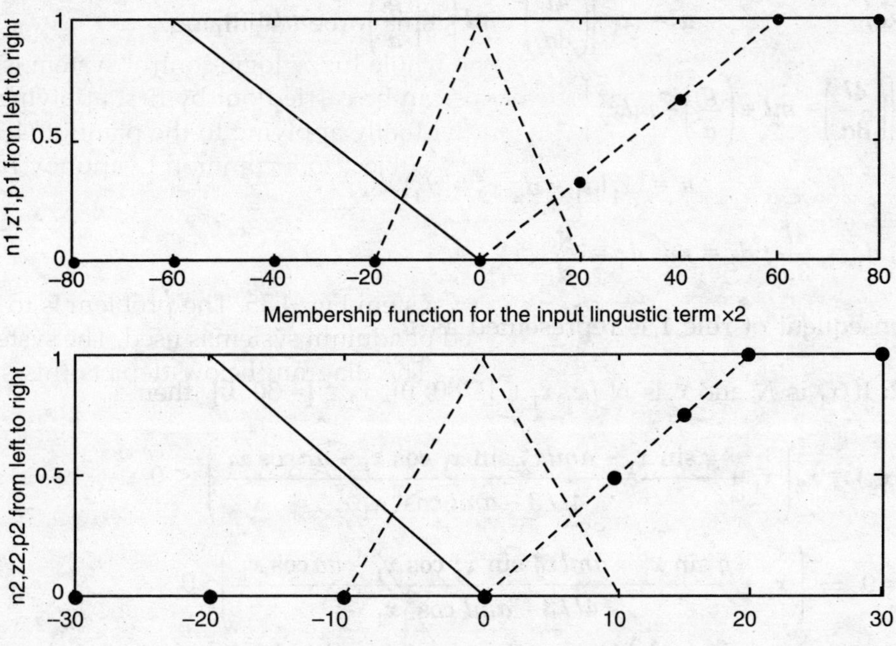

Fig. 4.36. Membership Function

Step 3: The Lyapunov function is: $V(x) = \dfrac{1}{2}(x_1^2 + x_2^2)$

Then, $\dot{V}(x_1, x_2) = x_1 \dot{x}_1 + x_2 \dot{x}_2 = x_2(x_1 + \dot{x}_2)$

$$\Rightarrow \quad \dot{V}(x_1, x_2) = x_2 \left(x_1 + \frac{g \sin x_1 - amlx_2^2 \sin x_1 \cos x_1 - ua \cos x_1}{4l/3 - aml \cos^2 x_1} \right)$$

$$\Rightarrow \quad \dot{V}(x_1, x_2) < 0$$

This is the required condition for stability.

Step 4: Analyzing the stability of the fuzzy subsystems individually, and determine the control signals u_i such that each fuzzy subsystems to be stabilizing.

Rule 1: If x_1 is P and x_2 is P *i.e.* $x_1 \in [0,80], x_2 \in [0,30]$, then

$$\dot{V}(x_1, x_2) = x_2\left(x_1 + \frac{g \sin x_1 - amlx_2^2 \sin x_1 \cos x_1 - ua\cos x_1}{4l/3 - aml \cos^2 x_1}\right) < 0$$

as $$x_2 \neq 0 \Rightarrow \left(x_1 + \frac{g \sin x_1 - amlx_2^2 \sin x_1 \cos x_1 - ua\cos x_1}{4l/3 - aml \cos^2 x_1}\right) < 0$$

Therefore the control variable must be satisfying the condition given below:

$$u > x_1\left\{\left(\frac{4l}{3a}\right)\sec x_1 - ml\cos x_1\right\} + \left(\frac{g}{a}\right)\tan x_1 - mlx_2^2 \sin x_1$$

We take, $$u = x_1\left\{\left(\frac{4l}{3a}\right) - ml\right\} + \left(\frac{g}{a}\right)x_1 - mlx_2^2\, x_1$$

$$\Rightarrow x_1\left\{\left(\frac{4l}{3a}\right) - ml + \left(\frac{g}{a}\right) - mlx_2^2\right\}$$

$$\Rightarrow \qquad u = x_1\{d_1 - d_2\, x_2^2 + d_3\}$$

where, $d_1 = \dfrac{4l}{3a}$; $d_2 = ml$; $d_3 = \dfrac{g}{a}$

The consequent of rule 1 is represented as u_1

Rule 2: If x_1 is N and x_2 is N *i.e.* $x_1\ \varepsilon\ [-80, 0]$, $x_2\ \varepsilon\ [-30, 0]$, then

$$\dot{V}(x_1, x_2) = x_2\left(x_1 + \frac{g \sin x_1 - amlx_2^2 \sin x_1 \cos x_1 - ua\cos x_1}{4l/3 - aml \cos^2 x_1}\right) < 0$$

as $$x_2 \neq 0 \Rightarrow \left(x_1 + \frac{g \sin x_1 - amlx_2^2 \sin x_1 \cos x_1 - ua\cos x_1}{4l/3 - aml \cos^2 x_1}\right) > 0$$

Therefore the control variable must be satisfying the condition given below:

$$u > x_1\left\{\left(\frac{4l}{3a}\right)\sec x_1 - ml\cos x_1\right\} + \left(\frac{g}{a}\right)\tan x_1 - mlx_2^2 \sin x_1$$

We take, $$u = x_1\left\{\left(\frac{4l}{3a}\right) - ml\right\} + \left(\frac{g}{a}\right)x_1 - mlx_2^2\, x_1$$

$$\Rightarrow x_1\left\{\left(\frac{4l}{3a}\right) - ml + \left(\frac{g}{a}\right) - mlx_2^2\right\}$$

$$\Rightarrow \qquad u = x_1\{d_1 - d_2\, x_2^2 + d_3\}$$

where, $d_1 = \dfrac{4l}{3a}$; $d_2 = ml$; $d_3 = \dfrac{g}{a}$

The consequent of rule 2 is represented as u_2

Rule 3: If x_1 is N and x_2 is N *i.e.* x_i *e* [0, 80], x_2 ε [– 30, 0] then the control variable must be u_3 same as given by rule 2.

Rule 4: $x_1 \left\{\left(\frac{4l}{3a}\right) - ml + \left(\frac{g}{a}\right) - mlx_2^2\right\}$

If x_1 is N and x_2 is P *i.e.*, x_i *e* [– 80, 0], x_2 ε [0, 30] then the control variable must be u_4 same as given by rule 1

Rule 5: If x_1 is P and x_2 is Z *i.e.* x_i *e* [0, 80], x_2 ε [– 5, 5], then to ensure $\dot{V}(x_1, x_2) < 0$, let it be equals to $-x_2^2$.

$$u = (x_1 + x_2)\left\{\left(\frac{4l}{3a}\right)\sec x_1 - ml\cos x_1\right\} + \left(\frac{g}{a}\right)\tan x_1 - mlx_2^2 \sin x_1$$

The consequent of rule 5 is represented as u_5.

Rule 6: If x_1 is N and x_2 is Z *i.e.* x_1 ε [– 80, 0], x_2 ε [– 5, 5], then the control variable must be u_6 same as given by rule 5.

Rule 7: If x_1 is Z and x_2 is P *i.e.*, x_1 ε [– 10, 10], x_2 ε [0, 30], then the control variable must be u_7 same as given by rule 1.

Rule 8: If x_1 is Z and x_2 is N *i.e.* x_1 ε [– 10, 10], x_2 ε [– 30, 0], then the control variable must be u_8 same as given by rule 2.

Rule 9: If x_1 is Z and x_2 is Z *i.e.* x_1 ε [– 10, 10], x_2 ε [– 5, 5], then the control variable must be u_9 same as given by rule 5.

Step 5: The fuzzy rules database is given in table 4.2.

The control signals u_i must be determinate by applying the theorem 2 where, $i = \overline{1, r}$ *i.e.* 1 to 9.

Table 4.2. Rule base matrix

x_1 / x_2	P	N	Z
P	u_1	u_3	u_5
N	u_4	u_2	u_6
Z	u_7	u_8	u_9

4.7.7. Results (and Discussion)

The designed FLC is applied to the process described by equation (4.4), and simulation is done in MATLAB. Results are on the last page for

$$m = 0.5 \text{ kg};\ M = 9 \text{ kg};\ g = 9.81 \text{ m/s}^2;\ 1 = 5 \text{ m};\ a = \frac{1}{m + M} = 0.1053;$$

For the initial states: $x_1(0) = 5$; $x_2(0) = -3$ Responses of x_1, x_2 and the control effort are shown in the Fig. 4.37 (*a*) and (*b*) respectively.

For the initial states: $x_1(0) = 10$; $x_2(0) = -3$; Responses of x_1, x_2 and the control effort are shown in the Fig. 4.38 (*a*) and (*b*) respectively.

For the initial states: $x_1(0) = 15$; $x_2(0) = -3$ Responses of x_1, x_2 and the control effort are shown in the Fig. 4.39 (*a*) and (*b*) respectively.

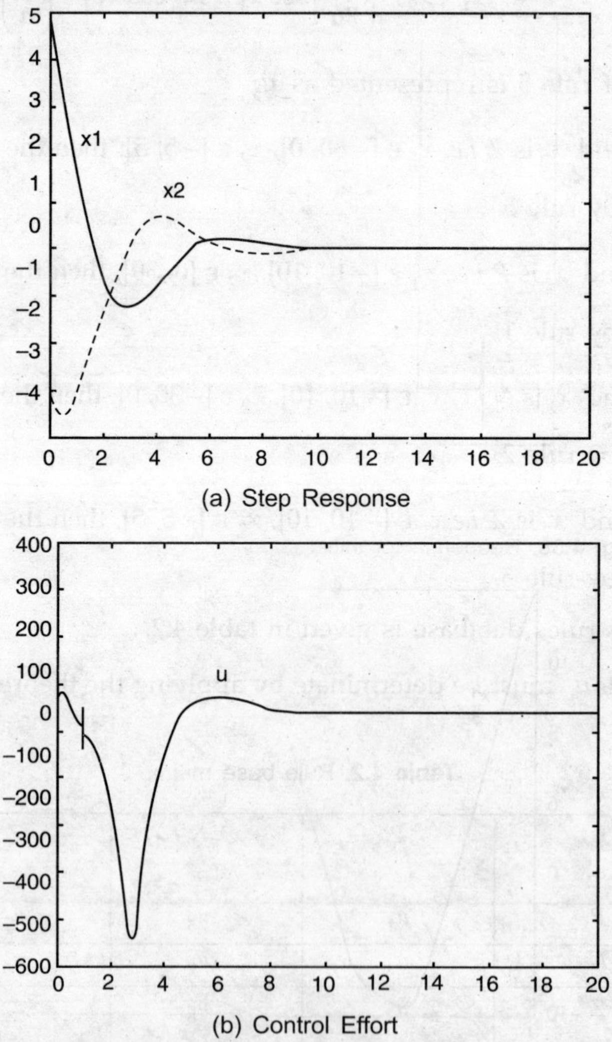

(a) Step Response

(b) Control Effort

Fig. 4.37. Response for initial condition $x_1(0) = 5$; $x_2(0) = -3$

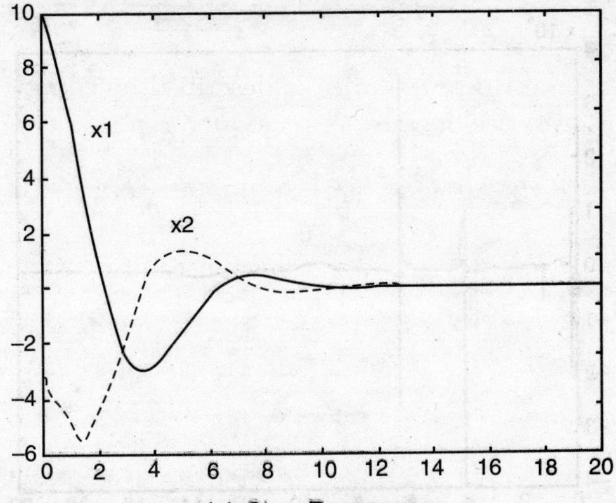

(a) Step Response

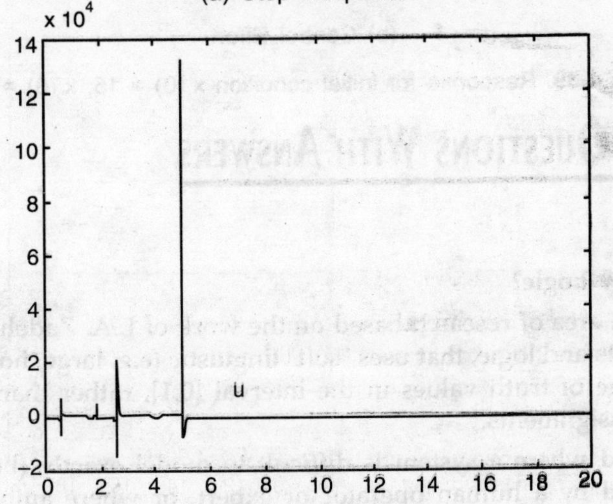

(b) Control Effort

Fig. 4.38. Response for initial condition $x_1(0) = 10$; $x_2(0) = -3$

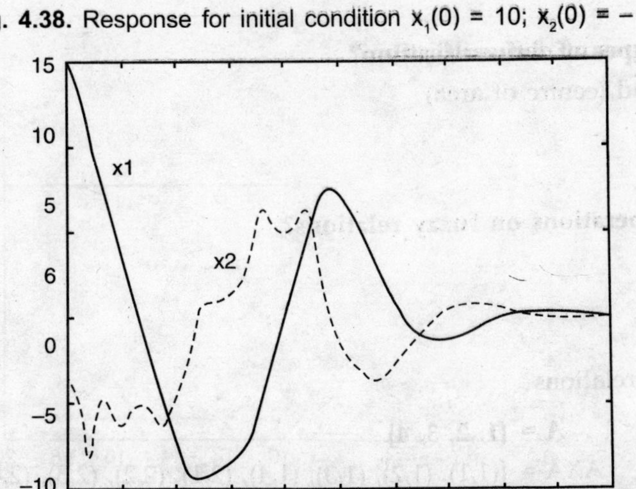

(a) Step Response

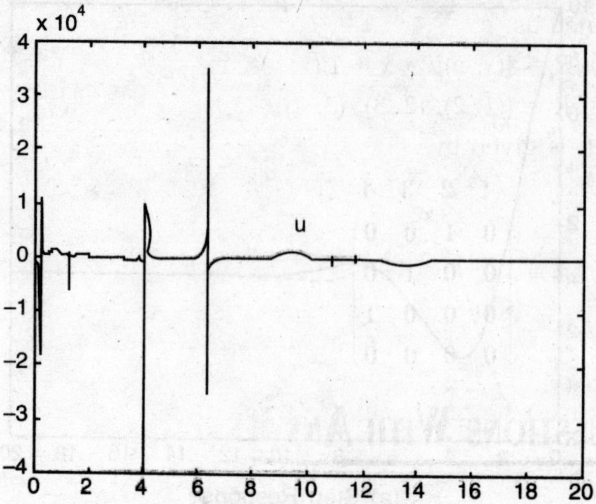

(b) Control Effort

Fig. 4.39. Response for initial condition $x_1(0) = 15$; $x_2(0) = -3$

Short Questions With Answers

Q.1. What about Fuzzy Logic?

Ans. Fuzzy Logic is an area of research based on the work of L.A. Zadeh. It is a departure from classical two-valued sets and logic, that uses "soft" linguistic (e.g. large, hot, tall) system variables and a continuous range of truth values in the interval [0,1], rather than strict binary (True or False) decisions and assignments.

Fuzzy logic is used where a system is difficult to model exactly (but an inexact model is available), is controlled by a human operator or expert, or where ambiguity or vagueness is common. A typical fuzzy system consists of a rule base, membership functions, and an inference procedure.

Q.2. What are the types of defuzzification?

Ans. Centroid method (centre of area)

Centre of sum

Mean of max

Q.3. What are the operations on fuzzy relations?

Ans. Union

Intersection

Complement

Compostion of relations

Q.4. Given, $A = \{1, 2, 3, 4\}$

Ans. AXA = {(1,1), (1,2), (1,3), (1,4), (2,1), (2,2), (2,3), (2,4), (3,1)(3, 2), (3, 3), (3, 4), (4, 1),(4, 2), (4, 3), (4, 4)}

Let relation R be defined as

$$R = \{(x, y)y = x + 1,(x, y) \varepsilon X\}$$

Then, $R = \{(1, 2), (2, 3), (3, 4)\}$

Then relation matrix R is given by

$$R = \begin{array}{c} \begin{matrix} \mathbf{1} & \mathbf{2} & \mathbf{3} & \mathbf{4} \end{matrix} \\ \begin{pmatrix} 0 & 1 & 0 & 0 \\ 0 & 0 & 1 & 0 \\ 0 & 0 & 0 & 1 \\ 0 & 0 & 0 & 0 \end{pmatrix} \end{array}$$

Long Questions With Answer

Q.1. What are different industrial applications of fuzzy logic?

Ans. Different applications of fuzzy logic are:

1. Two well-developed applied areas of fuzzy set theory they are directly relevant to industrial engineering are fuzzy control and fuzzy decision making fuzzy sets are convinie for estimating the service life of a given piece of equity for various conditions under which it operates.
2. **Robotics:** Robotics is a cross-dissiplinary subject is concerned with the design and construction of maximum that are capable of human-like behaviour to achieve human like behaviour an intelligent robot must have some key capable that distinguish human beings from classical non-intelligent machines. It is clear that the use of fuzzy set theory is essential for all these capabilities. The main subjects of fuzzy that are relevant include approximate reasoning, fuzzy systems, fuzzy controllers and fuzzy pattern recognition and fuzzy decision making.
3. **Medicine:** Medicine is one field in which the applicability fuzzy set theory was recognized itis the uncertainity four in process of diagnosis of diseases that has most frequently been the focus of applications of fuzzy set theory.
4. An application of genetic algorithms for altering members functions of fuzzy controllers on-line was developed in context of chemical engineering.
5. Among other areas connected with natural sciences, there fuzzy set theory has been observed in meteorology earthquake research.
6. An interesting application of fuzzy set theory is fuzzy questionnaires. This application is quite important, since questionnaires are fundamental tools for analyzing structures of informed.
7. One area in which fuzzy set theory has great potential is psychology.
8. Expert systems for medical diagnosis and treatment planner that are based on approximate resoaning is another application.

Q.2. Verify max-composition. With an example.

Ans. Given, the relation matrices of the relation R and S, the max-composition is defined as

$$T = R_0S$$

$$T(x, z) = \max(\min(R(x, y)), S(y, z)))$$

where $y \varepsilon Y$

For example:

Let R, S be defined on sets {1, 3, 5} × {1, 3, 5}.

$$R = \{x, y)/ \ y = x + 2\}$$

$S = \{(x, y)/ \ x < y\}$. Find max-min composition

$$RXS = \{(1, 1), (1, 3), (1, 5), (3, 1), (3, 3), (3, 5), (5, 1), (5, 3), (5, 5)\}$$

$$R = \{(1, 3), (3, 5)\}$$

$$S = \{(1, 3), (1, 5), (3, 5)\}$$

$$R = \begin{array}{c} \\ 1 \\ 3 \\ 5 \end{array}\begin{array}{c} \begin{array}{ccc} 1 & 3 & 5 \end{array} \\ \begin{bmatrix} 0 & 1 & 0 \\ 0 & 0 & 1 \\ 0 & 0 & 0 \end{bmatrix} \end{array} \quad S = \begin{array}{c} \\ 1 \\ 3 \\ 5 \end{array}\begin{array}{c} \begin{array}{ccc} 1 & 3 & 5 \end{array} \\ \begin{bmatrix} 0 & 1 & 1 \\ 0 & 0 & 1 \\ 0 & 0 & 0 \end{bmatrix} \end{array}$$

Using max-min composition

$$R_o S(1, 1) = \max\{\min(0, 0), \min(1, 0) \min(0, 0)\}$$

$$= \max(0, 0, 0) = 0$$

$$R_o S(1, 3) = \max(0, 0, 0) = 0$$

$$R_o S(1, 5) = \max(0, 1, 0) = 1$$

Similarly, $R_o S(3, 1) = 0$

$$RoS(3, 3) = RoS(3, 5) = RoS(5, 1) = RoS(5, 3) = RoS(5, 5) = 0$$

Hence, $R_o S = \begin{bmatrix} 0 & 0 & 1 \\ 0 & 0 & 0 \\ 0 & 0 & 0 \end{bmatrix} = S_o R$ **Ans.**

Q.3. What is fuzzification and defuzzification?

Ans. Defuzzification is a process of making crisp quantity fuzzy. We do this by simply recognizing that many of quantities we consider to be crisp and determkinistic are actually not deterministic at all. They carry consideral uncertainity. If the form of uncertainity happens to arise because of inmprecision, ambiguity or vagueness, then the variable is probably fuzzy and can be represented by a membership function. It is given by the expression.

$$\mu_{\text{large number}}(x) = \frac{1}{\left(1 + \frac{x}{F_2}\right)^{-F_1}}$$

where, $\mu(u)$ = fuzzyfying parameter.

F_1 = exponential fuzzifying parameter/number.

F_2 = denominational fozzifying parameter/number

$$\infty \equiv \int_X \frac{\frac{1}{\left(1+\left(\frac{x}{F_2}\right)\right)^{-F_1}}}{x}$$

where, A = a large number in continuous form

X = universe of discourse

Defuzzification: Defuzzification is the concession of a fuzy quantity to precise quantity, just as fuzzyfication is the conversion of a precise quantity to a fuzzy quantity. The output of a fuzzy process can be the logical union of two or more fuzzy membership functions defined on the universe of discourse of the output variable.

$$\mu_{(u)} \rightarrow u$$

where u = any crisp value. There are several types of defuzzification methods. These are → maximun membership principle → centroid method → weighted average method → mean of maxima method → centre of sum → centre of largest area → first (or last) of maximum.

1.1 Centroid (Centre of Area) COA: This procedure (also called centre of area, centre of gravity) is the most prevalent and physically appealing of all the defuzzicacion methods. In this method the crisp value u is taken to be the geometrical centre of output fuzzy value.

$\mu_{out}(u)$ is formed by the taking the union of the contribution of all the rules whose degree of Fuzziness is greater than zero.

$$u = \frac{\sum_{i=1}^{N} u_i \, \mu_{OUT}(u_i)}{\sum_{i=1}^{N} u_i \, \mu_{OUT}(u_i)} = \frac{\int \mu_{\underset{\sim}{c}}(z)\, z\, dz}{\int \mu_{\underset{\sim}{c}}(z)\, z\, dz}$$

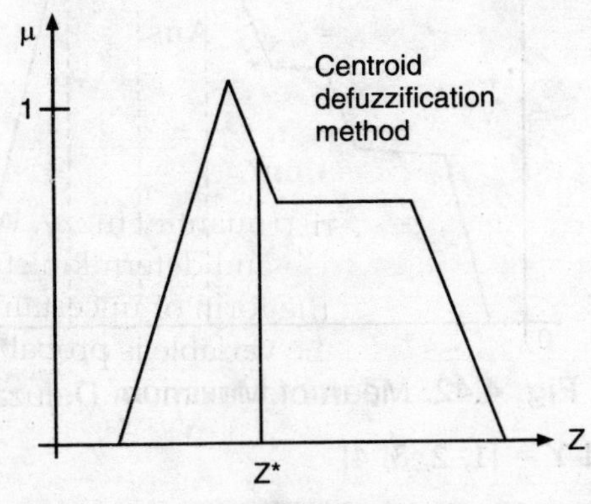

Fig. 4.40.

Where denotes ∫ an algebraic expression. The drawback of this method is that overlapped area is counted once.

2.1. Centre of Sums: This is faster than many defuzzification methdos that are press in use. This process involves the algebraic sum of individual output fuzzy sets, say $\underset{\sim}{C}_1$ and $\underset{\sim}{C}_2$ instead of their union. One drawba to this method is that the intresting areas are added twice. The defozzified value z is given by following equation

$$z^* = \frac{z\sum_{k=1}^{N} \mu_{\underset{\sim}{C}_k}(z)\,dz}{z\sum_{k=1}^{N} \mu_{\underset{\sim}{C}_k}(z)\,dz} = \frac{\sum_{k=1}^{N} u_i \sum_{k=1}^{N} \mu_k(u_i)}{\sum_{k=1}^{N} \sum_{k=1}^{N} \mu_k(u_i)}$$

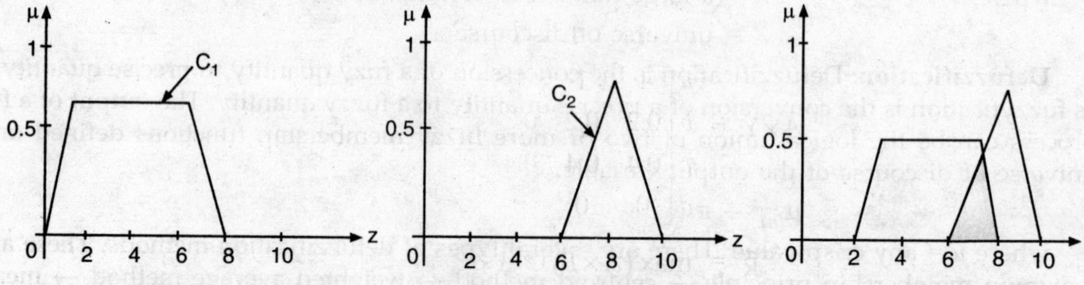

(a) First Membership Function (b) Second Membership Function (c) Defuzzification Step

Fig. 4.41. Centre of Sum Method

3.1. Mean of maxima method: This method (also called middle of maxima) expect that the location of the maximum membership can be non-unique (r. e. the maximum membership can be a plateau rather than a single point). This method is given by the expression.

$$Z^* = \frac{a+b}{z} = \sum_{m=1}^{M} \frac{u_m}{M}$$

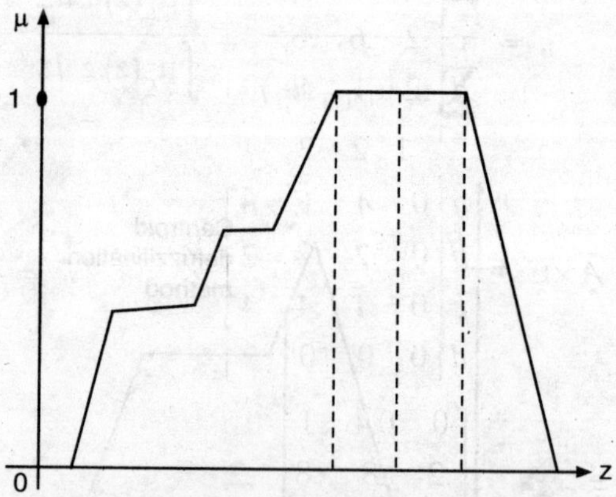

Fig. 4.42. Mean of Maximum Defuzzificant

Q.4. Let $X = \{a, b, c, d\}$ and $Y = \{1, 2, 3, 4\}$

and $\underset{\sim}{A} = \{(a, 0), (b, 0.8), (c, 0.6), (d, 1)\}$

$\underset{\sim}{B} = \{(1, 0.2), (2, 1), (3, 0.8), (4, 0)\}$

$\underset{\sim}{C} = \{(1, 0), (2, 0.4), (3, 1), (4, 0.8)\}$

Determine th e implication relation.

(*i*) If X is $\underset{\sim}{A}$ then Y is B

(*ii*) If X is $\underset{\sim}{A}$ the Y is $\underset{\sim}{B}$ else Y is C.

Ans. (*i*)

$$\underset{\sim}{A}\times\underset{\sim}{B} = \begin{array}{c} \\ a \\ b \\ c \\ d \end{array}\begin{array}{c} \begin{array}{cccc} 1 & 2 & 3 & 4 \end{array} \\ \begin{bmatrix} 0 & 1 & 0 & 0 \\ 0.2 & 0.8 & 0.8 & 0 \\ 0.2 & 0.6 & 0.6 & 0 \\ 0.2 & 1 & 0.8 & 0 \end{bmatrix} \end{array}$$

$$\bar{\underset{\sim}{A}}\times Y = \begin{array}{c} \\ a \\ b \\ c \\ d \end{array}\begin{array}{c} \begin{array}{cccc} 1 & 1 & 1 & 1 \end{array} \\ \begin{bmatrix} 1 & 1 & 1 & 1 \\ 0.2 & 0.2 & 0.2 & 0.2 \\ 0.4 & 0.4 & 0.4 & 0.4 \\ 0 & 0 & 0 & 0 \end{bmatrix} \end{array}$$

$$\tilde{R} = \max(\underset{\sim}{A}\times\underset{\sim}{B}) \cup \bar{\underset{\sim}{A}}\times Y$$

$$\tilde{R} = \begin{bmatrix} 1 & 1 & 1 & 1 \\ 0.2 & 0.8 & 0.8 & 0 \\ 0.4 & 0.6 & 0.6 & 0.4 \\ 0.2 & 1.0 & 0.8 & 0 \end{bmatrix}$$

(*ii*)

$$\underset{\sim}{A}\times\underset{\sim}{B} = \begin{array}{c} \\ a \\ b \\ c \\ d \end{array}\begin{array}{c} \begin{array}{cccc} 1 & 2 & 3 & 4 \end{array} \\ \begin{bmatrix} 0 & 0 & 0 & 0 \\ .2 & .8 & .8 & 0 \\ .2 & .6 & .6 & 0 \\ .2 & 1 & .8 & 0 \end{bmatrix} \end{array}$$

$$\bar{\underset{\sim}{A}}\times\underset{\sim}{C} = \begin{array}{c} \\ a \\ b \\ c \\ d \end{array}\begin{array}{c} \begin{array}{cccc} 1 & 2 & 3 & 4 \end{array} \\ \begin{bmatrix} 0 & .4 & 1 & .8 \\ 0 & .2 & .2 & .2 \\ 0 & .4 & .4 & .4 \\ 0 & 0 & 0 & 0 \end{bmatrix} \end{array}$$

$$(\bar{\underset{\sim}{A}}\times B) \cup (\bar{\underset{\sim}{A}}\times\underset{\sim}{C}) = \begin{bmatrix} 0 & 0.4 & 1 & 0.8 \\ .2 & .8 & .8 & .2 \\ .2 & .6 & .6 & .4 \\ .2 & 1 & .8 & 0 \end{bmatrix}$$

Q.5. Discuss the following with example.

Ans. (*i*) **Fuzzy logic:** Logic is the study the methods and principles of reasoning in all its possible forms fuzzy systems implement fuzzy logic which uses sets and predicate of the kind: values true, not true, very true, not very true more 0 less true, not very false, false etc. The main concepts of fuzzy logic are fuzzy sets, linguistic variables possibility distribution and fuzzy IF-THEN rules.

(*ii*) **Fuzzy propositions:** Fuzzy proposition is a statement which accquires a fuzzy truth value. Therefore, $\underset{\sim}{P}$ is a fuzzy proposition, $T(\underset{\sim}{P})$ represents the truth value (0 – 1) attach to $\underset{\sim}{P}$. The fuzzy membershhip value associated with the fuzzy set $\tilde{A}$ for $\tilde{p}$ is treated as the fuzzy truth value $T(\tilde{P})$.

i.e.,
$$T(\tilde{P}) = \mu(x) \text{ where } 0 \le \mu_A(x) \le 1$$
$$T(\tilde{P}) = \text{multivalued truth value.}$$

***(iii)* Fuzzy connectives:** Fuzzy uses connectives to describe the operation between two propositions. Fuzzy support the following connectives.

Symbol	Connective	Usage	Definition
$-$	Negative	$\bar{\underset{\sim}{P}}$	$1-T(\underset{\sim}{P})$
$\vee$	Disjunction	$\underset{\sim}{P} \vee \underset{\sim}{Q}$	$\max(T(\underset{\sim}{P}), T(\underset{\sim}{Q}))$
$\wedge$	Conjunction	$\underset{\sim}{P} \wedge \underset{\sim}{Q}$	$\min(T(\underset{\sim}{P}), T(\underset{\sim}{Q}))$
$\Rightarrow$	Implication	$P_n \Rightarrow \underset{\sim}{Q}$	$\bar{\underset{\sim}{P}} \vee \underset{\sim}{Q} = \max(1-T(\underset{\sim}{P}), T(\underset{\sim}{Q}))$

***(iv)* Fuzzy quantifiers:** Fuzzy quantifiers are fuzzy numbers that take part in fuzzy propositions. They are of two kinds fuzzy quantifiers of the first kind are defined on R and characterize linguistic terms such as baout in ???? of the second kind are defined on [0, 1] and characterize line terms such as almost all, about half, most and so on. Two basic for of propositions that contains fuzzy quantifiers of first kind are

→ P: There are in Q_{is} I such that $\nu(i)$ is F.

→ P: There are Q_{is} in I such that $\nu(i)$ is F1 and $\nu_2(i)$ is F_2.

Fuzzy propositions with quantifiers of the second kind have general form

→ P: Among is in I such that $\nu_i(i)$ is F_1 there are Q_{is} in I such that $\nu_2(i)$ F_2. Where ν is a variable that for each individual i in a given set I assumes a value $\nu(i)$, F is a fuzzy set defined on set of values of variable ν.

Fuzzy inference also called approximate reasoning refers to computational procedure used for evaluating or finding out linguistic description there are two procedures for it.

***(i)* Generalized Modus Ponems:**

$$\text{If } x \text{ is } \underset{\sim}{A} \text{ then y is } \tilde{B} = \frac{x \text{ is } \underset{\sim}{A}'}{y \text{ is } \underset{\sim}{B}'}$$

$$\underset{\sim}{B}' = \underset{\sim}{A}' o R(x, y) \Rightarrow \mu'_B(y) = \max(\min(\mu_{\underset{\sim}{A}'}(x), \mu_{\underset{\sim}{R}}(x, y)))$$

***(ii)* Generalized modus Tollens:**

If x is $\underset{\sim}{A}$ then y is $\underset{\sim}{B}$

$$\frac{x \text{ is } \underset{\sim}{B}'}{y \text{ is } \underset{\sim}{A}'} \quad \underset{\sim}{A}' = \underset{\sim}{B}'_o R(x, y) \Rightarrow \mu_{A'}(x) = \max(\min(\mu_{\underset{\sim}{B}'}(y), \mu_{\underset{\sim}{R}}(x, y))).$$

Q.6. Discuss with an example, fuzzy rule based system.

Ans. Fuzzy is used as a calculus for the interpretation of natural language. The use of natural language in the expression of knowledge form is known as rule based systems.

For example: Suppose we define a specific atomic term in the universe of natural language, x is element α, and a fuzzy set A in the universe of interpretation or meaning y, as a specific meaning for the term α. Then natural language can be expressed as a mapping, M from a set of atomic terms in x to a corresponding set of interpretation defined on universe y which can be denoted by membership function as $\mu \underset{\sim}{M}(x, y) = \mu \underset{\sim}{A}(y)$

As an example, suppose we have the atomic term young (x) and we want to interpret this linguistic interms of age, y, by membership function. This can be expressed as

$$A = \text{"young"} = \int_0^{25} \frac{1}{y} + \int_{25}^{100} \frac{1}{y}\left[\left(\frac{y-25}{5}\right)^2\right]^{-1}$$

$$\mu_{\underset{\sim}{M}}(\text{young}, y) = \left[\left(1+\left(\frac{y-25}{5}\right)^2\right)\right]^{-1} \quad y > 25 \text{ year} \quad y \le 25 \text{ year}$$

SUMMARY

This Chapter goes into the details of Fuzzy Membership functions . First we define a membership function and function with smooth transitions. Next, different ways of defining fuzziness as uniform distributed fuzziness, point fuzziness, three point fuzziness and non uniform distributed fuzziness are discussed. These interference in fuzzy logic help us to reach crisp relation at the output. Next, we discuss the fuzzy if-then rules, Fuzzy implications and Fuzzy algorithms. Major Fuzzy implications being: Larsen, Lukasiewicz, Standard Strict,Kleeene-Dienes-Luk and to top it all, the most famous Mamdani. Along with the definitions of Modifiers and partitioning, we complete our tour in the fuzzy world. Last stop is to convert the fuzzy instructions into crisp instruction using defuzzification process. Three methods are discussed here with example and they are: Centroid Method, Centre of Sums Method and Mean of Maxima Method. Next we discuss the industrial applications of Fuzzy Logic in form of fuzzy controller. We discuss the Classification of conventional control methods namely Proportional Integral and Derivative Control methods, jointly known as PID Controller. Next we discuss fuzzy controller as Proportional Integral and Derivative Control methods, jointly known as PID Fuzzy Controller. Next we go into the details of Takagi-Sugeno Fuzzy controller model where we do the fuzzy modeling, derive the mathematical model of Fuzzy Logic Controller and algorithm therein. A very interesting example of fuzzy control of inverted pendulum is also discussed. First, Control System then analysis and design, and other Industrial applications.

REVIEW QUESTIONS

1. What are the definitions and properties of a membership function?
2. Discuss about the possible ways of the definition of Fuzzy information .
3. Discuss various fuzzification rules applied in Fuzzy Inference Machine.
4. Taking an example, discuss the max/min and max prod inference.

5. Write a short note on Fuzzy-if-then rules.
6. Discuss Fuzzy implications.
7. Discuss Modifiers and partitioning in Fuzzy domain.
8. Compare and contrast between different defuzzification methods, namely, Centroid method, Centre of Sums method, Mean of Maxima method.
9. Show the block diagram of a conventional controller. Upgrade it to make a fuzzy controller.
10. What are the different modes of control. Use hybrid controllers also for better working.
11. Classify the Fuzzy Controller in view of the control actions.
12. Use PI and PD controller to show the block diagram of control action of a reversing truck.
13. Draw the logic block diagram of TS controller.

NUMERICAL PROBLEMS

1. Write predicate logic for
   (*a*) Ram likes all games
   (*b*) Sita likes to watch all games Ram likes
   (*c*) Laxman like those liked by Sita and Ram
   (*d*) Mehul John likes some of what Ram likes.
2. Consider the following fuzzy sets

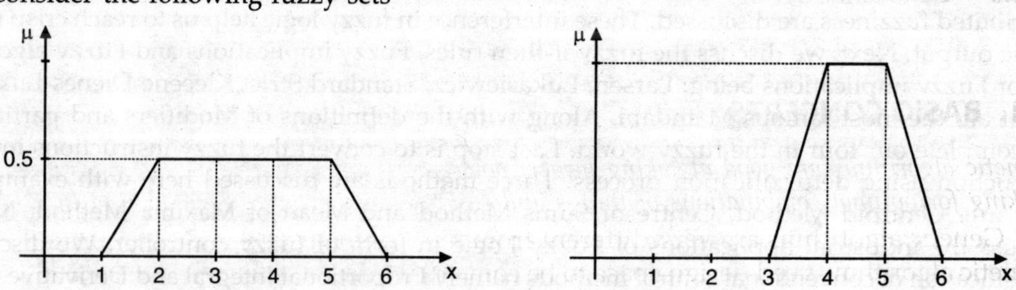

Fig. 4.43.

   (*a*) Draw the agreegate fuzzy diagram
   (*b*) Find out centroid
   (*c*) Find out centre of sums
   (*d*) Find out mean of maxima
   (*e*) Compare all the defuzzification outputs in one graph only.

5 CHAPTER FIVE

GENETIC ALGORITHM (GA)

CHAPTER OUTLINE

5.1 Basic Concepts
5.2 Working Principle
5.3 Procedures of GA
5.4 Flow chart of GA
5.5 Genetic Representations
5.6 Encoding Intialization and Selection
5.7 Genetic Operators
5.8 Mutation
5.9 Genetional Cycle
5.10 Applications

5.1. BASIC CONCEPTS

Genetic algorithms are good at taking larger, potentially huge, search spaces and navigating them looking for optimal combinations of things and solutions which we might not find in a life time.

Genetic algorithms are very different from most of the traditional optimization methods. Genetic algorithms need design space to be converted into genetic space. So, genetic algorithms work with a coding of variables. The advantage of working with a coding of variable space is that coding discretizes the search space even though the function may be continuous. A more striking difference between genetic algorithms and most of the traditional optimization methods is that GA uses a population of points at one time in contrast to the single point approach by traditional optimization methods. This means that GA processes a number of designs at the same time. As we have seen earlier, to improve the search direction in traditional optimization methods, transition rules are used and they are deterministic in nature but GA uses randomized operators. Random operators improve the search space in an adaptive manner.

NOTE Three most important aspects of using GA are:

1. Definition of objective function
2. Definition and implementation of genetic representation
3. Definition and implementation of genetic operators.

Once these three have been defined, the GA should work fairy well beyond doubt. We can, by different variations, improve the performance, find multiple optima (species if they exist) or parallelize the algorithms.

Genetic Algorithms (GAs) are adaptive heuristic search algorithm premised on the evolutionary ideas of natural selection and genetic. The basic concept of GAs is designed to simulate processes in natural system necessary for evolution, specifically those that follow the principles first laid down by Charles Darwin of survival of the fittest. As such they represent an intelligent exploitation of a random search within a defined search space to solve a problem.

- Genetic algorithm is based on Darwinian Paradigm
- Intrinsically a robust search and optimization mechanism

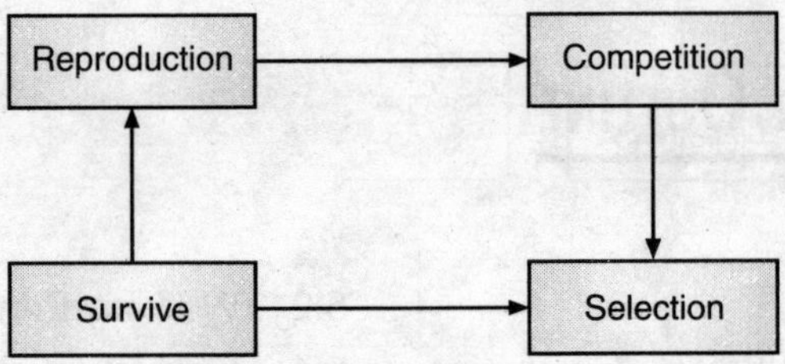

Fig. 5.1. Representation of genetic algorithm

Some points regarding genetic algorithm (GA)

- A class of probabilistic optimization algorithms
- Inspired by the biological evolution process
- Uses concepts of "Natural selection" and "Genetic Inheritance" (Darwin 1859)
- Originally developed by John Holland (1975)
- Particularly well suited for hard problems where little is known about the underlying search space
- Widely-used in business, science and engineering

Inspiration

- Inspired by natural evolution
- Population of individuals
- Individuals is feasible solution to problem
- Each individual is characterized by a Fitness function
- Higher fitness is better solution
- Based on their fitness, parents are selected to reproduce offspring for a new generation
- Filter individuals have more chance to reproduce
- New generation has same size as old generation; old generation dies
- Offspring has combination of properties of two parents

If well designed, population will converge to optimal solution.

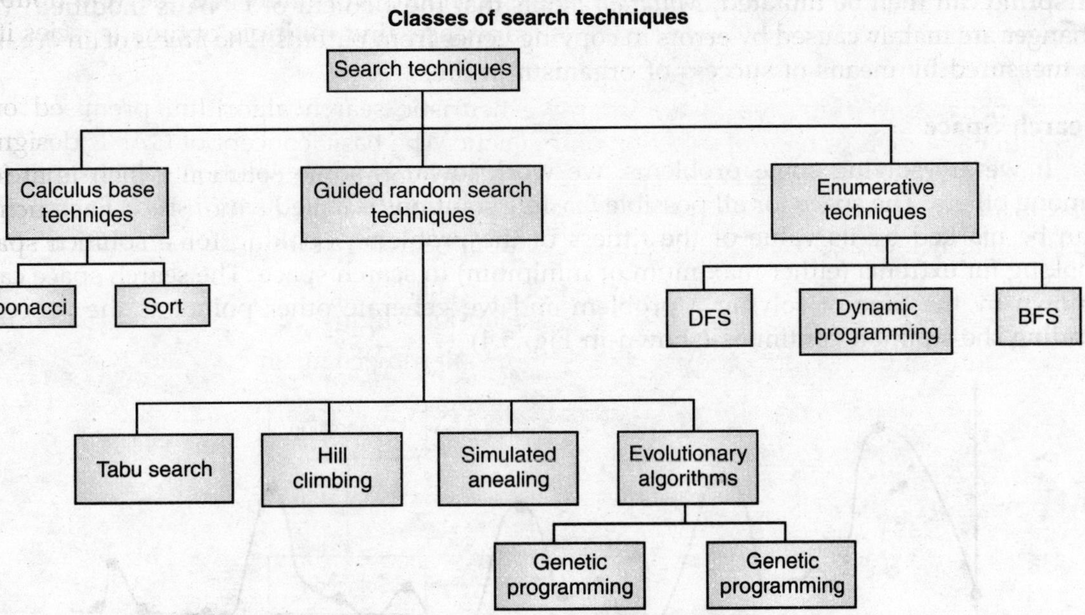

Fig. 5.2. Classification of genetic algorithm

Biological Background

All living organisms consist of cells. In each cell, there is a set of chromosomes which are strings of DNA and serve as a model for the whole organism. A chromosome consists of genes on blocks of DNA as shown in Fig. 5.3 Each gene encodes a particular pattern. Basically, It can be said that each gene encodes a trait, e.g. colour of eyes. Possible settings of traits (bluish brown eyes) are called *alleles;* each gene has its own position in the chromosome search space. This position is called *locus.* Complete set of genetic material is called *genome* and a particular set of genes in genome is called *genotype.* The genotype is based on organism's phenotype development after birth, its physical and mental characteristics such as eye colour, intelligence and so on.

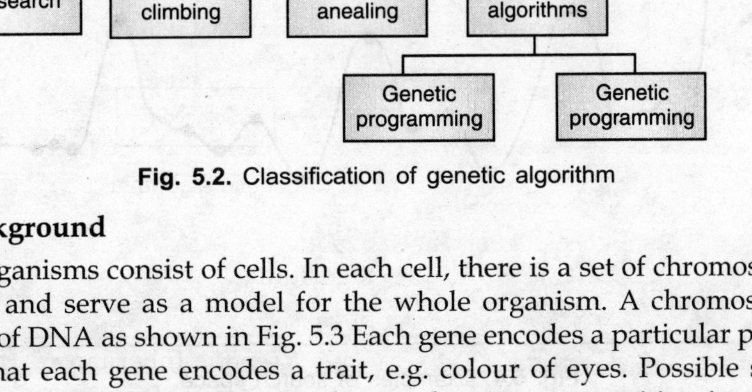

Fig. 5.3. Genome consisting of chromosomes

Creation Of Offsprings

During the creation of offspring, recombination occurs (due to cross over) and in that process genes from parents form a whole new chromosome in some way. The new created offspring can then be mutated. *Mutation means* that the element of DNA is modified. These changes are mainly caused by errors in copying genes from parents. The *fitness* of an organism is measured by means of success of organism in life.

Search Space

If we are solving some problems, we work towards some solution which is the best among others. The space for all possible feasible solutions is called *search space*. Each solution can be marked by its value of the fitness of the problem. 'Looking for a solution' means looking for extrema (either maximum or minimum) in search space. The search space can be known by the time of solving a problem and we generate other points as the process of finding the solution continues (shown in Fig. 5.4).

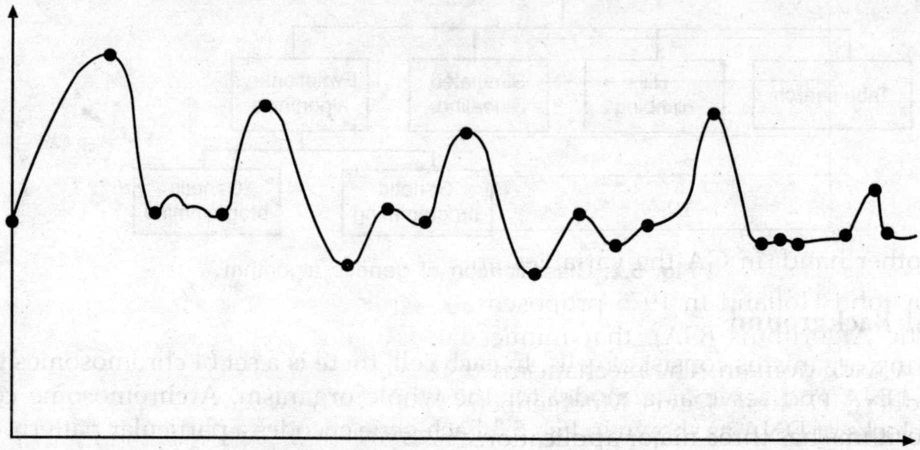

Fig. 5.4. Examples of search space.

The problem is that, search space is complicated and one does not know where to look for the solution or where to start from and this is where genetic algorithm is useful. Gas are inspired by *Darwinian theory of the survival of the fittes.* Algorithm is started with a set of solutions (represented by chromosomes) called *populations.* Solutions for one population are taken and used to form a new population. This is motivated by a hope that new population will be better than the old one. Solutions, which are selected to form new population (offspring), are selected according to their fitness. The more suitable they are, the more chances they have to reproduce. This is repeated until some conditions (number of populations) for improvement of best solution are satisfied.

5.2. WORKING PRINCIPLE

To illustrate the working principle of GA, we first consider unconstrained optimization problem. Later, we shall discuss how GA can be used to solve a constrained optimization problem. Let us consider the following maximization problem.

Maximize $f(x)$ $\quad X_i^{(L)} \le X_i \le X_i^{(U)}$ for $i = 1, 2, ..., N$ ...(5.1)

If we want to minimize $f(X)$, for, for, for $f(x) > 0$, then we can write the objective function as

$$\text{Maximize } \frac{1}{1 + f(X)} \quad ...(5.2)$$

If $f(X) < 0$ instead of minimizing $f(X)$, maximize $\{-f(X)\}$. Hence, both maximization and minimization problem can be handled by GA.

If the same problem is solved by multiple regression analysis, given independent variables, for regressing the dependent variable $2^{(k+1)}$ including the intercept which are given in Table 5.1.

Table 5.1. Subsets for regression analysis

Variable	Subsets
2	7
3	15
–	–
–	–
9	1023
–	–
19	10, 48, 574

On the other hand, in GA the variables are coded.

Professor John Holland in 1975 proposed an attractive class of computational models, called Genetic Algorithms (GA), that mimic the biological evolution process for solving problems in a wide domain. The mechanisms under GA have been analyzed and explained later by Goldberg, De Jong, Davis, Muehlenbein, Chakraborti, Fogel, Vose and many others. Genetic Algorithms has three major applications, namely, intelligent search, optimization and machine learning. Currently, Genetic Algorithms is used along with neural networks and fuzzy logic for solving more complex problems. Because of their joint usage in many problems, these together are often referred to by a generic name: ";soft-computing". A Genetic Algorithms operates through a simple cycle of stages:

(*i*) Creation of a "population" of strings,
(*ii*) Evaluation of each string,
(*iii*) Selection of best strings and
(*iv*) Genetic manipulation to create new population of strings.

The cycle of a Genetic Algorithms is shown in Fig. 5.4. presented below

Each cycle in Genetic Algorithms produces a new generation of possible solutions for a given problem. In the first phase, an initial population, describing representatives of the potential solution, is created to initiate the search process. The elements of the population are encoded into bit-strings, called chromosomes. The performance of the strings, often called fitness, is then evaluated with the help of some functions, representing the constraints of the problem. Depending on the fitness of the chromosomes, they are selected for a subsequent genetic manipulation process. It should be noted that the selection process is mainly responsible for assuring survival of the best-fit individuals. After selection of the population strings is over, the genetic manipulation process consisting of two steps is carried out. In the first step, the crossover operation that recombines the bits (genes) of each two selected strings (chromosomes) is executed. Various types of crossover operators are found in the literature. The single point and two points crossover operations are illustrated.

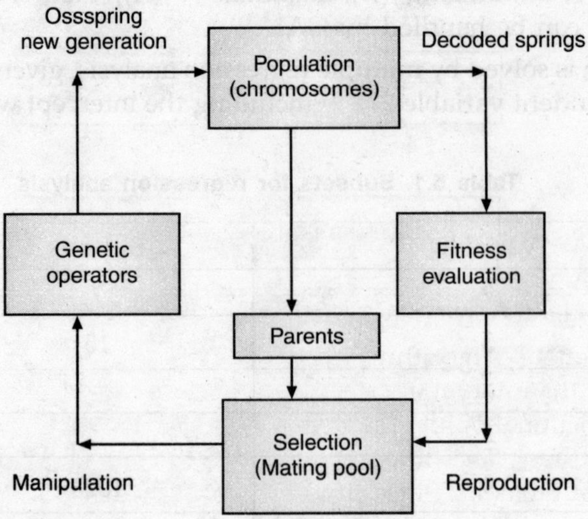

Fig. 5.5. G A Cycle

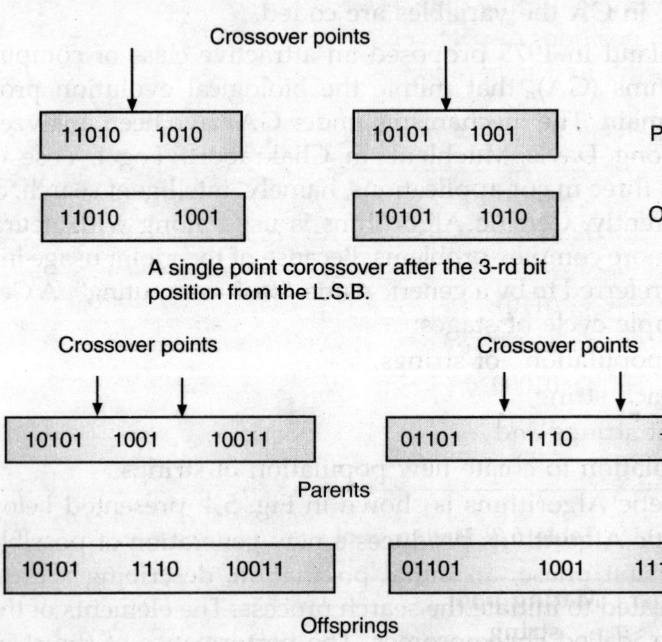

Fig. 5.6. Operators of GA

The crossover points of any two chromosomes are selected randomly. The second step in the genetic manipulation process is termed mutation, where the bits at one or more randomly selected positions of the chromosomes are altered. The mutation process helps to overcome trapping at local maxima. The offsprings produced by the genetic manipulation process are the next population to be evaluated.

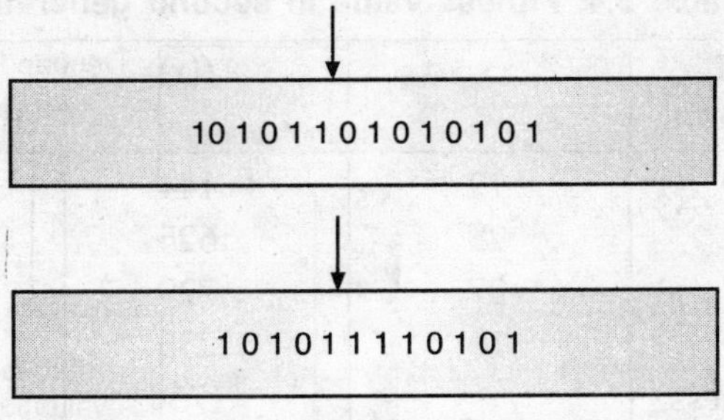

Fig. 5.7. Mutation of a chromosome at the 5th bit position

Example: The Genetic Algorithms cycle is illustrated in this example for maximizing a function $f(x) = x_2$ in the interval $0 = x = 31$. In this example the fitness function is $f(x)$ itself. The larger is the functional value, the better is the fitness of the string. In this example, we start with 4 initial strings. The fitness value of the strings and the percentage fitness of the total are estimated in table 5.2. Since fitness of the second string is large, we select 2 copies of the second string and one each for the first and fourth string in the mating pool. The selection of the partners in the mating pool is also done randomly. Here in table 5.3, we selected partner of string 1 to be the 2-nd string and partner of 4-th string to be the 2nd string. The crossover points for the first-second and second-fourth strings have been selected after o-th and 2-nd bit positions respectively in table *B*. The second generation of the population without mutation in the first generation is presented in table 5.4.

Table 5.2. Initial population and their fitness values

string number	initial population	x	$f(x)$	strength fitness (% of total)
1.	01101	13	169	14.4
2.	11000	24	576	49.2
3.	01000	08	64	5.5
4.	10011	19	361	30.9
Sum-fitness			1170	100.00

Table 5.3. Mating pool strings and crossover

string number	Mating pool string	Mates	Swapping	New population
1.	01101	2	0110[1]	01100
2.	11000	1	1100[0]	11001
3.	11000	4	11[000]	11011
4.	10011	2	10[011]	10000
Sum-fitness				

Table 5.4. Fitness value in second generation

Initial population	x	$f(x)$	strength (% of total)
01100	12	144	8.2
11001	25	625	35.6
11011	27	729	41.5
10000	16	256	14.7
Sum-fitness		1754	100.00

A Schema (or schemata in plural form) / hyperplane or similarity template is a genetic pattern with fixed values of 1 or 0 at some designated bit positions. For example, $S = 01?1??1$ is a 7-bit schema with fixed values at 4-bits and don't care values, represented by ?, at the remaining 3 positions. Since 4 positions matter for this schema, we say that the schema contains 4 genes.

Deterministic Explanation of Holland's Observation

To explain Holland's observation in a deterministic manner let us presume the following assumptions.

(i) There are no recombination or alternations to genes.

(ii) Initially, a fraction f of the population possesses the schema S and those individuals reproduce at a fixed rate r.

(iii) All other individuals lacking schema S reproduce at a rate s < r.

Thus with an initial population size of N, after t generations, we find $Nf\ r\ t$ individuals possessing schema S and the population of the rest of the individuals is $N(1 - f)$st. Therefore, the fraction of the individuals with schema S is given by–

$$= \frac{(N\ f\ r^t)}{[N(1-f)\ s^t + Nfr^t]} = \frac{f(r/s)^t}{[1 + f\ \{r/s\}^t - 1\}]}$$

For small t and f, the above fraction reduces to $f\ (r\ /\ s)\ t$, which means the population having the schema S increases exponentially at a rate $\left(\frac{r}{s}\right)$. A stochastic proof of the above property will be presented shortly, vide a well-known theorem, called the fundamental theorem of Genetic algorithm.

Stochastic Explanation of Genetic Algorithms

For presentation of the fundamental theorem of Genetic Algorithms, the following terminologies are defined in order.

Definition: The order of a schema H, denoted by O(H), is the number of fixed positions in the schema. For example, the order of schema H = ?001?1? is 4, since it contains 4 fixed positions.

For example, the schema ?1?001 has a defining length d(H) = 4 - 0 = 4, while the $d(H)$ of ???1?? is zero.

Definition: The schemes defined over L-bit strings may be geometrically interpreted as hyperplanes in an L-dimensional hyperspace (a binary vector space) with each L-bit string representing one corner point in an n-dimensional cube.

The Fundamental Theorem of Genetic Algorithms

Schema theorem: Let the population size be N, which contains mH (t) samples of schema *H* at generation *t*. Among the selection strategies, the most common is the proportional selection. In proportional selection, the number of copies of chromosomes selected for mating is proportional to their respective fitness values. Thus, following the principles of proportional selection, a string is selected with probability.

$$\frac{f_i}{\sum_{i=1}^{n} f_1}$$

where f_i is the fitness of string *i*. Now, the probability that in a single selection a sample of schema *H* is chosen is described by:

$$= \frac{\sum_{i=1}^{m_H(t)} f_i}{\sum_{i=1}^{n} f_i} = \frac{m_H(t)\, F_H}{\sum_{i=1}^{n} f_i}$$

where *f H* is the average fitness of the mH (*t*) samples of *H*.

Thus, in a total of N selections with replacement, the expected number of samples of schema H in the next generation is given by–

$$m_H\,(t+1) = N\, m_H\,(t)\ \frac{f_H}{\sum_{i=1}^{n} f_i} = \frac{m_H(t)\, f_H}{f_{av}}$$

where $$f_{av} = \frac{\sum_{i=1}^{n} f_i}{N}$$

where f_{av} is the population average fitness in generation *t*. The last expression describes that the Genetic algorithm allocates over an increasing number of trials to an above average schema. The effect of crossover and mutation can be incorporated into the analysis by computing the schema survival probabilities of the above average schema. In crossover operation, a schema H survives if the cross-site falls outside the defining length d(H). If pc is the probability of crossover and L is the word-length of the chromosomes, then the disruption probability of schema H due to crossover is

$$= \frac{p_e d(H)}{(L-1)}$$

A schema H, on the other hand, survives mutation, when none of its fixed positions is mutated. If pm is the probability of mutation and O(H) is the order of schema H, then the probability that the schema survives is given by $(1 - p_m)^{O(H)}$.

Therefore, under selection, crossover and mutation, the sample size of schema H in generation (*t* + 1) is given by– $m_H\,(t+1) \geq (m_H\,(t)\, f_H \,/\, f_{av})\,[1 - p_e\, d\,(H) \,/\, (L-1) - p_m\, O(H)]$

The above theorem is called the fundamental theorem of GA or the Schema theorem. It is evident from the Schema theorem that for a given set of values of d(H), O(H), L, pc and pm , the population of schema H at the subsequent generations increases exponentially when $f_H > f_{av}$. This, in fact, directly follows from the difference equation:

$$m_H\ (t+1) - m_H\ (t) \geq (f_H / f_{av} - 1/K)\ K\ m_H\ (t)$$

where, $K = 1 - p_e\ d\ (H)\ /\ (L-1) - p_m\ O(H)$

$$\Rightarrow \Delta m_H\ (t) \geq K\ (f_H\ /\ f_{av} - 1/\ K)\ m_H(t).$$

Replacing Δ by $(E - 1)$, where E is the extended difference operator, we find

$$(E - 1 - K_1)\ m_H\ (t) \geq 0$$

where, $K_1 = K(f_H\ /\ f_{av} - 1/\ K)$.

Since $m_H(t)$ in equation $(E\text{-}1\text{-}K1)m_H(t) \geq 0$ is positive, $E \geq (1 + K_1)$. Thus, the solution of is given by $m_H\ (t) \geq A\ (1 - K_1)^t$

where A is a constant. Setting the boundary condition at $t = 0$, and substituting the value of K_1 by therein, we finally have: $m_H\ (t) \geq m_H\ (0)(K\ f_H\ /\ f_{av})^t$

Since K is a positive number, and $f_H\ /\ f_{av} > 1$, $m_H\ (t)$ grows exponentially with iterations. The process of exponential increase of $m_H\ (t)$ continues until some iteration r, when f_H approaches f_{av}. This is all about the proof of the schema theorem.

The Markov Model for Convergence Analysis

For the sake of understanding, let us now consider the population size = 3 and the chromosomes are 2-bit patterns, as presumed earlier. The set S now takes the following form.

S = {(00, 00, 00), (00, 00, 01), (00, 00, 10), (00, 00, 11),
(00, 01, 00), (00, 01, 01), (00, 01, 10), (00, 01, 11),
....
....
(11, 11, 00), (11, 11, 01), (11, 11, 10), (11, 11, 11) }

It may be noted that the number of elements of the last set S is 64. In general, if the chromosomes have the word length of m bits and the number of chromosomes selected in each Genetic Algorithm cycle is n, then the cardinality of the set S is 2 ^ mn. The Markov transition probability matrix P for 2-bit strings of population size 2, thus, will have a dimension of (16 × 16), where the element pij of the matrix denotes the probability of transition from i-th to j-th state. A clear idea about the states and their transitions can be formed from figure.

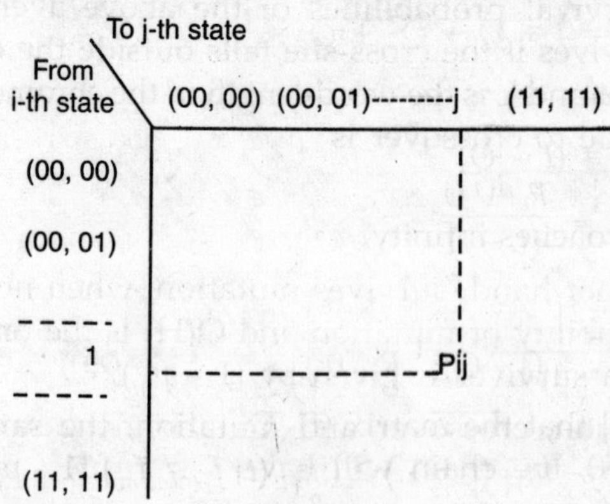

Fig. 5.8.

It needs mention that since from a given i-th state, there could be a transition to any 16 j-th states, therefore the row sum of P matrix must be 1. Formally,

$$\sum_{\forall j} P_{ij} = 1$$

for a given *i*.

Now, let us assume a row vector πt, whose k-th element denotes the probability of occurrence of the k-th state at a given genetic iteration (cycle) *t*; then $\pi t + 1$ can be evaluated by

$$\pi_{i+1} = \pi_t P$$

Thus starting with a given initial row vector $\pi 0$, one can evaluate the state probability vector after n-th iteration π_n by

$$\pi_n = \pi_0 . P^n$$

The behavior of Genetic Algorihms without mutation can be of the following three types.

(i) The GA may converge to one or more absorbing states (i.e., states wherefrom the GA has no transitions to other states).

(ii) The GA may have transition to some states, wherefrom it may terminate to one or more absorbingstates.

(iii) The GA never reaches an absorbing state.

Taking all the above into account, we thus construct P as a partitioned matrix of the following form:

$$P = \begin{pmatrix} I & 0 \\ R & Q \end{pmatrix}$$

where *I* is an identity matrix of dimension $(a \times a)$ that corresponds to the absorbing states; *R* is a $(t \times a)$ transition sub-matrix describing transition to an absorbing state; *Q* is a $(t \times t)$ transition sub-matrix describing transition to transient states and not to an absorbing state and 0 is a null matrix of dimension $(t \times t)$.

It can be easily shown that P^n for the above matrix *P* can be found to be as follows.

$$P^n = \begin{pmatrix} 1 & 0 \\ N_n R & Q^n \end{pmatrix}$$

where the n-step transition matrix Nn is given by

$$N_n = I + Q + Q^2 + Q^3 + \ldots + Q^{n+1}$$

As n approaches infinity,

$$\underset{n \to \infty}{Lt} N_n = (I - Q)^{-1}$$

Consequently, as n approaches infinity,

$$\underset{n \to \infty}{\text{Lt}} P^n = \begin{pmatrix} I & 0 \\ (1-Q)^{-1} R & 0 \end{pmatrix}$$

Goodman has shown [8] that the matrix (I - Q) -1 is guaranteed to exist. Thus given an initial probability vector p 0, the chain will have a transition to an absorbing state with probability 1. Further, there exists a non-zero probability that ab sorbing state will be the globally optimal state . We now explain: why the chain will finally terminate to an absorbing state. Since the first 'a' columns for the matrix Pn, for $n \to \infty$, are non-zero and the remaining columns are zero, therefore, the chain must have transition to one of the absorbing states. Further, note that the first 'a' columns of the row vector p^n for $n \to \infty$ denote the probability of absorption at different states, and the rest of the columns denote that the probability of transition to nonabsorbing states is zero. Thus probability of transition to absorbing states is one. Formally,

$$= \sum_{i=1}^{a} \underset{n \to \infty}{Lt} \; (\pi_n)_i \; = \sum_{i=1}^{a} \underset{n \to \infty}{\text{Lt}} \; (\pi_0 P^n)_i$$

$$= \sum_{i=1}^{a} \underset{n \to \infty}{\text{Lt}} \left(\pi_0 \cdot \left(\frac{I}{(I-Q)^{-1} \; R} \right) \right)_i$$

To summarize working principle of genetic algorithm:

- Coding of Representation
  - String with all parameters
- Fitness function
  - Parent selection
- Reproduction
  - Cross over
  - Mutation
- Convergence

  When to stop
- An individual is characterized by a set of parameters: **Genes**
- The genes are joined into a string: Chromosome
- The chromosome forms the genotype
- The genotype contains all information to construct an organism: the phenotype
- Reproduction is a "dumb" process on the chromosphere of the **genotype**
- Fitness is measured in the real world ('struggle for life') of the phenotype.

5.3. PROCEDURE OF GENETIC ALGORITHM

The GA consists four main stages:

1. evaluation,
2. selection
3. crossover and
4. mutation

The *evaluation* procedure measures the fitness of each individual solution in the population and assigns to it a score. The *selection* procedure randomly selects individuals of the current population for development of the next generation. The *crossover* procedure takes two selected individuals and combines them about a crossover point thereby creating two new individuals.

The **mutation** procedure randomly modifies the genes of an individual subject to a small mutation factor, introducing further randomness into the population.

The iterative process continues until one of the possible termination criteria is met: if a known optimal or acceptable solution level is attained; or if a maximum number of generations have been performed; or if a given number of generations without fitness improvement occur.

5.4. GENETIC ALGORITHM FLOWCHART

The genetic algorithm flowchart consists of following four steps:

1. Initialization of generation chromosomes
2. Calculating the scores
3. To select parent chromosomes
4. To perform genetic operations flow chart consists of

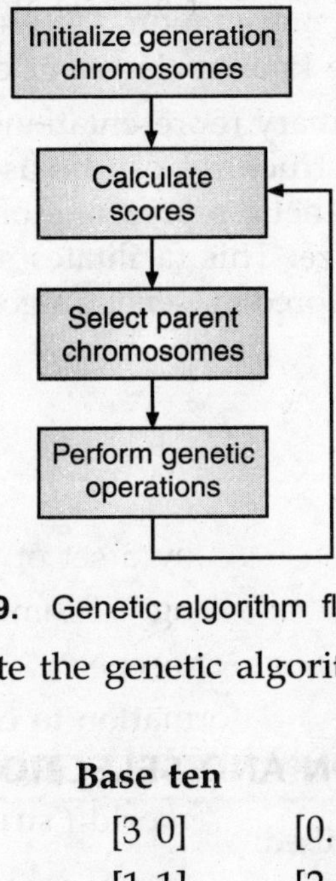

Fig. 5.9. Genetic algorithm flowchart

Following examples demonstrate the genetic algorithm:

Evaluation

Binary		Base ten		Scores
[0 0 0 1 1 0 0 0 0 0]		[3 0]		[0. 0 3 3 1]
[0 0 0 0 1 0 0 0 0 1]		[1 1]		[2. 4 3 3 8]
[0 0 1 0 1 1 1 1 0 0]	⇔	[5 28]	→	[0]
[1 1 1 1 0 0 0 1 1 0]		[30 6]		[0]
[0 0 0 1 0 0 0 0 1 1]		[2 3]		[0. 0 0 5 7]

Selection:

[0 0 0 1 1 0 0 0 0 0]	[3 0]	[0. 0 3 3 1]
[0 0 0 0 1 0 0 0 0 1]	[1 1]	[2. 4 3 3 8]
[0 0 1 0 1 1 1 1 0 0]	[5 28]	[2. 4 3 3 8]
[1 1 1 1 0 0 0 1 1 0]	[3 0]	[0. 0 3 3 1]
[0 0 0 1 0 0 0 0 1 1]	[2 3]	[0. 0 0 5 7]

Crossover:

[0 0 0 1 1 0 0 0 0 0] [0 0 0 1 1 0 0 0 0 1]
[0 0 0 0 1 0 0 0 0 1] [0 0 0 0 1 0 0 0 0 0]
[3 0] [3 1]
[1 1] [1 0]

Crossover:

[0 0 0 0 1 0 0 0 0 1] [0 0 0 0 1 0 1 0 0 1]
[1 1] ⟶ [1 9]

5.5. GENETIC REPRESENTATION

Genetic representation is a way of representing solutions/individuals in evolutionary computation methods. Genetic representation can encode appearance, behaviour, physical qualities of individuals. Designing a good genetic representation that is expressive and evoluble is a hard problem in evolutionary computation. Difference in genetic representation is one of the major criteria drawing a line known classes of evolutionary computation.

Genetic algorithms use linear binary representations. The most standard one is an array of bits. Arrays of other types and structures can be used in essentially the same way. The main property that makes these genetic representations convenient is that their parts are easily aligned due to their fixed size. This facilitates simple crossover operation. Variable length representations were also explored in Genetic algorithms, but crossover implementation is more in this case.

Common genetic representation

- binary
- binary tree
- genetic tree
- natural language
- parse tree

5.6. ENCODING, INITIALIZATION AND SELECTION

Many techniques are used as described:

5.6.1. Binary Encoding

Binary encoding is the most common, mainly because first works about GA used this type of encoding.

In binary encoding every chromosome is a string of bits, 0 or 1.

Arithmetic crossover: Some arithmetic operation is performed to make a new offspring.

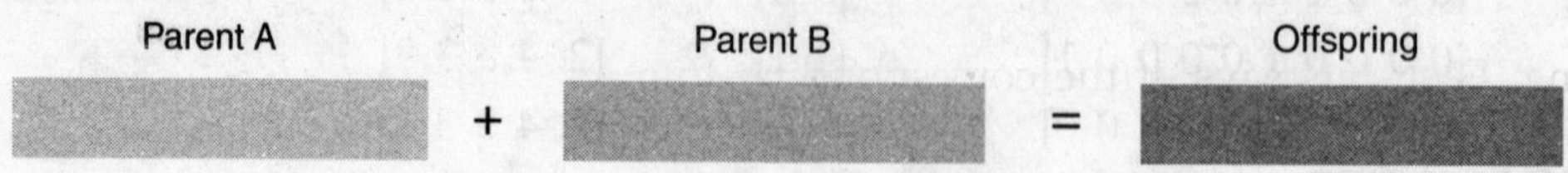

11001011 + 11011111 = 11001001 (AND)

Mutation

Bit inversion: Selection bits are inverted

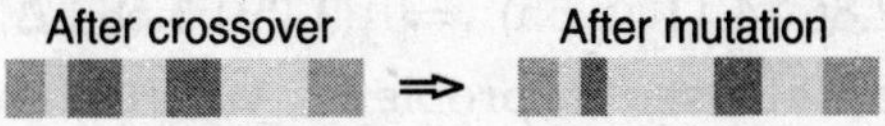

11001001 → 10001001

Crossover

Single point crossover: One crossover point is selected, binary string from beginning of chromosome to the crossover point is copied from one parent, the rest is copied from the second parent.

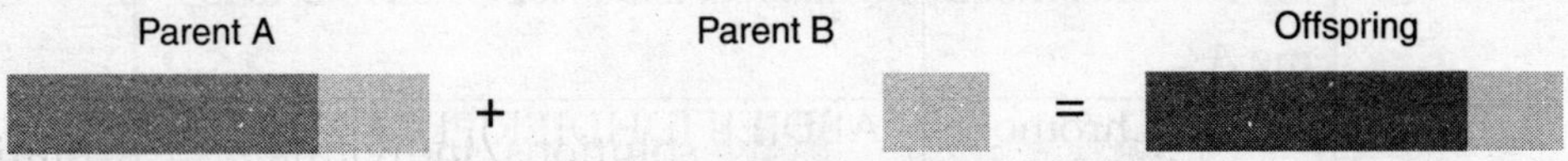

11001011 + 11011111 = 11001111

Two point crossover: Two croissover pointn are selected, binary string from beginning of chromosome to the first crossover point is copied from one parent, the part from the first to the second crossover point is copied from the second parent and the rest is copied from the first parent.

110010011 + 11011111 = 11011111

Uniform corossover-bits are randomly copied from the first or from the second parent.

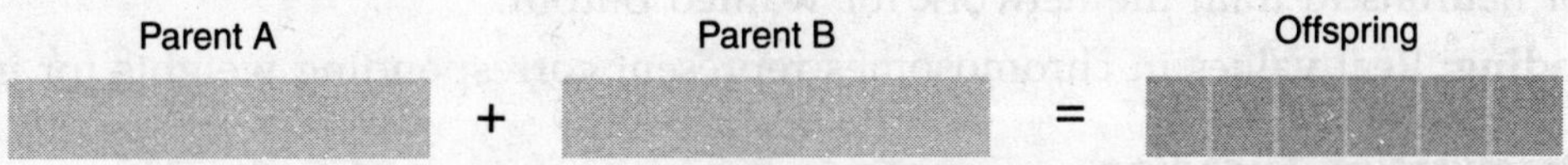

11001011 + 11011101 = 11011111

Chromosome *A*	101100101100101011100101
Chromosome *B*	111111100000110000011111

Example of chromosomes with binary encoding

Binary encoding gives many possible chromosomes even with a small number of alleles. On the other hand, this encoding is often not natural for many problems and sometimes corrections must be made after crossover and/or mutation.

Example of problem: Knapsack problem

The problem: There are things with given value and size. The knapsack has given capacity.

Encoding: Each bits says, if the corresponding thing is knapsack.

5.6.2. Value Encoding

Crossover: All crossovers from binary encoding can be used

Mutation: Adding a small number (for real value encoding) to selected values is added (or subtracted) a small number

(1.29 5.68 2.86 4.11 5.55) ⇒ (1.29 5.68 2.73 4.22 5.55)

Direct value encoding can be used in problems, where some complicated value, such as real numbers, are used. Use of binary encoding for this type of problems would be very difficult.

In value encoding, every chromosome is a string of some values. Values can be anything connected to problem, form numbers, real numbers or charts to some complicated objects.

Chromosome *A*	1.234 5.3243 0.4556 2.3293 2.4545
Chromosome *B*	ABDJEIFJDHDIERJFDLDFLFEGT
Chromosome C	(back), (back), (right) (forward), (left)

Example of chromosome with value encoding

Value encoding is very good for some special problems. On the other hand, for this encoding is often necessary to develop some new crossover and mutation specific for the problem.

Example of Problem: Finding weights for neural network.

The problem: There is some neural network with given architecture. Find weights for inputs of neurons to train the network for wanted output.

Encoding: Real values in chromosomes represent corresponding weights for inputs.

5.6.3. Permutation Encoding

Crossover

Single point crosover: One crossover point is selected, till this point the permutation is copied from the first then the second parent is scanned and if the number is not yet in the offspring it is added.

Note: there are more ways how to produce the rest after crossover point.

(1 2 3 4 5 6 7 8 9) + (4 5 3 6 8 9 7 2 1) = (1 2 3 4 5 6 8 9 7)

Mutation

Order changing: Two numbers are selected and exchanged

(1 2 3 4 5 6 8 9 7) → (1 8 3 4 5 6 2 9 7)

Permutation encoding can be used in ordering problems, such as travelling sales man problem or task ordering problem.

In **permutation encoding**, every chromosome is a string of numbers, which represents number in a **sequence.**

Chromosome *A*	1 5 3 2 6 4 7 9 8
Chromosome *B*	8 5 6 7 2 3 1 4 9

Example of chromosomes with permutation encoding

Permutation encoding is only useful for ordering problems. even for this problem for some types of crossover and mutation corrections must be made to leave the chromosome consistent (*i.e.*, have real sequence in it).

Example of problem: Travelling salesman problem (TSP) all of them, but he does not to travel very much. Find a sequence of cities to minimize travelled distance.

Encoding: Chromosome says order of cities, in which salesman will visit them.

5.6.4. Tree Encoding

Crossover

Tree crossover: In both parent one crossover point is selected, parents are divided in that point and exchange part below crossover point to produce new offspring.

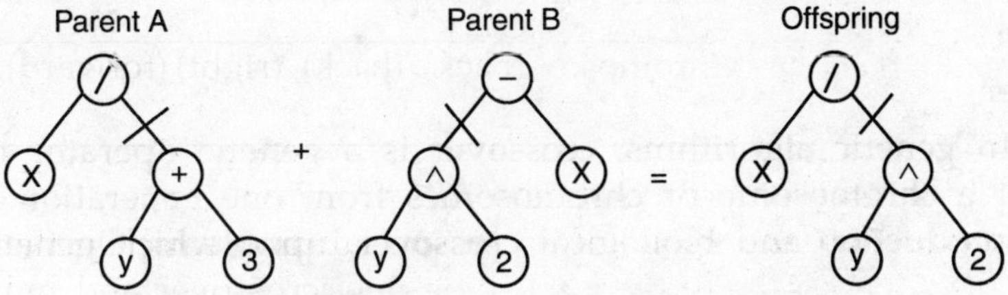

Fig. 5.10. Generation of offspring by crossover

Mutation

Changing operator, number: Selected nodes are changed

Tree encoding is used mainly for evolving programs or expressions, for genetic programming.

In tree encoding every chromosome, is a tree of some objects, such as functions or commands in programming language.

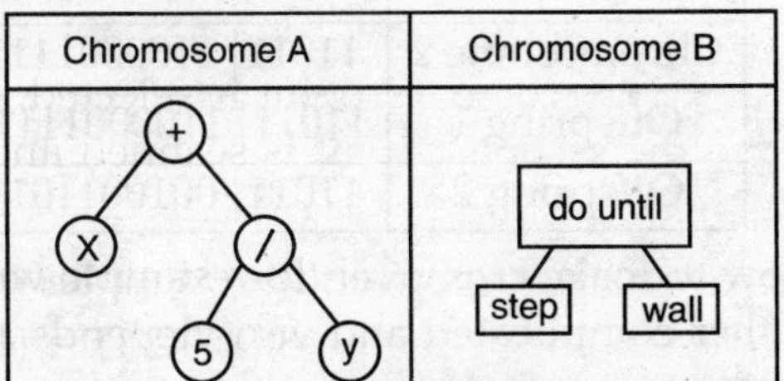

Fig. 5.11.

(+ x (/ 5 y))	(do_until step wall)

Example of chromosome with tree encoding

Tree encoding is good for evolving programs. Programming language LISP is often used to this, becuase programs in it are represented in t his form and can be easily parsed as a tree, so the crossover and mutation can be done relatively easily.

Example of problem: Finding a function from given values.

The problem: Some input and output values are given. Task is to find a function, which will give the best (closet to wanted) output to all inputs.

Encoding: Chromosome are functions represented in a tree.

Generation cycle: A genetic Altogithms operates through a simple cycle of stages:

(*i*) Creation of a "population" of strings,

(*ii*) Evaluated of each string,

(*iii*) Selection of best strings and

(*iv*) Genetic manupulation to create new population of strings.

5.7. GENETIC OPERATORS

A genetic operator is an operator used in genetic algorithms to maintain genetic diversity, known as Mutation and to combine existing solutions into others, Crossover The main difference between them is that the mutation operators operate on one chromosome, that is they are unary, while the crossover operators are binary operators.

Types of Operators

1. Mutation
2. Crossover

Crossover: In genetic algorithms, crossover is a genetic operator used to vary the programming of a chromosome or chromosomes from one generation to the next. It is analogous to reproduction and bioilogical crossover, upon which genetic algorithms are based.

Crossover: After we have decided what encoding we will use, we can make a step to crossover. Crossover selects genes from parent chromosomes and creates a new offspring. The simplest way how to to this is to choose randomly some crossover point and everything before this point copy from a first parent and then everything after a crossover point copy from the second parent.

Crossover can then look this (| is the crossover point):

Chromosome 1	11011 | 00100110110
Chromosome 2	11011 | 11000011110
Offspring 1	11011 | 11000011110
Offspring 2	11011 | 00100110110

There are other ways how to make crossover, for example we can choose more crossover points. Crossover can be rather complicated and very depends on encoding of the encoding of chromosome. Specific crossover made for a specific problem can improve performance of the genetic algorithm.

Mutation: After a crossover is performed, mutation take place. This is to prevent falling all solutions in population into a local optimum of solved problem. Mutation changes randomly the new offspring. For binary encoding we can switch a few randomly chosen bits from 1 to 0 or from 0 to 1. Mutation can then be following ;

Original offspring 1	1101111000011110
Original offspring 2	1101100100110110
Mutated	1100111000011110

offspring 1	
Mutated offspring 2	1101101100110110

The mutation depends on the encoding as well as the crossover. For example when we are encoding permutations, mutation could be exchanging two genes.

5.8. MUTATION

After cross over, the strings are subjected to mutation. *Mutation* of a bit involves flipping it, changing 0 to 1 and vice versa with a small mutation probability P_m. The bit-wise mutation is performed bit-by flipping a coin with a probability of P_m. Flipping a coin with a probability of P_m is simulated as follows.

A number between 0 to 1 is chosen at random. If the random number is smaller than P_m then the outcome of coin flipping is true, otherwise the outcome is false. If at any bit, the outcome is true then the bit is altered, otherwise the bit is kept unchanged. The bits of the string are independently muted, that is, the mutation of a bit does not affect the probability of mutation of other bits. A simple genetic algorithm treats the mutation only as a secondary operator with the role or restoring lost genetic materials. Suppose, for example, all the strings in a population have conveyed to a zero at a given position and the optimal solution has a one as that position. Then cross over cannot regenerate a one at that position while a mutation chold. The mutation is simply an isurance policy against the irreversible loss of genetic material. The mutation operator introduces new genetic structures in the population by randomly modifying some of its building blocks. It helps the search algorithm to escape from local minima traps since the modification is not related to any previous genetic structure of the population. It creates different structure representing other sections of the search space. The mutation is also used to maintain diversity in the population. For example consider the following population having four eight-bit strings.

0111 011

0011 1101

0001 0110

0111 1100

Notice that all four strings have a zero in the leftmost bit position. If the true optimum solution requires a one in that position, then neither reproduction nor cross over operator described above will be able to create one in that position. The inclusion of mutation introduces some probability (N_{pm}) of turning zero to one as

0110 1011

0011 1101

0001 0110

1111 1100

Mutation for real numbers can be done as

Before (1.38 – 69.4 326.44 01)

After (1.38 – 67.5 326.44 0.1)

Hence, mutation causes movement in the search space (local or global) and restores lost information to the population.

5.8.1 Mutation Rate P_m

Mutation rate is the probabililty of mutation which is used to calculate number of bits to be muted. The mutation operator preserves the diversity among the population which is also very important fo the search. Mutation probabilities are smaller in natural populations leading us to conclude that mutation is appropriately considered a secondary mechanism of genetic algorithm adoption. Typically, the simple genetic algorithm uses the population size of 30 to 200 with the mutation rates vrying from 0.001 to 0.5.

The cycle of a Genetic Algorithms is presented below

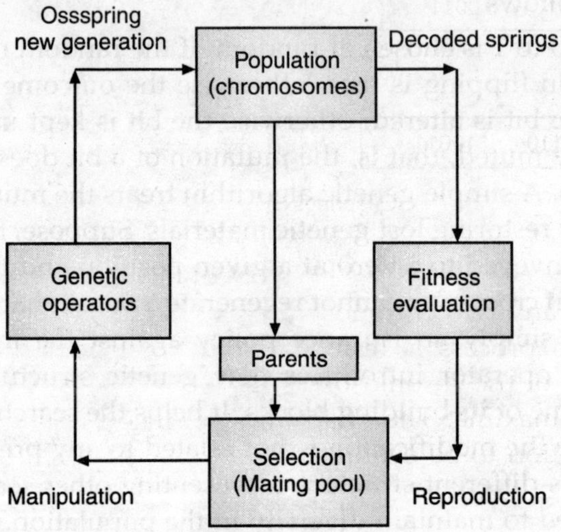

Fig. 5.12. G A Cycle

Each cycle in Genetic Algorithms produces a new generation of possible solutions for a given problem. In the first phase, an initial population, describing representatives of the potential solution, is created to initiate the search process. The elements of the population are encoded into bit-strings, called chromosomes. The performance of the strings, often called fitness, is then evaluated with the help of some functions, representing the constraints of the problem. Depending on the fitness of the chromosomes, they are selected for a subsequent genetic manipulation process. It should be noted that the selection process is mainly responsible for assuring survival of the best-fit individuals. After selection of the population strings is over, the genetic manipulation process consisting of two steps is carried out. In the first step, the crossover operation that recombines the bits (genes) of each two selected strings (chromosomes) is executed. Various types of crossover operators are found in the literature. The single point and two points crossover operations are illustrated.

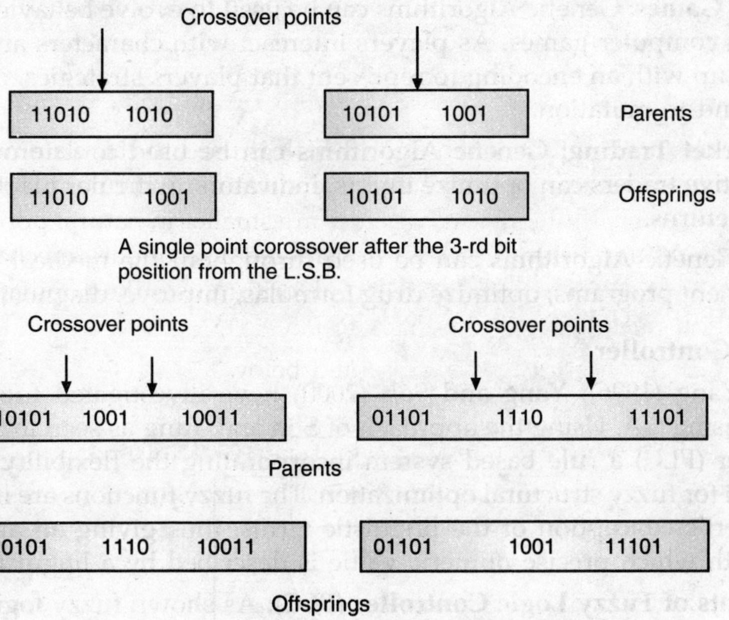

Fig. 5.13. Operators of GA

The crossover points of any two chromosomes are selected randomly. The second step in the genetic manipulation process is termed mutation, where the bits at one or more randomly selected positions of the chromosomes are altered. The mutation process helps to overcome trapping at local maxima. The offsprings produced by the genetic manipulation process are the next population to be evaluated.

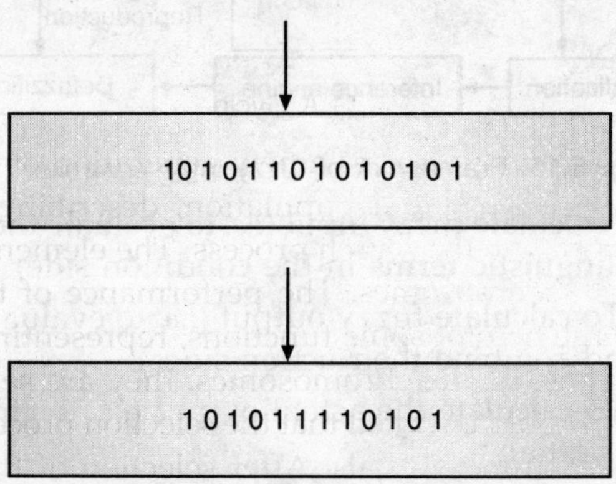

Fig. 5.14. Mutation of a chromosome at the 5th bit position

5.9. COMMON APPLICATIONS OF GENETIC ALGORITHM

Scheduling: Genetic Algorithm (GA) cna be used for numerous scheduling problems. Using a GA for difficult scheudling problems enables relatively arbitary constraints and objectives to be incorporated painlessly into a single optimization method.

Computer Games: Genetic Algorithms can be used to evolve behaviors of characters and events with in computer games. As players intereact with characters and events the games GA will come up with an encoding to represent that players strategies, one that is amenable to crossover and to mutation.

Stock Market Trading: Genetic Algorithms can be used to improve trading systems. Using GAs, active traders can optimize inputs, indicators, and rules for their trading systems and increase returns.

Medical: Genetic Algorithms can be used throughout the medical field. GAs can help develop treatment programs, optimize drug formulas, improve, diagnostics, and much more.

Fuzzy Logic Controller

Soh and Yang (1996), Yang and Soh (2000) have investigated fuzzy based structural optimization using GA. Using the approach of Soh and Yang as seen in Section 15.4, a fuzzy logic controller (FLC) a rule based system incorporating the flexibility of human decision making is used for fuzzy structural optimization. The fuzzy functions are intended to represent a human expert's conception of the linguistic terms, thus giving an approximation of the confidence with which precise numeric value is described by a linguistic label.

Components of Fuzzy Logic Controller (FLC): As shown fuzzy logic controller process is divided into three stages.

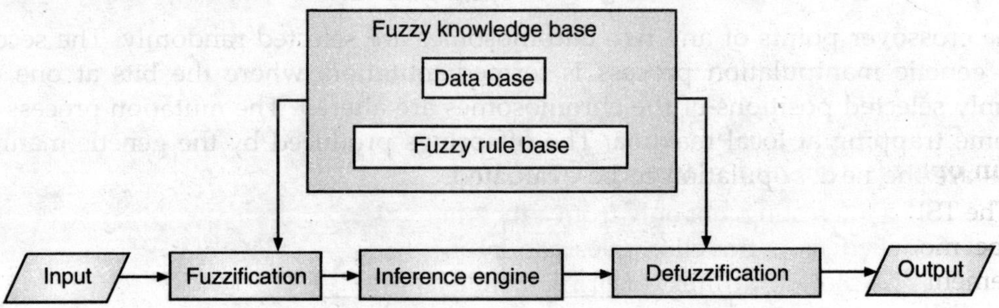

Fig. 5.15. Framework of fuzzy logic controller (FLC)

(*a*) **Fuzzificataion:** To calculate fuzzy input (*i.e.* to evaluate the input variable with respect to corresponding linguistic terms in the condition side).

(*b*) **Fuzzy inference:** To calculate fuzzy output (*i.e.* to evaluate the activation strength of every rule base and combine their action sides).

(*c*) **Defuzzification:** To calculate the actual output (*i.e.* to convert the fuzzy output into precise numerical value).

FLC-GA Based Structural Optimization

First, coding scheme is to be defined and the initial population is produced. The computation with genetic operators is used to evaluate fitness function with respect to the objective function. Figure 5.16 shows the FLC-GA Based Optimization Procedure. Using FLC we can get the expert's experience in fuzzy rule base of FLC. Hence, the search can react optimum solution quickly. As a result, computing time is very much reduced. The predefined probability and fuzzy representation of design constraints causes FLC to reduce the risk of premature problem solution caused by improper rule.

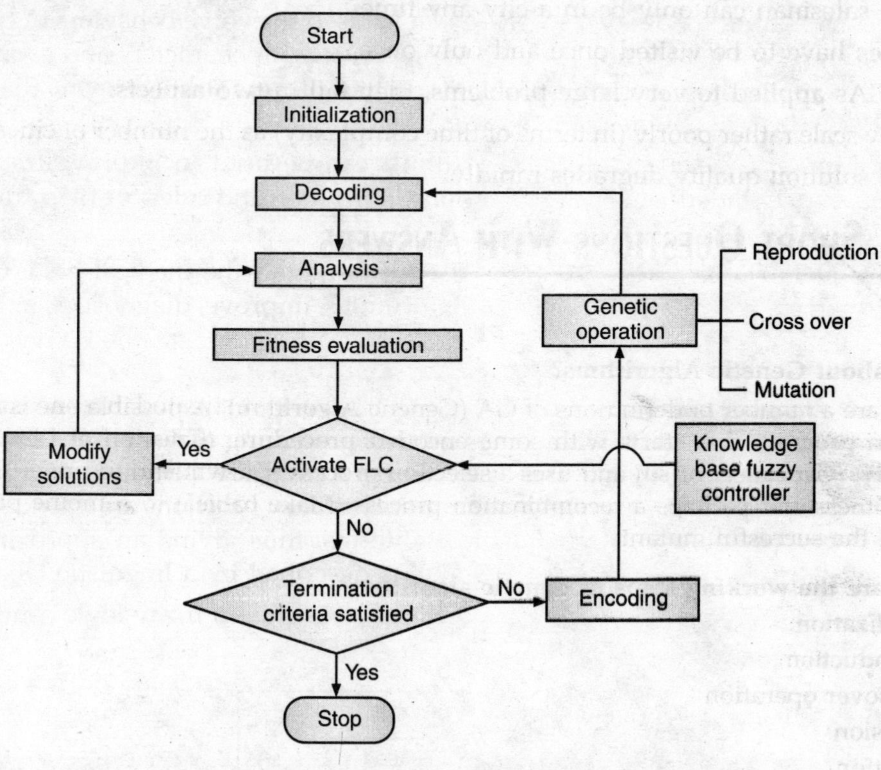

Fig. 5.16. Flow chart of FLC GA based optimization

GA an optimisation and planning: Travelling Salesman Problem

The TSP is interesting not only from a theoretical point of view, many practical applicatins, can be modelled as a travellin salesman problem or as variants of it, for example, pen movement of a plotter, drilling, of printed circuit boards (PCB), real-world routing of school buses, airlines, delivery, trucks and postal carriers. Researchers have tracked TSPs to study biomolecular pathways, to route a computer network parallel processing to advance cryptography, to determine the order of thousands of exposures needed in X-ray crystallography and to determine routes searching for forest fires (which is a multiplesalesman problem partitioned into single TSPs). Therefore, there is a tremendous need for algorithms.

In the last two decades an enormous has been made with respect to solving travelling salesman problems to optimality which of course, is the ultimate goal of every researcher. one of landmarks in the search for optimal solutions in a 3038-city problem. This problem is only party due to the increasing hardware power of computers. Above all, it was made possible by the development of mathematical theory and of efficient algorithms. Here, the GA approach is discussed.

There are strong relations, between the constraints of the problem, the representation adopted and the genetic operators that can be used with it. The goal of travelling Salesman problem is to devise a travel plan (a tour) which minimise the total distance travelled. TSP is is NP-hard (NP stands for non-deterministic polynomial time) – it is generally believed cannot be solved (exactly) in time polynomial. The TSP is constrained.

- The salesman can only be in a city any time
- Cities have to be visited once and only once.

When GAs applied to very large problems, they fail in two aspects:

1. They scale rather poorly (in terms of time complexity) as the number of cities increases.
2. The solution quality degrades rapidly.

Short Questions With Answers

Q.1. What about Genetic Algorithms?

Ans. There are a number of definitions of GA (Genetic Algorithm). A possible one is A GA is an optimization program that starts with some encoded procedure, (Creation of Life) mutates it stochastically, (Get cancer or so) and uses a selection process (Darwinism) to prefer the mutants with high fitness and perhaps a recombination process (Make babies) to combine properties of (preferably) the succesful mutants.

Q.2. What are the working steps of genetic algorithm?

Ans. Initialization
Reproduction
Crossover operation
Inversion
Mutation

Q.3. What are the different operators in genetic algorithm?

Ans. Reproduction/selection
Crossover
Mutation

Summary

In this chapter we covered the Basics concepts of Genetic Algorithm(GA)and its algorithm lying between reproduction or creation of a population of strings, competition or evaluation of each string, selection of best string and survival or genetic manipulation to create the new population of strings. We discuss the Biological Background and inspiration therein. We also discuss the classification of genetic algorithm and working principle. Thereafter we discuss the deterministic explanation of Holland's Observation and Stochastic Explanation of Genetic Algorithms. The fundamental theorem of Genetic Algorithm known as Schema theorem is discussed to show the convergence of GA process In very short, we discussed the Markov model as this is the backbone of all futuristic procedures. Next we discuss the procedures of GA, flow chart of GA and Genetic representations. For Encoding, Initialization and selection; Binary encoding Value Encoding, Permutation Encoding and Tree Encoding are discussed. In all these methods the Crossover and mutation process is discussed. Genetic operators, like Crossover and Mutation, with mutation rate and its affect on GA cycle are discussed. Finally, Common applications of Genetic Algorithm are discussed.

REVIEW QUESTIONS

1. What are the most important aspects of using Genetic Algorithm ?
2. Draw the classification diagram of Genetic Algorithm.
3. Discuss the biological motivation for using Genetic Algorithm as Artificial Intelligence tool.
4. Discuss the working principle of Genetic Algorithm.
5. What are the different operators of Genetic Algorithm? Write short notes about them. What is the affect of mutation rate?
6. What is Holland's Observation? Evolve it to derive the fundamental theorem of Genetic Algorithm, namely, the schema theorem.
7. What is Hidden Markov Model?
8. Discuss with flow chart the procedure of Genetic Algorithm.
9. Discuss Mutation and crossover in Binary encoding, value encoding, permutation encoding, and tree encoding.
10. What are common applications of Genetic Algorithm in our day to day life.

NUMERICAL PROBLEMS

1. The following variable bounds are to be solved with Genetic Algorithm with three variables x, y, z:

$$-6 < x \leq 12$$
$$0.002 \leq y \leq 0.004$$
$$104 \leq z \leq 10^5$$

   What should be the length of minimum string of any point in binary string to achieve two significant digits. Do the same exercise to achieve the three significant digits.

2. Use Genetic Algorithm to solve the following nonlinear problem:

$$\text{minimize } (x_1 - 2.5)^2 + (x_2 - 5)^2$$
$$\text{subject to } 5.5\, x_1 + 2x_2^2 - 18 \leq 0$$
$$0 \leq x_1,\ x_2 \leq 5$$

   How many bits are required to give three and two decimal places of accuracy for variables x_1, x_2.

□□□

ANNEXURES

Annexure–A: History and Summary (For your Leisure Reading)

Annexure–B: More ANN Networks (For more Enthusiastic Students)

Annexure–C: Research Areas in Soft Computing (For more Ambitious Students)

Annexure–D: ANN Applications

1 ANNEXURE ONE

HISTORY AND SUMMARY (FOR YOUR LEISURE READING)

CHAPTER OUTLINE

A.1.1. History and Summary

A.1.1. HISTORY AND SUMMARY

Introduction

This introduction presents the underlying principles of neural networks so one can understand the strengths and weaknesses of each type and see how they relate to each other. Consequently, this introduction is generally non-mathematical with an emphasis on network diagrams. What follows are the significant neural network designs seeking to mimic how memory and pattern recognition might work in the brain. Other neural network designs not covered here include those for classical conditioning, motor control in the cerebellum, and rhythmic motor generation.

The First Neural Network Theory

The first neural network was presented by Alexander Bain (1818 - 1903) of the United Kingdom in his 1873 book entitled "Mind and Body. The Theories of Their Relation". His work was no doubt inspired by the then new findings of neuroanatomists who using the newly discovered neural stains of carmine (by Gerlach in 1858), methylene blue (by Nissl in 1858), and hematoxylin (by Waldeyerin 1863 - stains axon fibers) were able for the first time to see the true extent of the interlocking fibers of the brain. The state of knowledge before these stains is given by the 1853 Manual of Human Histology written by Rudolf Kolliker which states that there were various forms of nerve cells and "besides these there are a good many fine pale fibers, like the processes of cells, only more extended of which nothing more can be said as to whether they are nerve tubes or are to be referred to as the processes of cells." (McHenry - 1969)

Bain first described memory as a set of nerve currents weaker than that produced by the original stimulus.

"If we suppose the sound of a bell striking the ear, and then ceasing, there is a certain continuing impression of a feebler kind, the idea or memory of the note of the bell; and it would take some very good reason to deter us from the obvious inference that the continuing impression is the persisting (although reduced) nerve currents from the past - the remembrance of the former sound of the bell" (1873, p 90 as reprinted in Wilkes and Wade - 1997).

The ability to recall a specific memory requires that an association (grouping) first be made with some other memory, sensation, or motor action via some kind of neural growth.

"for every act of memory, every exercise of bodily aptitude, every habit recollection, train of ideas, there is a specific grouping, or coordination, of sensations and movements, by virtue of specific growths in cell junctions" (1873, p. 91 as reprinted in Wilkes and Wade - 1997)

Bain next applied these general ideas in a more concrete fashion suggesting an early form of threshold decision making as shown in Fig. 1 where each line represents an idea which sums (co-operates) with other ideas.:

"If each set of sensory fibres had one definite connection with motory or outcarrying fibres, we should have always the same movement answering to the stimulation of the same nerves, as in the reflex system; the fibre a could do nothing but affect the movement x. It is necessary to the variety and flexibility of our acquirements, that the fibre a should at one time take part in stimulating x, and at another time take part in stimulating y, the circumstances being different. The stroke of the clock will stimulate us at one time to set in one direction, and at another time in another direction, according to the ideas that it co-operates with."

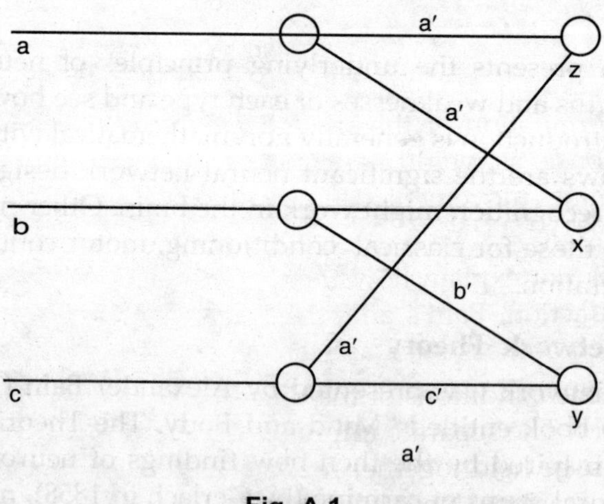

Fig. A.1.

Yet Bain realized that if his theory were true every possible association or grouping would have to be hardwired into the brain:

"...the number of fibres and cells brought into action, before the grouping can converge upon some one set of cells definitely connected with an out-going motor arrangement, or with some other internal grouping, - must be very great indeed; and but for the vast number of fibres and cells, demonstrably present in the& brain, the separate embodiment of every separate impression and idea would seem impractical." (1873, p. 113, as reprinted in Wilkes and Wade - 1997)

In an attempt to reduce the degree of hardwiring needed Bain suggested that a signal attenuation strategy through the brain network would add greater flexibility. In Fig. 2 each node has some resistance so signal strength is proportional to path length.:

"... a more energetic current necessarily takes a more extended sweep, and affects a number of cells and fibres that are left quiescent under a feebler current. The cells being viewed as crossings - where a current in one circuit induces a current in an adjoining circuit - there is, at each crossing, a certain resistance to overcome, and the feebler current is sooner exhausted and stops short of the distance reached by the stronger." (1873, pp. 113-114, as reprinted in Wilkes and Wade - 1997)

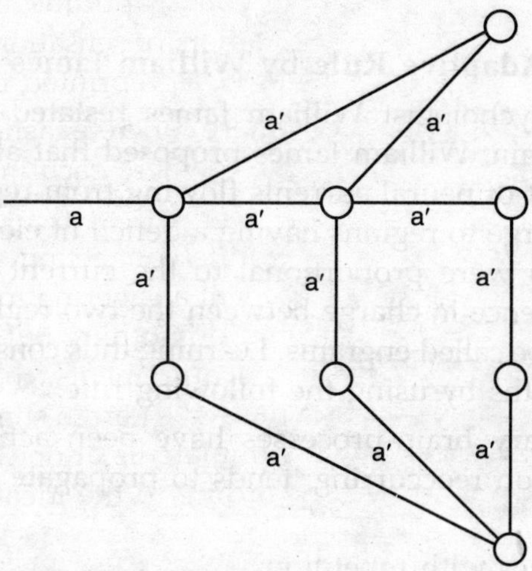

Fig. A.2.

" ... the degree of stimulation of the same fibres will determine not merely a greater energy of the same response, as would happen in reflex stimulation, but a totally different response: a, weak, determines movement x; a, strong, determines y." (1873, p.109, as reprinted in Wilkes and Wade - 1997).

Notice that Bain does not explain how the the stronger signal does not simultaneously activate the weak signal action. Bain's adaptive rule for associations (groupings) precedes that of later, better known authors such as Hebb:

"We know what are the conditions of making an acquirement, or of fixing two or more things together in memory. The separate impression must be made together, or flow in close succession; and they must be held together for a certain length of time, either, either on one occasion or on repeated occasions. Now to each impression, each sensation or thought, there corresponds physically a group or series of nerve currents; when two impressions concur, or closely succeed one another, the nerve currents; find some bridge or place of continuity, better or worse, according to the abundance of nerve matter available for the transition. In the cells or corpuscles where the currents meet and join, there is, in consequence of the meeting, a strengthened connexion or diminished obstruction - a preference track for that line over lines where no continuity has been established. (1873, p. 117, as reprinted in Wilkes and Wade - 1997)

Most of the ideas for Bain's book Mind and body were prepared 10 years earlier for three lectures to the Philosophical Society in Aberdeen, Scotland. Yet the idea that most of the brain had to be hardwired apparently did not appeal to most scientists. Even his student, David Ferrier who published his own classical work The Functions of the Brain (1876) never mentioned the details of Bain's work although the work itself is mentioned. In the greatest irony of all even Bain came to doubt his own idea's:

"The hypothesis was a legitimate one; but subsequent reflection led to the belief that the number of psychical elements, although run up to hundreds of thousands, was still inadequate." (1904, p 313).

Not until 1943 with McCulloch and Pitts at the dawn of computer age were brain theorists of equivalent caliber to be found who went beyond language based general descriptions to actual network drawings.

Restatement of Bain's Adaptive Rule by William James (1890)

In 1890 American psychologist William James restated Bain's adaptive rule without apparently knowing of Bain. William James proposed that all thoughts and bodily actions were produced as a result of neural currents flowing from regions (brain-processes) having an excess of electrical charge to regions having a deficit of electrical charge. The intensity of the thoughts and actions were proportional to the current flow rate which in turn was proportional to the difference in charge between the two regions. Later these reverberating electrical currents would be called engrams. Learning thus consisted of changing these current paths or forming new paths by using the following rule:

"When two elementary brain-processes have been active together or in immediate succession, one of them, on reoccurring, tends to propagate its excitement into the other." (James 1890, p566)

Such learning improves with repetition:

"... nerve currents propagate themselves easiest through those tracts of conduction which have been already most in use." (James 1890, p563)

This also implies that paths not used will tend to close producing forgetting.

Actual characterization of these nerve currents as a burst of action potentials traveling in one direction down a single celled neuron was only established between 1890 and 1910. This quickly lead to the standard neuron model in which a neuron sums all its facilitory and inhibitory inputs from other connecting neurons and if the neural charge exceeds a threshold then an action potential is produced. The greater some sensory stimulus intensity the greater is the frequency of the action potentials with a typical maximum frequency typically being about 100 pulses per second.

According to the theory of William James a neuron's transmission efficiency between stimulus and its output should increase with use as the skill is learned but the first experimental test found just the opposite. In 1898 England's Charles Sherrington found that spinal neurons in cats reduce their efficiency with low intensity use instead of increasing their efficiency as expected. He had found a phenomena known today as habituation.

First Neural Logic (1938 to 1943)

Yet no theory arose to challenge the holistic, reverberatory scheme of William James until 1938 when N. Rashevsky proposed that the brain could be organized around binary logic operations since action potentials could be viewed as binary 1 (true) values. He even presented the circuit shown in Fig. 3 showing how a binary logic EXCLUSIVE OR operation could be implemented using addition and subtraction operations. He did not formulate the other logical operations in analog terms so his formulation remained incomplete.

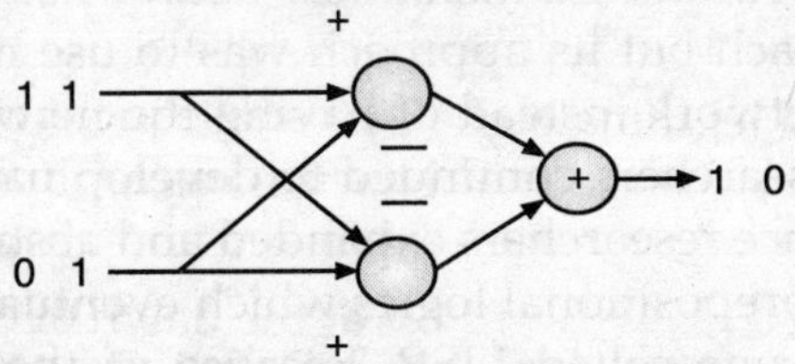

Fig. A.3. Rashevsky's EXCLUSIVE OR

In 1943 Warren McCulloch and Walter Pitts realized that the natural consequence of the standard neuron model's threshold in combination with binary action potentials produced another type of logic called threshold logic. Since each action potential pulse is an all or nothing binary event a threshold value of 2 defines an AND operation as shown in Fig. 4. Likewise, a threshold value of 1 defines an INCLUSIVE OR operation since only one action potential on either line is sufficient to produce an output. They also described the use of the subtraction operation in their paper but never explicitly associated it with the logical CONDITIONAL operation.

Threshold logic was expanded by adding more input lines so that the output of the summation node would become analog since this seemed to fit in more with the standard neuron model. Yet by using more than two inputs the operations ceased to be logical, that is, they ceased to define the basic set of state symmetry testers used as building blocks in larger circuits. Instead these operations became parameter classifiers which classified their input patterns based upon the one parameter of signal magnitude given by the summation of the binary inputs.

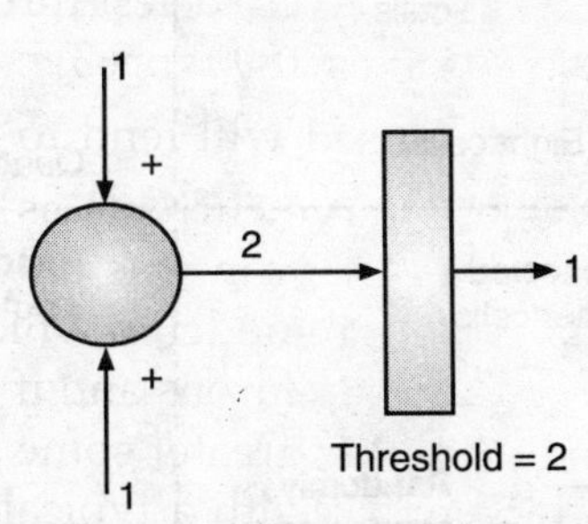

Fig. A.4. Threshold Logic AND

The very existence of an analog value for the first time allowed these operations to become adaptive using multiplication factors called weights. Exactly how that adaptability was implemented defines the many different neural networks described below. Yet modifications to a single parameter can never fully characterize the state (the pattern or distribution) of a system's input. Some ambiguity is always the result yet at the time this ambiguity was hailed as a great achievement of neural networks for many saw it as meaning that neural networks could work with partial knowledge even though this ambiguity could never be precisely defined and controlled.

So the achievement of adaptability in the neural networks of this time came at the cost of loosing the decision making resolution of logic. By 1963 this split brain researchers into the separate groupings of neural network researchers and artificial intelligence researchers. The 1963 paper by R.O. Winder seems to have been the last major paper which still attempted to combine the weight based and mathematical neural network approach with the logical artificial intelligence approach but its approach was to use mathematical techniques to find the desired weights of a network instead of having the network learn them. The result was that the neural network researchers continued to develop parameter based classifier circuits while the artificial intelligence researchers expanded and abstracted the binary logic approach to form propositional and prepositional logics which eventually resulted in the development of the programming language called LISP. Because of the precision of logic the field of artificial intelligence enjoyed great initial success. Yet its high level algorithmic and sequential approach eventually proved to have serious limitations in terms of adaptability and the handling of uncertainty.

The First Randomly Connected Reverberatory Networks (1954)

Neural network research really only became possible at the dawn of the computer age when ideas could be validated by simulation on various types of electronic calculators. The impetus for these simulations was provided by Donald Hebb of McGill University in Canada who in 1949 proposed this unidirectional variation of the Bain / James learning rule:

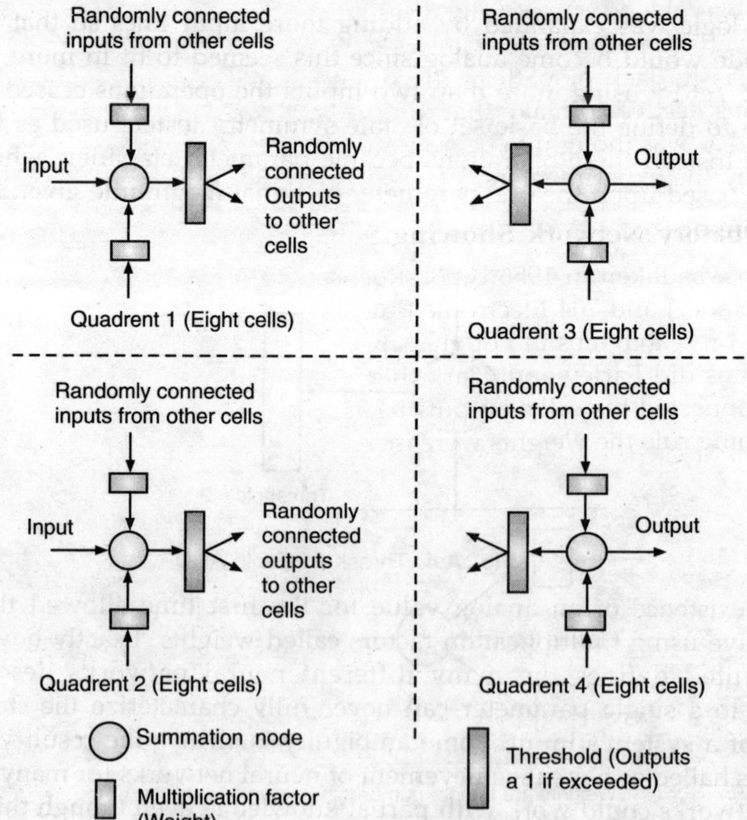

Fig. A.5. The Farlay and Clark Network

"Let us assume then that the persistence of repetition of a reverberatory activity (or trace) tends to induce lasting cellular changes that add to its stability. The assumption can be precisely stated as follows: When an axon of cell A is near enough to excite cell B and repeatedly or persistently takes part in firing it, some growth process or metabolic change takes place on one or both cells so that A's efficiency as one of the cells firing B is increased" (Hebb 1949, p62)

The first such Hebbian inspired network was simulated by Farley and Clark in 1954 on an early digital computer at M.I.T. Their network shown in Fig. 5 consisted of nodes representing neurons randomly connected with each other by unidirectional lines having multiplication factors (weights). Each node summed its inputs and produced an output of one if the sum of its inputs exceeded its threshold value. In order to recognize patterns the network was divided into quarters with the inputs from one pattern entering the front top quarter while the inputs from the other pattern entered the front bottom quarter. Only one pattern was presented at a time so each tended to produce its own unique flow path through the network. A pattern was considered to be recognized if it consistently produced more activity in one of the back network quarters than in the other.

Yet in order to get this network to work Farley and Clark had to modify Hebb's learning rule. This new rule required that the activity level of each network line be examined at each instant in time so that if the line value changed in the desired *direction*then all of its nodal input weights would be incremented (increased). If the change was in the wrong direction then the nodal input weights would be decremented (decreased). With this new directional rule the network was able to successfully discriminate between two widely differing patterns as long as they were presented alternately (see Farley, 1960). The major problem with this type of network is its lack of pattern discrimination resolution combined with the high number of neurons needed to make that discrimination. Also the cells in the network were not self-assembled as was thought to be necessary at the time but were assigned to assemblies (the quadrants) by the experimenters.

The First Reverbatory Network Showing Self-Assembly 1956)

The next step was taken in 1956 by N. Rochester and friends using an early IBM Type 701 (2 Kbytes memory) and 704 Electronic Calculator (as computers were often called then) located at the IBM research labs in Poughkeepsie, New York. Instead of connecting the cells randomly overall as did Farley and Clark they organized their cells into a single layer and then randomly connected the cell's outputs back to other cell's inputs (Fig. 6). In accordance with Hebb's learning rule the weights were increased with use and they had this to say about it:

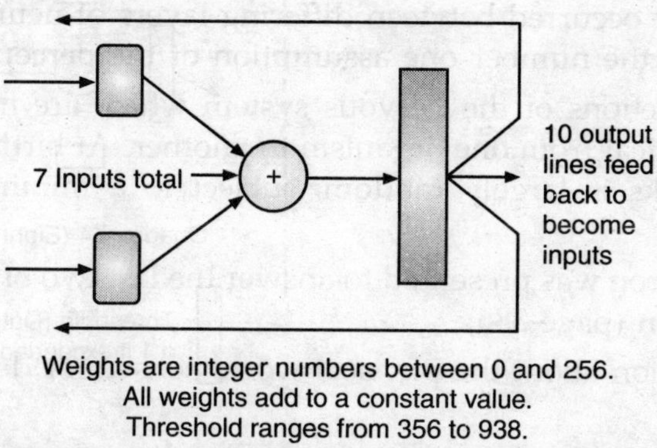

Weights are integer numbers between 0 and 256.
All weights add to a constant value.
Threshold ranges from 356 to 938.

Fig. A.6. One Cell of the Self-Assembly Network

"No process of just this sort has been observed in living tissue. However, it has not been possible to demonstrate, by measurement, that the Hebb postulate is false. Nothing else has been observed that could account for learning and memory in a plausible way."

Despite this they were forced to modify Hebb's rule by normalizing all the weight values so that they always added up to some constant value. The reason was that with use all the weights would eventually increase to their maximum value. What was desired was that the more heavily used weights be larger than the lessor used weights so if any weight is increased the other weights must be decreased by an equivalent overall amount. (Significantly, they did not discuss how this process could be accomplished neurally).

They also tried neural habituation in their network which occured when a rapidly firing neuron increased it threshold so as to become less responsive. The result was a network which did not form any assemblies but which did show a continuing wave-like reverberation after the inputs ceased.

Meanwhile, Peter Milner, an associate of Donald Hebb, suggested that cell assemblies could form if most of the synapses within a cell assembly were exitory (positive) while those between cell assemblies were inhibitory (subtractive). So they changed the network weight ranges from 0 to 256 to -1 to 1. In order to speed up the simulation they ceased using binary pulses on the network lines and instead had the lines represent analog firing frequencies which in their model ranged from 0 to 15. They called this the F.M. model. The result was that cell assemblies built up around the input regions but only around those regions. They did not appear any where else in the network. In response to this partial success they said the following:

"This kind of investigation cannot prove how the brain works. It can, however, show that some models are unworkable and provide clues as to how to revise the models to make them work."

The First Pattern Regularity Detector - The Original Perceptron (1958)

Random neural networks were not having much success yet many researchers were stuck on the idea that the neural connections in the brain were mostly random. So a slight compromise was made in 1958 by Frank Rosenblatt who initiated a new phase in neural network research by abandoning the idea of self-forming cell assemblies. Instead memory was simply a change in the relation between some input and some output which changed with regular use. This meant that those patterns which regularly and consistently occurred would be learned. This type of network for unsupervised learning is called an auto-learning network or a pattern regularity detector. Consequently, the randomness of neural connections was not overall but only occurred between differing layers of neural cells. On page 388 of his 1958 paper he states the number one assumption of the perceptron's design as:

"The physical connections of the nervous system which are involved in learning and recognition are not identical from one organism to another. At birth, the construction of the most important networks is largely random, subject to a minimum number of genetic constraints."

The Original Perceptron was presented to answer the last two of these three fundamental questions about the brain (page 386):

1. How is information about the physical world sensed, or detected, by the biological system?

2. In what form is information stored, or remembered?
3. How does information contained in storage, or in memory, influence recognition and behavior?

The simplicity and random connections of the Original Perceptron and the later Perceptrons made them a fascinating subject for mathematical analysis using probability theory. Rosenblatt's preference for this approach as opposed to the developing algorithmic approach of artificial intelligence is shown by this passage in his 1960 paper.

"Simulation should not, in general, be attempted without a theoretical analysis of the nerve net in question, sufficient to indicate questions of theoretical interest. The examination of arbitrary networks in the hope that they will yield something interesting, or the simulation of networks which have been specially designed to compute a particular function by a definite algorithmic procedure seem to be about equally lacking in value."

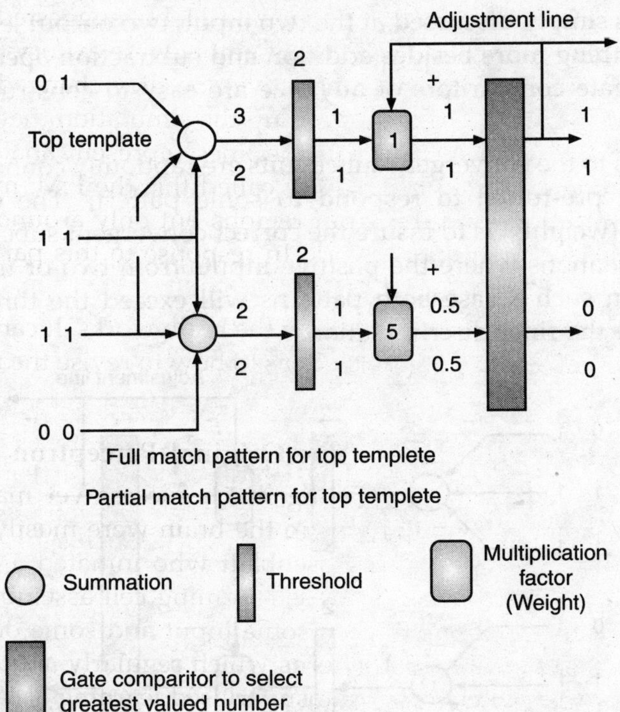

Fig. A.7. The Original Perceptron – Positive Inputs

The original Perceptron circuit shown in Fig. 7 consisted of many convergent type subcircuits feeding into a decision making element which only passed its greatest valued input (gate comparitor). A convergent subcircuit is characterized by several input lines feeding into some central operation or series of operations. In this case the central operation is a summation node connected to a threshold unit which in turn connects to a multiplication factor (weight). Rosenblatt called these subcircuits "A" units for association units. Each received a certain number of binary (0 or 1) positive inputs and a lessor number of negative (subtractive) inputs connected in a random manner. If the sum of these inputs exceeded the threshold value then the threshold would produce an output value of 1. This value was then modified by the weight valued between 0 and 1 which Rosenblatt called the value of the "A" unit.

The output of each convergent subcircuit was next sent to the gate comparitor which selected its largest input value, passed it on and fed it back to its convergent subcircuit's ("A" unit's) weight to increment it by a certain amount. In this way the most used template would would increase its transmission efficiency in conformity to the idea's of Donald Hebb mentioned above. Any neural network using a gate comparitor is known as a competitive network because the different convergent subcircuits compete to get their signal selected.

The irony here is that a gate comparitor seems to be impossible to form beyond the two input, two output level using only the mathematical operations of addition and subtraction (in such a case the circuit is equivalent to the EXCLUSIVE OR circuit of Fig. 1 without the final summation node). In all the competitive type of neural networks this function is either left vague of is simply stated as a mathematical function. In Rosenblatt's paper the gate comparitor function is simply discussed at the two input, two output level. This should have been a clue that something more besides addition and subtraction operations were required in neural networks (gate comparitors of any size are easy to construct using multivalued logic operations).

Because the inputs to the convergent subcircuits are randomly connected every subcircuit is already somewhat pre-tuned to respond to some pattern. The only purpose of the multiplication factors (weights) is to insure the correct convergent subcircuit selection by the gate comparitor in situations where the positive inputs from two or more patterns overlap as shown in Fig. 7. In such a case both patterns will exceed the threshold value leaving weight value to make the final discrimination.

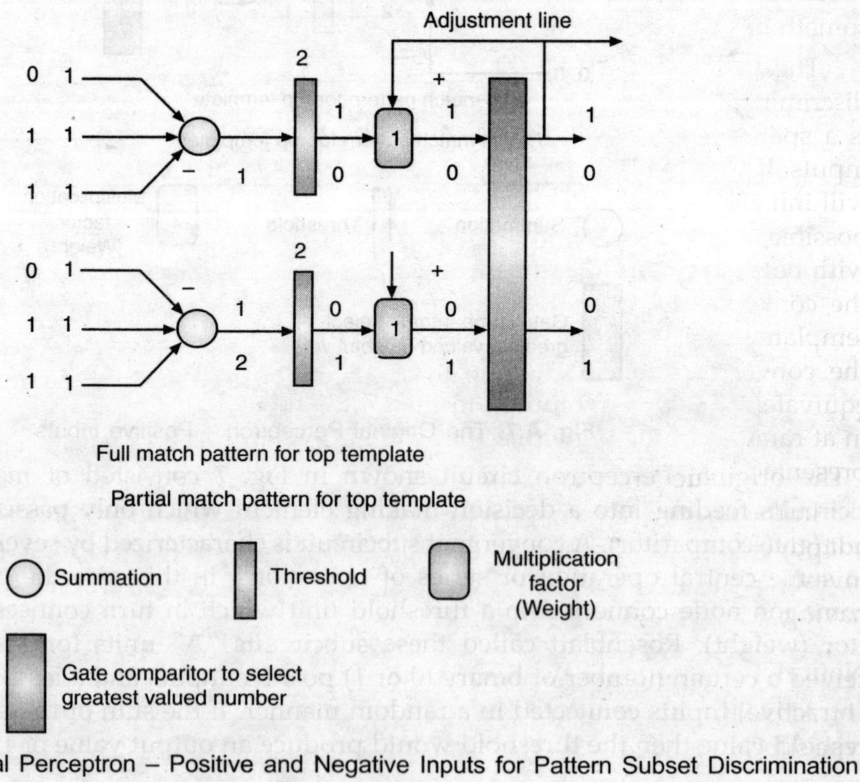

Fig. A.8. The Original Perceptron – Positive and Negative Inputs for Pattern Subset Discrimination

The major test of any neural network's ability to classify patterns is the subset test in which the neural network must discriminate a pattern which is smaller and completely contained within another larger pattern. A variation of this test is the so called "exclusive or" test in which two different smaller patterns are contained within the larger pattern and all three must be discriminated. The Original Perceptron is not always able to pass the subset test if it only uses positive inputs (Fig. 7) yet it can if negative inputs are used (Fig. 8). The ability of the all positive input Original Perceptrons to discriminate between a full pattern and its subset (partial) pattern depends on the order of pattern presentation and on luck. If the partial match pattern is presented first then both the top and bottom convergent subcircuits of Fig. 7 would respond equally given equal initial weights. Yet in true marketing fashion, this lack of discrimination ability was touted as the significant brain-like characteristic of generalization in which the pattern does not have to be an exact match in order for it to be classified! Still, the lack of control over this process is a weakness of the original positive input only perceptron.

This subset pattern discrimination weakness in the all positive input Original Perceptron can be remedied to a certain degree if the total multiplication factor values (weights) of the network remain constant (if they add up to one this requirement is called normalization). Thus whenever a weight is incremented a certain amount all the other weights together must be decremented (reduced) the same overall amount. This biases the response towards the more recently presented and larger patterns. Thus in Fig. 7 if the the full pattern for the top subcircuit has been presented most recently then the presentation of the partial pattern will also activate the top subcircuit. Instead, if the full pattern for the middle subcircuit has been presented most recently then the presentation of the partial pattern will activate the middle subcircuit. While this is an interesting phenomena weight normalization does not seem to be something that real neuronal circuits could accomplish.

The use of negative inputs as shown in Fig. 8 partly overcomes the subset pattern discrimination problem by having larger patterns inhibit the smaller subset patterns. A template is a spatial filter, passing only that pattern (or image) which exactly matches its positive inputs. If any part of that pattern exceeds (overlaps) a template's border then negative inputs will inhibit the activation of the convergent subcircuit. A partial template leaves part of the possible pattern border undefined (no inhibition). Another way to characterize templates with border inhibition is by hardness and softness. A hard template will completely inhibit the convergent subcircuit if any part of the pattern overlaps the negative inputs. A soft template requires a large degree of overlap onto some inhibitory region in order to suppress the convergent subcircuit. The network template matching function has an algorithmic equivalent called convolution. In the Original Perceptron the negative inputs were just put in at random in the hope that they might accomplish the partial template function for some presented set of patterns.

In the Original Perceptron overlapping patterns could also have been discriminated by adaptively changing the values of the thresholds but this does not seem to have been investigated. For example, a threshold value of 3 can easily discriminate between a pattern having 3 positive binary inputs and one having 2 positive binary inputs.

Neural Pattern Classification Arrives - ADALINE (1960)

ADALINE is an acronym for ADAptive LINear element. For the first time a convergent type subcircuit having weights before the summation node is used to formally classify patterns.

Adaline was developed by Bernard Widrow and Marcian Hoff for the purpose of recognizing binary patterns so that it could predict the next bit in a flowing stream of bits. As such it was based on the ideas of R.L. Mattson, then a student at MIT. Consequently, the paper Widrow and Hoff published is an engineering paper concerned with computing which makes no mention of the brain or brain memory. Perhaps because of this Adalines do not use a threshold.

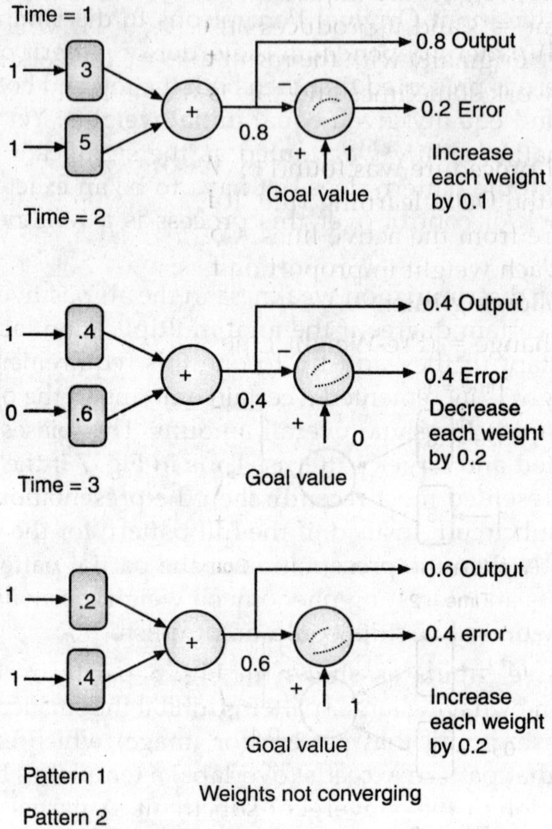

Fig. A.9. Adaline – No Convergence of Weights to a Stable Value

Adalines are based on the use of an attractive (as opposed to repulsive) goal seeking learning procedure in which a convergent subcircuit output value is defined as the goal for each pattern. Consequently, the subcircuit must learn the proper weight values to produce that goal value for any given set of input patterns. This is in contrast to the later perceptrons which were repulsive driven meaning that a specified goal value is not defined for each subcircuit, only misclassification errors. Even less complete error information leads to reinforcement learning which only provides the information that an error or success has occurred without the magnitude or direction (type) of that error. Positive reinforcement is the attractive form while negative reinforcement is the repulsive form.

Figure 9 shows an adaline using a learning procedure which does not work while Fig. 10 shows a procedure which does work. The patterns chosen for this example again demonstrate the subset problem in which one pattern is completely contained within another pattern. This type of discrimination is the most difficult for convergent type sub-circuits to discriminate.

Learning begins in Fig. 9 by first choosing a goal value which the pattern presented to the Adaline is supposed to match. In Widrow and Hoff's paper these were 1, 0, or -1. Pattern One is supposed to produce a 1. It is presented at time = 1 and the error between the goal value and the actual Adaline value is noted. The error value is then divided among the weights which thus eliminates that error. Next Pattern Two is presented which is supposed to produce a 0 and the weights are adjusted again to eliminate the error. Next pattern 1 is presented again at time = 3 and it produces an error of 0.2 which will change the weights back to what they were originally with the result that the network is not learning the patterns. Yet this method will work sometimes and this is the procedure used in Widrow and Hoff's first paper.

A more successful procedure was found by Widrow in 1962 and it is called the Widrow-Hoff learning rule or the Delta learning rule. It is based on the realization that the greatest sources of the error are from the active lines. Consequently, the Widrow-Hoff learning rule changes the value of each weight in proportion to its pre-weight line value (in this case 1 or 0) according to the following rule:

Weight Change = (Pre-Weight Line Value) * (Error / (Number of Inputs)).

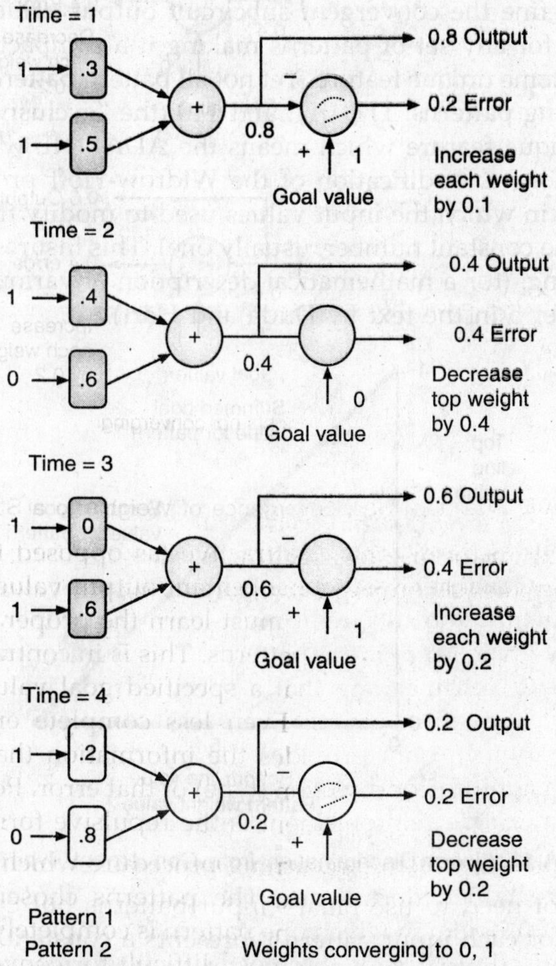

Fig. A.10. Adaline – With Convergence

While this is the official Delta Rule notice that it does not conserve the error value in accordance with Widrow and Hoff's first paper. With the error value per weight divided by the number of inputs not all of it may be used. (although repeated presentations will gradually reduce this error). Yet if the error is conserved as is done in Fig. 10 so that all of it is distributed to the weights then the error is totally eliminated (see time = 2 of Fig. 10). Using error conservation increases the rate of learning.

Instead of just using the pre-weight line value to indicate the greatest source of error a variation would be to include the weight values as well since the post-weight line value give a more accurate measure of error assignment. The trade-off is a slightly more complex learning rule and slower learning if the error value is not conserved since smaller numbers are the result. In fact, this variation is the rule used in the Back Propagation networks discussed below. (This is part of a larger problem in neural network research known as the credit assignment problem. As networks get more complex determining the source of any resultant error also becomes more complex.)

What happens with the Widrow-Hoff procedure is that all those input lines representing features which are common to the patterns will tend to zero. This leaves the non-overlapping pattern features to define the convergent subcircuit output value. Consequently, only one ADALINE is needed for any set of patterns making it a compact solution for a pattern set having patterns with some unique feature. Yet not all pattern pattern sets exhibit this property. Consider the set having patterns: 111, 011, and 110 (the "exclusive or" problem again). No one pattern has a unique feature which means the ADALINE will never converge to zero error. In these situations a modification of the Widrow-Hoff procedure is used called the Relaxation Procedure in which the input values used to modify the weights are normalized (that is add up to some constant number, usually one). This insures that large patterns do not overly bias the learning. (for a mathematical description of various other less used learning procedures see chapter 5 in the text by Duda and Hart).

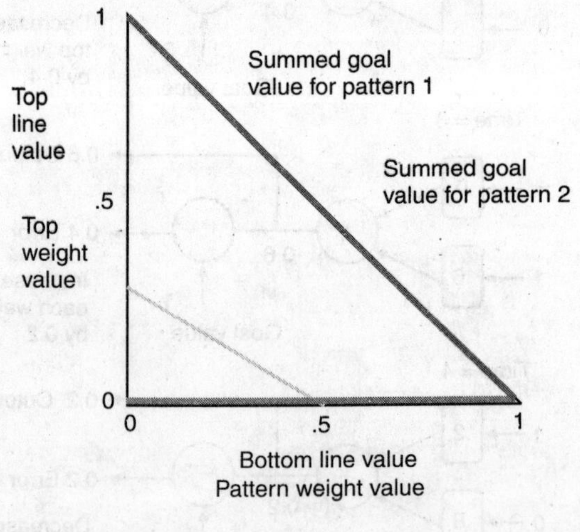

Fig. A.11. Linear Discrimination from Convergent Type Subcircuits

ADALINES do not need to use binary input patterns but can also use analog inputs. Then the goal value for each input pattern represents a constant valued line relative to the inputs as shown in Fig. 11. The learning procedure then shifts the line so that it tends to connect the two goal lines. This shows that weight and summation node convergent subcircuits always produce straight lines, in other words they are linear. Any analog pattern must lie on a straight line in order for it to be completely learned by any ADALINE with zero error convergence. This is an important limitation of these kinds of circuits. Not realized at the time was the significance of the error value used in the ADALINEs. Logical negation of error values produces multivalued and fuzzy logic certainty values which allows downstream circuits to work with a "degree of matching" signal.

The Classic Perceptrons (1962)

In 1962 Frank Rosenblatt published a book which combined the concepts of his original perceptron with those of ADALINE to come up with the classic perceptron design shown in Fig. 1. In contrast to ADALINE, perceptrons are based on repulsive learning in which only the weights on the non-active lines are changed in response to an error. In other words the weights change only in response to a misclassification. Thus the weight values are not pulled towards some defined goal but are pushed away from non-goals. Consequently each subcircuit can represent a whole class of patterns.The adaptive multiplication factors (weights) are now placed before the summation node like ADALINE instead of after the node as in the original perceptron. In addition all convergent subcircuits now share a common set of inputs instead of having randomly connected inputs (although the initial values of the weights may be randomized which would effectively accomplish the same thing). These changes allowed the input pattern to dispense with the binary line signal requirement in favor of analog signals which could represent the frequency of an action potential pulse or the ionic charge on a neuron. Yet, in order for patterns to be reliably discriminated by perceptrons the pattern inputs had to be normalized, that is the numbers in each pattern had to add up to the same value - usually one. Using analog values (and thus analog equations) also required that the binary threshold be replaced with a subtractive threshold.

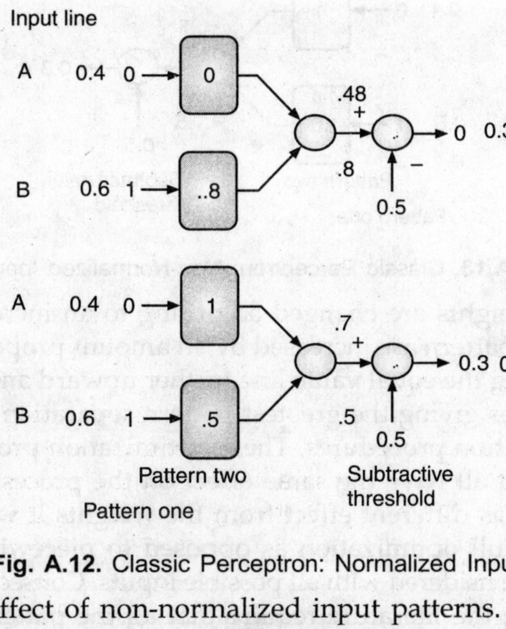

Fig. A.12. Classic Perceptron: Normalized Inputs

Figure 2 shows the effect of non-normalized input patterns. The values of pattern one add up to 1.0 yet the values of pattern two add up to 1.2. No combination of weight values or threshold values will allow each of these patterns to have their own unique convergent subcircuit output.. However, if an additional weight is placed after the summation operation then classifications of non-normalized patterns are possible. Yet this does not seem to have been done for the manipulation of post summation node value is not easily incorporated (mathematically) into the learning procedures used to find the pre-summation operation weight values.As was seen with ADALINE changing convergent subcircuit weights only shifts the angle and height of the equal value lines but since the perceptron uses repulsive learning that equal value line now becomes the basis for defining the border between pattern classes. The equal value graph for the Fig. 1 example is shown in Fig. 3. The axis's of the graph list the values of the pattern input lines which will only produce an output from the subcircuit's threshold if they are above or to the right of their equal value line. Thus the value on input "B" must be above .625 in order for the top subcircuit (represented by the red line) to produce an output. Since the top subcircuit has a zero valued weight on the "A" line it can be any value. In contrast the input values for the bottom subcircuit must be above the blue line for it to produce an output. Since the greatest valued output is the one selected the input with the greatest effective distance from its equal value line is selected. Consequently, the perceptron has the same linear limitation as the ADALINE although in this case it is called linear separability.

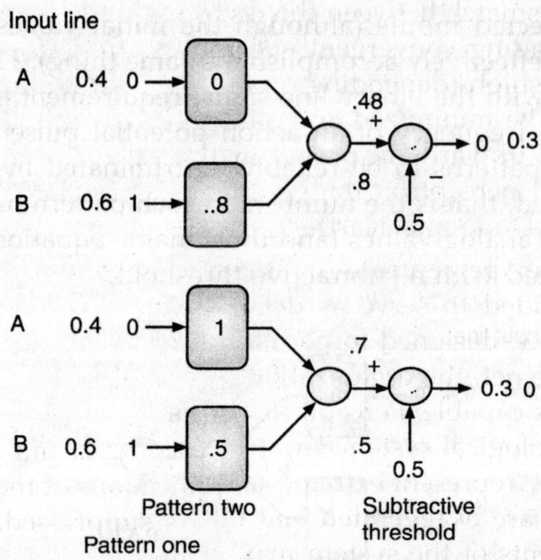

Fig. A.13. Classic Perceptron: Non-Normalized Inputs

During learning the weights are changed according to an increment rule such that the weights on a misclassified pattern are increased by an amount proportional to the total error. This has the effect of moving the equal value line further upward and rightward. The process of finding the weight values giving the greatest pattern separation is complex and requires the use of various optimization procedures. These optimization procedures work best when the changes they command all have the same effect on the process being optimized. Since the subtractive threshold has different effect from the weights it was generally set to some fixed value and ignored. Full optimization as opposed to piecewise optimization requires that the whole problem be considered with all possible inputs. Consequently, all the perceptron optimization procedures in the literature require that all the patterns be known before the optimization procedure begins which really limits their use.

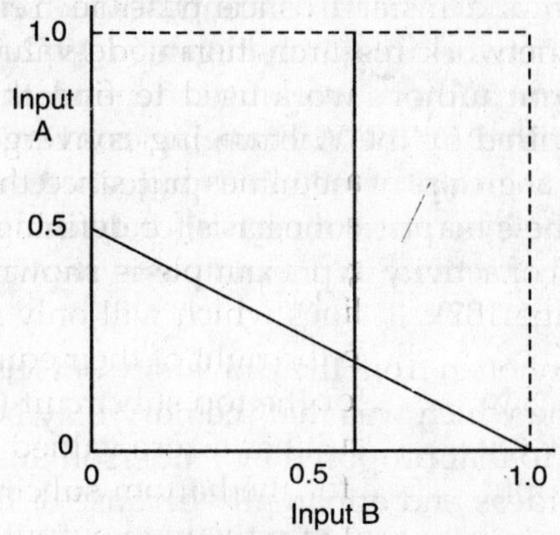

Fig. A.14. Pattern Separation Space for Fig. A.12

These optimization procedures work by initializing the weights to some low value, not necessarily random. All the patterns are presented and for those patterns which are incorrectly classified, the degree of mismatch between the response of the correct convergent subcircuit and the incorrectly responding convergent subcircuit is noted. All these mismatches are added together in some fashion (depending on the type of optimization procedure) to give the total error which is to be minimized and all the weights on the incorrectly responding templates are incremented upwards in proportion to the total error. (for a comprehensive mathematical review of all perceptron improvements centered around linear discriminant functions see chapter 5 of Duda and Hart).

Rosenblatt summed up perceptrons in this passage from his 1962 book (page 28): "Perceptrons are not intended to serve as detailed copies of any actual nervous system. They're simplified networks, designed to permit the study of lawful relationships between the organization of a nerve net, the organization of its environment, and the 'psychological' performances of which it is capable. Perceptrons might actually correspond to parts of more extended networks and biological systems; in this case, the results obtained will be directly applicable. More likely they represent extreme simplifications of the central nervous system, in which some properties are exaggerated and others suppressed. In this case, successive perturbation and refinements of the system may yield a closer approximation.

"The 1960's also saw the growth of artificial intelligence techniques based mostly on net search techniques and higher order logic languages known as propositional and predicate calculus. The aim of artificial intelligence (A.I.) researchers was to simulate intelligent processes at a level more abstract than that of the direct neural level and they were having good success at the time. They could see what was required of any intelligent machine so when the supporters of perceptrons began to oversell its potential one of A.I.'s founders, Marvin Minsky (who had started out with neural networks), with Seymour Papert were inspired in 1969 to write a book describing the Perceptron's inherent limitations. Since the perceptron was the most sophisticated neural network idea at the time that book essentially ended neural network research in the United States for a period of time.

The Association Networks of Kohonen and Anderson (1972)

Still enamored by the idea that brain memory is based upon distributed pathways the next stage in neural network research was to avoid the pattern classification problems as represented by the perceptron and instead concentrate on how these memories might be formed. This stage in neural network research began in 1972 with the publication of two papers on the subject by different author's working independently of each other. One was by James Anderson who was inspired by the William James - Donald Hebb model of memory for he states in his paper:"If a group of neurons projects to another we shall show that strengthening or weakening the synaptic connection between the two groups according to a simple multiplicative function of activity in pre and postsynaptic cells automatically generates an interactive memory ..." (page 182).

In a similar way, Teuveo Kohonen from the University of Helsinki in Finland was inspired by an idea prevalent at the time which was that memory may be holographic in nature. The result was a network identical to that proposed by Anderson. Interestingly, both used matrix mathematics to describe their ideas and apparently because of this did not realize that what they produced was an array of analog ADALINE circuits.

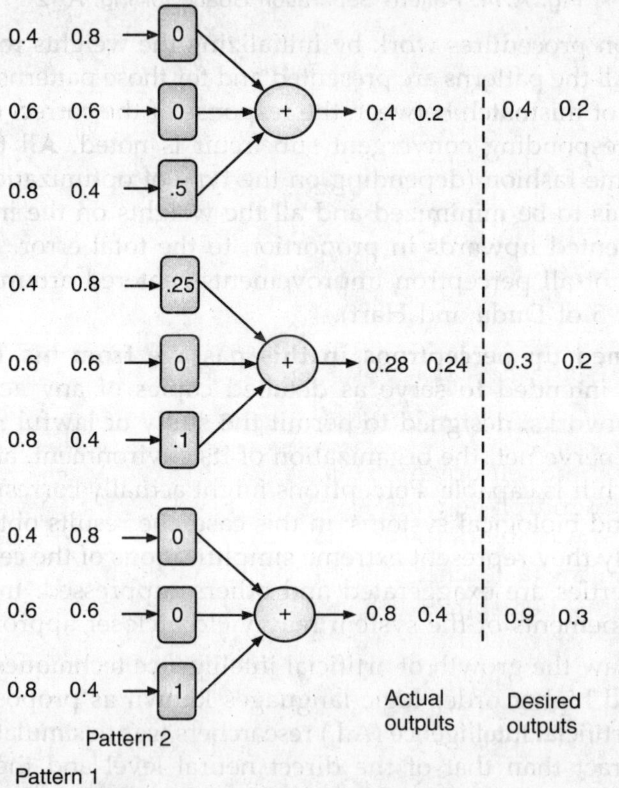

Fig. A.15. An Association Network

Fig. 4 shows such a network, the purpose of which is to associate or correlate an input pattern (vector) with some desired output pattern (vector). The weights for each convergent subcircuit represent a row in a matrix such that the calculations done by each convergent subcircuit itself is the inner (dot) product between the input vector and the weight vector. Consequently, figure 15 represents a square 3 row by 3 column matrix. An association network is different from perceptron like networks in that a presented pattern is supposed to activate a set of output lines instead of just one. Consequently, a single convergence for a decision is not required in association networks.

Producing an an output pattern which matches exactly the desired output pattern requires that the input patterns don't overlap thus avoiding the subset problem. In vector terminology: this is the same thing as saying the input pattern vectors need to be orthogonal. Alternatively, a successful matches can also occur if all input patterns have differing overall magnetudes (distances) that are proportional to their desired output patterns. Figure 4 shows the failure of pattern association with overlapping and normalized (adding up to 1.8 in this case and thus having the same overal magenetude) patterns using only positive inputs. Yet such a correlation is possible using a mix of negative inputs (represented by negative valued weights) and positive inputs and that the Widrow-Hoff procedure for learning the weight values will work in most cases.

The Widrow-Hoff procedure has trouble when both high valued patterns are mixed with low valued patterns. In this case the weight values for higher valued inputs will tend to oscillate wildly while the weights associated with the low valued inputs would change very little thus preventing any convergence. In order to minimize this problem the Widrow-Hoff procedure is often modified for use in these association network by making the weight changes proportional to the product of the input values and the desired output values instead of just proportional to the input values. This tends to average out the value variations at the cost of longer learning times. Still, to be absolutely sure of convergence where such solutions are possible, other more complex algorithmic (not likely to be realizable in a biological neural setting) optimization procedures are be needed. Solutions are even less likely to occur if the input patterns are not normalized which results in more situations in which the convergent subcircuits have identical values for most of their input values leaving less "room" for the remaining weights to adjust themselves to achieve the desired output values. For an extensive mathematical review of this type of neural network with many its many variations and weight change rules see the 1984 book by Kohonen.

The Cognitron - First Multilayered Network (1975)

In 1975, inspired by the self-organization ability of the brain, Kunihiko Fukushima from Japan introduced the Cognitron network as an extension of the Perceptrons. Like the Original Perceptron the Cognitron is a pattern regularity detector meaning it is able to learn patterns without some mechanism (a teacher) to indicate the success or non-success of a pattern match. Unlike the original Perceptron the Cognitron is better able to handle (but not perfectly) the pattern subset problem in which one pattern is completely contained within the other. It does this by using a special inhibitory input to the convergent subcircuit node which tends to counteract the effects of larger patterns. Also unlike the original Perceptron the Cognitron can discriminate to some degree between analog patterns although binary patterns are usually presented to the first layer.

A basic unit (section) of the Cognitron having two convergent subcircuits is shown in Fig. 5. It has four input lines labeled A through D. Notice lines B and C are common to both convergent subcircuits. It learns by increasing the weights on the the active convergent subcircuit lines of the subcircuit selected by the gate comparitor as having the greatest output. Like the original perceptron, the best match is simply strengthened. The rule for adjusting each positive line weight in a selected convergent subcircuit is:

Facilitory Weight Increment (Proportionality Constant) * (Positive Line Value) / (Number of Subcircuit Inputs).

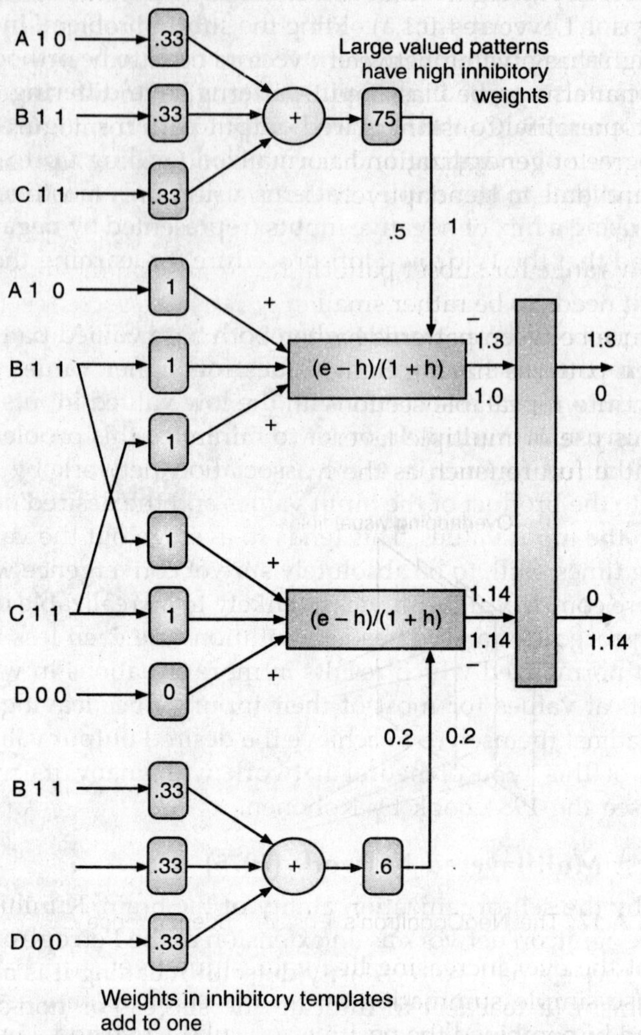

Fig. A.16. The Repeatable Unit of the Cognitron

In the Cognitron the weights can increase without limit but this is balanced by increasing the weights on the inhibitory inputs at the same time. The rule for adjusting the inhibitory weight is:

Inhibitory Weight Increment = (Pattern Generality Constant) * [(Sum of all positive inputs into the subcircuit node) / (Total Pattern Value)].

The Pattern Generality Constant in the original paper was 1/2 and it is needed to help define the dynamic equilibrium of the network between the positive and negative line values feeding into the subcircuit node. This dynamic equilibrium in turn defines the degree of pattern discrimination versus pattern generality.

Dynamic equilibrium effects are best seen in the example shown in figure 5 which represents a Cognitron section at a particular moment in time in which the positive weights have a value of 1 and the top inhibitory weight has a value of 0.75 while the bottom subcircuit has an inhibitory weight value of 0.6. These weights allow a subset pattern discrimination (such as 1,1,1 verses 0,1,1). Yet this discrimination is only possible if the bottom inhibitory weight has a value between 0.75 and 0.45. Any weight value outside that range forces the two patterns to be classified as belonging to the same general class. The choice of the Pattern Generality Constant is what ever works for no analytical derivation as to its value for any degree of generalization has yet been devised. Also one would think that it would be a prime candidate to be adaptively determined itself but no method has yet been devised for that either.

With such a narrow range for subset pattern discrimination the number of input lines of a convergent subcircuit needs to be rather small in order to preserve resolution. For example, the percentage difference between patterns having 20 and 21 binary values is not as great as the difference between patterns having 3 and 4 binary values. Consequently, Fukushima divided the Cognitron into repeatable sections and to connect the sections he was forced to use several layers. This use of multiple layers was to inspire other multilayered yet quite different networks in the future (such as the hybrid network below).

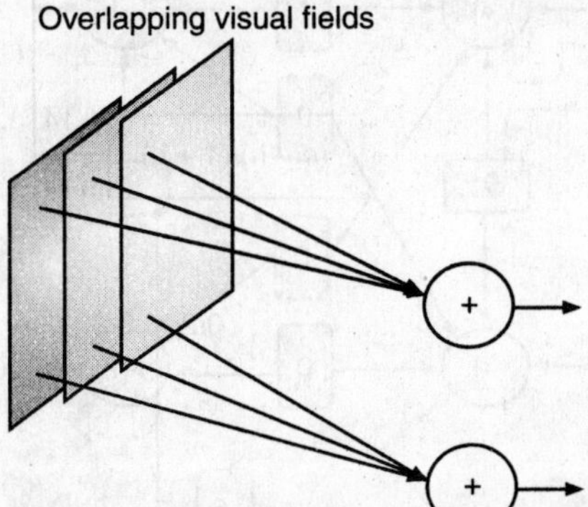

Fig. A.17. The NeoCognitron's Position Independence Strategy

In order to combat the ever increasing line values due to ever increasing weight values Fukushima did not use simple summation and subtraction operations for the convergent subcircuit node. Instead he combined the positive and subtractive nodal inputs with a formula which slows the growth of the output value. The exact equation is $(e - h)/(1 + h)$ where e is the exitory or additive input and h is the inhibitory or subtractive input.

The many layers and sections of the Cognitron allowed it to be modified so that it could respond in the same way (having the same final output) if the same object moved around in a visual field. This modification was called the Neocognitron by Fukushima who published it in 1980. All that was done was to add a final set of summation nodes after a layer's gate comparitor which summed all the outputs from all the convergent subcircuits in the same location of each section. (see figure 6). If a feature pattern was moved it would be in the same location in some new section as it had it had been in its old previously learned section. Consequently it would activate the same final summation node as before to effect position independence which is limited only by the degree of overlap between sections (if the sections do not overlap very much then the pattern would have a low probability of being in its exact relative location in the new section).

The Hopfield Association Network - Revival of the Reverberatory Networks (1982)

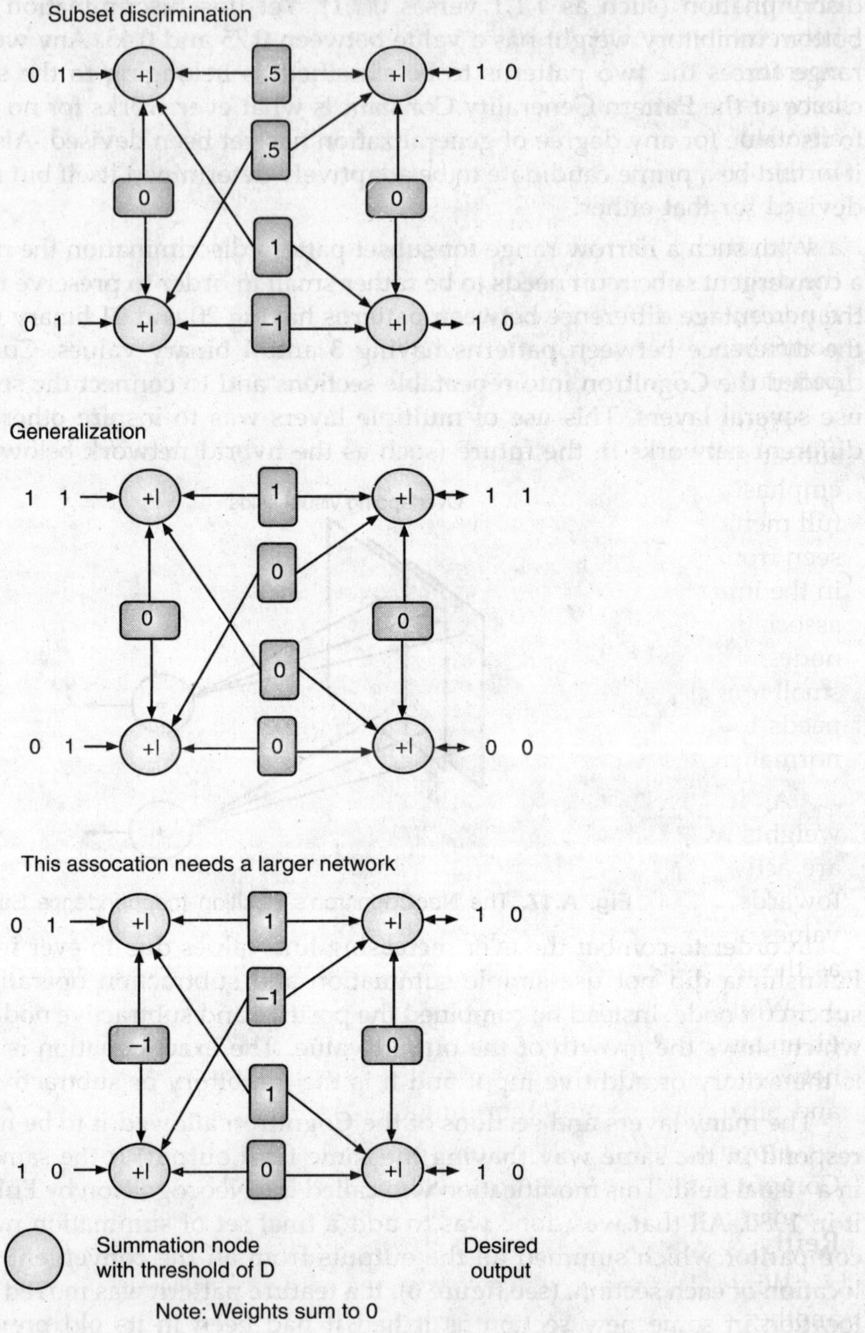

Fig. A18. Hopfield Network Examples

In 1982, John Hopfield revived interest in neural networks in the United States with the introduction of a new type of reverberatory network of the association type. It differed from the earlier versions by using bi-directional lines (equivalent to two reciprocal unidirectional lines) between summation nodes instead of unidirectional lines and emphasized individual cells (nodes) instead of cell assemblies. The summation nodes have a threshold of zero and only produce an output value of one if that threshold is met or exceeded. The weight values between the nodes can be any number between – 1 and 1 so both exitory and inhibitory operations are represented. The general learning idea behind the network is that the weights between the active nodes (those producing an output of 1) will increment while those between all other nodes will decrease. This is usually accomplished by normalizing all the weights. In the original paper by Hopfield all the weights added up to zero.

So in order to learn a correlation (association) between some input and output pattern a separate training session must be held in which the input pattern is presented along with the desired output pattern. In the original paper the weights were assigned initial values at random. The weights between active nodes are then incremented by some amount and the normalization process then decrements the remaining weights. The process is repeated for all patterns until the weights stop changing (or nearly so).

Figure 7 shows some of the characteristics of the Hopfield association network. Both subset discrimination and generalization are possible. A use for this generalization ability emphasized by Hopfield is this network's use as a content addressable memory in which the full memory (pattern) would be retrievable by providing only partial information. As can be seen from the example this mostly amounts to creating a pathway for the common elements in the input patterns. Yet not all associations are possible. In practice the number of different associations which can be learned and recalled is only about 15% of the number of summation nodes. This lack of association formation is shown in the bottom example in Fig. 18. The small four node network cannot learn the associations presented in the example. The network needs to be larger or the normalization value must be larger than zero. Yet increasing the normalization value will decrease the pattern discrimination ability of the network.

As learning proceeds in a Hopfield network with initially random weight values the weights will change in a manner so that fewer paths will be active. So the fewer paths which are active the closer is the network to fully learning its pattern set. This general tendency towards minimizing active pathways can be measured by summing together the output values of all the weights. Hopfield calls this the energy level of the network which is minimized as the network learns.

While the Hopfield association network can discriminate between almost any pattern given sufficient size the number of connections increases exponentially with the size of the network meaning it will get very complicated very fast. Also the requirements for normalization and binary inputs are other limitations.

Hopfield, J.J. (1982). Neural Networks and Physical Systems with Emergent Collective Computational Abilities, Proceedings of the National Academy of Sciences, 79 : 2554-2558.

Reilly & Cooper's Hybrid Network - The First Hybrid Network (1982)

While the Hopfield network got most of the press during the neural network revival another significant type of network, called at this site a hybrid network, was presented by by Doug Reilly, Leon Cooper and Charles Elbaum. This is a two layered network with each layer accomplishing (although imperfectly) a different strategy (thus the hybrid name). The first (input) layer of convergent subcircuits is responsible for generalization while the second (output) layer of convergent subcircuits is responsible for specification. This turns out to be an important and powerful pattern classification strategy.

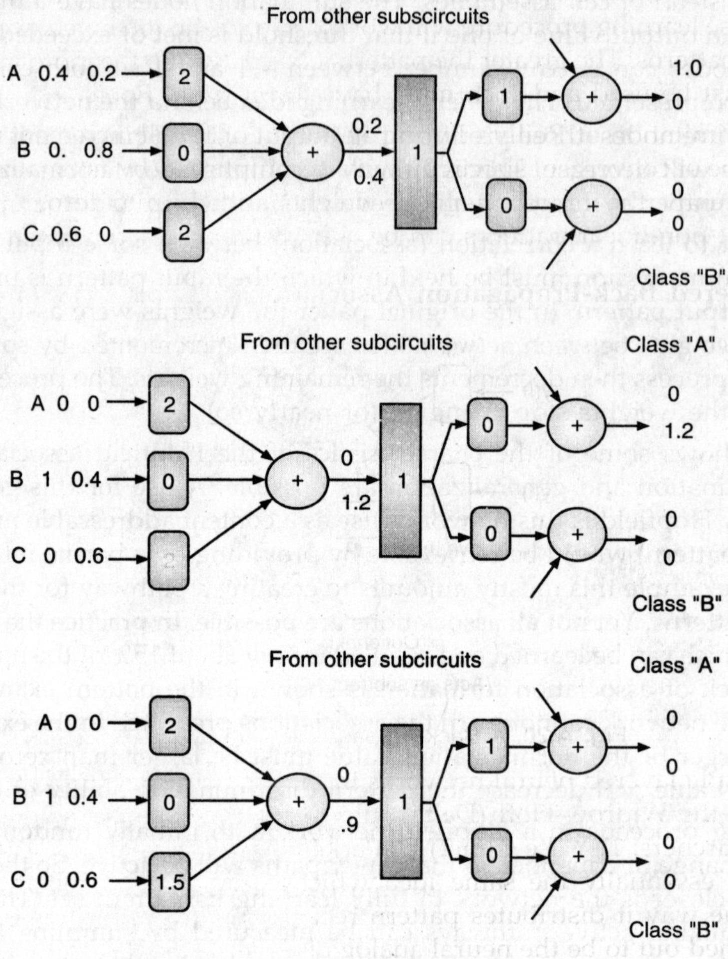

Fig. A.19. Two Layer Hybrid Network

Figure 8 shows the basic workings of a section Reilly and Cooper's hybrid network having two pattern classification outputs and three pattern feature inputs. A network would consist of many such sections. The top output line of this section is activated for class "A" patterns while the bottom output line is activated for class "B" patterns.

If a pattern is presented which does not produce an output then the weights on the active lines of of the front layer convergent subcircuits are incremented to some value greater than one. Since the threshold has a value of one this insures that the pattern will produce an output from the front layer. Next the weight on a branch of that output line (which is now active) in the second layer which has been assigned to represent that class of patterns is now incremented to one resulting in an output.

If a pattern (the blue "A" class pattern) is then presented which is supposed to belong to the same class as the red pattern yet it is so different as to not produce an output then the above procedure is repeated. Another possibility is that a pattern belonging to a different class (the blue "B" class pattern) will falsely produce an output from the "A" class output. In such a case called confusion, the front layer of the "A" class convergent subcircuit must be trimmed back by reducing the weight values on the active lines until an output is no longer produced. These learning procedure will work as long as no too much overlap occurs between the presented patterns. The greater the pattern feature overlap the more front layer convergent subcircuits must be used and each must have fewer input lines.

The main limitation of Reilly's hybrid network is that it has all the limitations found in the Adaline type of convergent subcircuit with its multiplication factor weights and summation node. Consequently, the input patterns need to be normalized (adding up to 1 here). Its great strength is that non-linear patterns can be classified and discriminated.

The Multilayered Back-Propagation Association Networks (1986)

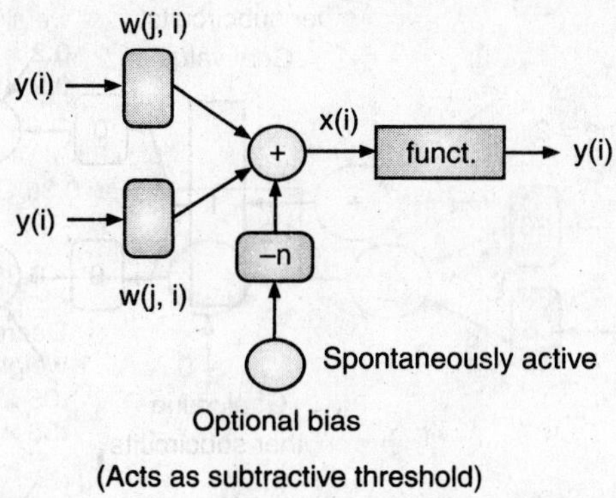

Fig. A.20. A Back Propagation Convergent Subcircuit

With multiple layered neural networks in the news the question was what was the best way to extend the Widrow-Hoff (Delta) rule to multiple layers. In 1986 three independent groups of researchers: 1) Y. Le Cun 2) D. Parker 3) D. Rumelhart, G. Hinton, & R. Williams came up with essentially the same idea which came to be called the Back Propagation network for the way it distributes pattern recognition errors throughout the network. Yet their ideas turned out to be the neural analog of the a steepest descent algorithm discovered by Paul Werbos in 1974. (I have not seen this paper so I am taking Stephen Grossberg's word on this).

The basic repeatable unit (convergent subcircuit) used in the Back-Propagation network as described by Rumelhart, Hinton, and Williams is shown in Fig. 9. The weights, represented by w(i,j), can be any positive or negative value but they start out as small, randomly chosen numbers (– 0.3 and 0.3 were used). The squashing function (funct.) is generally used to keep the output of the convergent circuit 1 or less. The error is the difference between the desired output and the actual output.

The strategy used in back-propagation is to determine and use the; error contribution made by each layered pathway through the network. In a three layered network the learning rule for each layer is (red indicates the additions relative to the last layer):

Weight value change for last layer = (proportionality constant)(error)(change in output value from last layer's function)(pre-weight line value) (weight of last layer) Weight value change for middle layer = (proportionality constant)(error)(change in output value from last layer's function) (value change from middle layer function) (pre-weight line value) (weight of last layer) (weight of middle layer)

Weight value change for front layer = (proportionality constant)(error)(change in output value from last layer's function)(value change from middle layer function)(value change from first layer function) (pre-weight line value)(weight of last layer) (weight of middle layer)(weight of first layer)

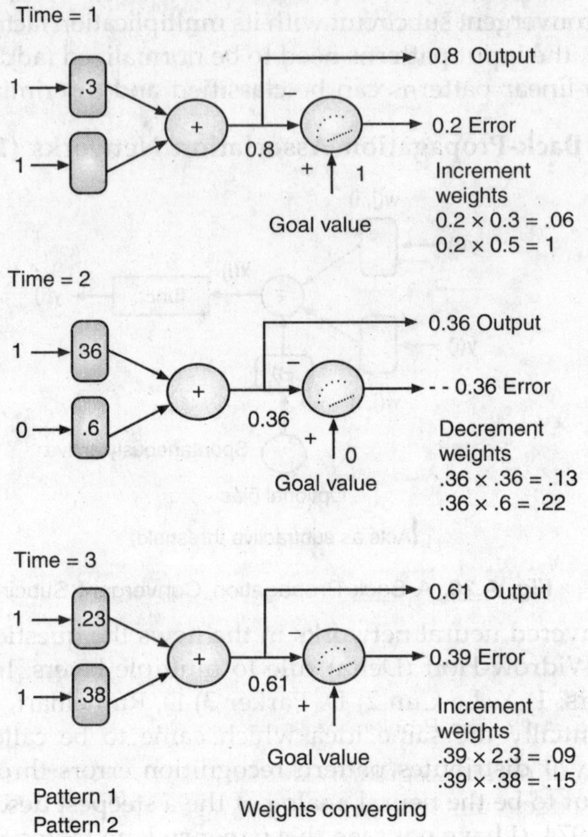

Fig. A.21. Learning Rate Example

Notice that the rule for changing the weights in the last layer is the variation of the Widrow-Hoff (Delta) rule used first in their Adaline network but making use of the more accurate but time consuming post-weight line value given by (pre-weight line value) (weight value). The other change is the incorporation of the value change due to the function used to keep the line values between layers less than or equal to 1. If the network is strictly connected layer to layer without any convergent subcircuits bypassing a layer the function component can be left out of the weight change rule for they would no longer produce any possible difference in the error contribution.

So as one moves forward in the network towards the inputs the weight change rules include all the weight values of the more rearward layers in its error path. This produces smaller weight changes nearer to the front. The assumption here is that the weights near the back of the layer are more responsible for the error than those near the front in accordance with The Fundamental Pattern Classification Strategy. Whereas Hybrid networks use just two layers to accomplish this strategy the Back-Propagation networks spread it over several.

The result is slow learning often needing as many as thousands of iterations to learn a set of patterns.An example of how including weight values in the weight change rule is given in figure 10. Weights less than 1 produce simply produce small weight changes.

The First Regularity Detector with Dynamic Allocation - The Adaptive Resonance (ART) Networks (1987)

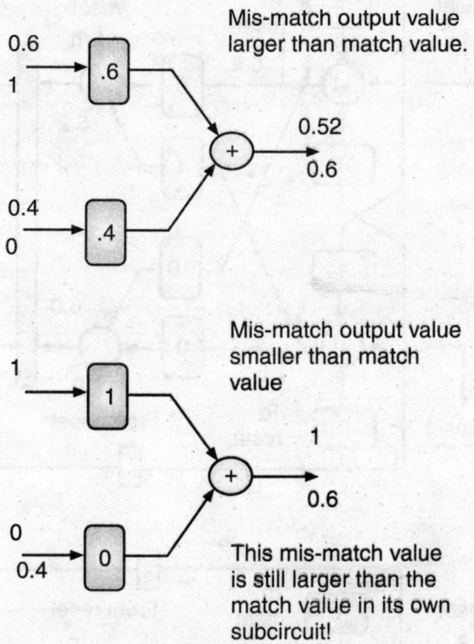

Fig. A.22. The Problem of Dectecting a Pattern MisMatch

The first generation of pattern regularity detectors for unsupervised learning were based upon using convergent subcircuits having random input connections. Those input connections which happened to match some presented pattern were then strengthened. Those networks run into the combinatorial explosion problem as the number and size of the presented patterns increased since the number of average random connections had to increase even more.

The second generation of regularity detectors was introduced in 1987 by Gail Carpenter and Stephen Grossberg of Boston University. Their network using binary input patterns was called ART 1 (1987a) while their network using analog inputs was called ART 2 (1987b). They were the first networks to use the convergent subcircuit dynamic allocation strategy which assigns a new convergent subcircuit to the presented pattern if no other convergent subcircuit has a sufficient match.

The problem here is determining when a convergent subcircuit does not match its pattern. The easiest approach is simply to assume that the greater the output of the subcircuit the better the match so one can simply use a Gate Comparitor subcircuit to pick the greatest output value. Yet this is not always the case as shown in the top subcircuit of figure 11 in which its matching pattern (the superset pattern) produces a lower output value than a mismatched subset pattern. Yet the bottom convergent subcircuit does produce an output which is larger for its matching pattern when the matching pattern is the subset pattern. If these two subcircuits were in the same network and the superset pattern was presented, the bottom subcircuit would still produce the greater output even though the pattern is not matched to it! This competitive mismatch problem is inherent in all convergent subcircuits based on summation nodes even when they use normalized inputs as is done in Fig. 11.

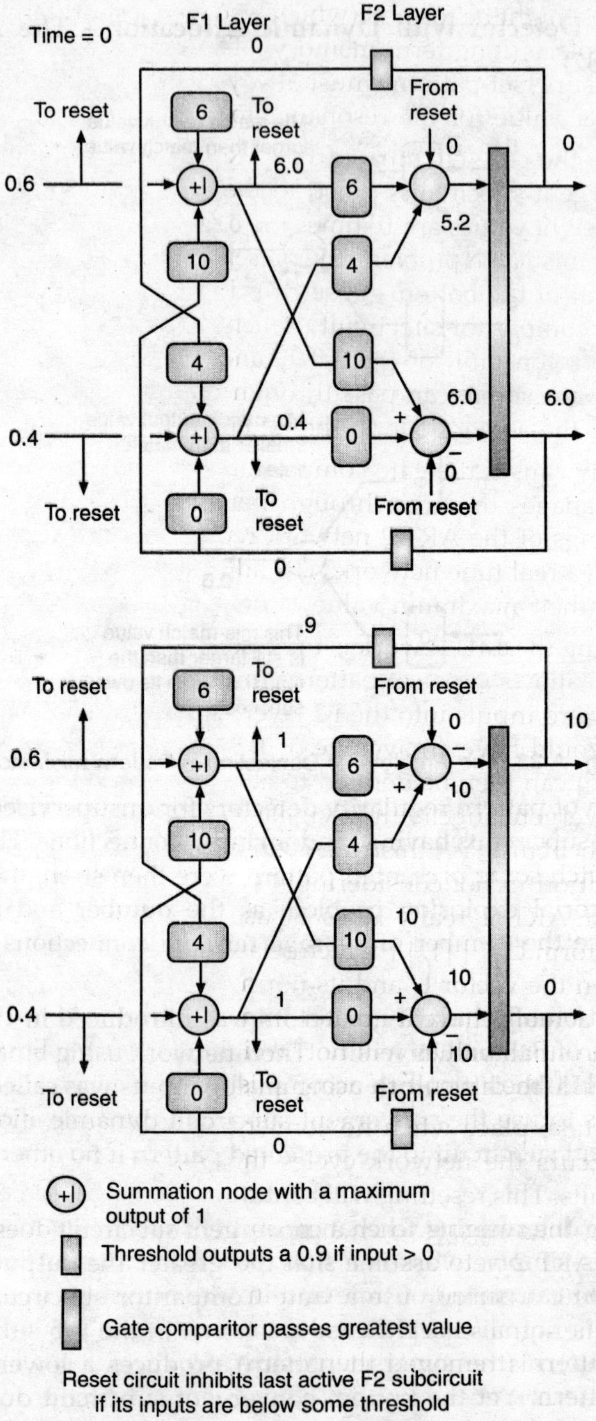

Fig. A.23. The Core of the ART 2 Network

The competitive mismatch problem of summation node convergent subcircuits can be mostly solved by using the positive feedback approach of the ART networks. The positive feedback is used to enhance the superset signals in a process called Adaptive Resonance, a term which was first coined by Stephen Grossburg in 1976 as he investigated how stable reverberation circuits might be formed (which he called a short term memory) so that weight changes could take place (long term memory). The limitation of adaptive resonance is that during learning the superset patterns must always be presented first so that the weights can be set to their proper value for the resonance to work.

Figure 12 shows the core circuitry of the ART 2 network (analog inputs). The F2 layer contains two convergent subcircuits tuned to the same patterns which were used in figure 11 except here the weight values are 10 times the pattern values. At time = 0 and in accordance with the competitive mismatch problem, the presented superset pattern of 0.6 and 0.4 produces the largest output out of the bottom subset subcircuit (tuned to the pattern 1, 0). This value is selected by a gate comparitor subcircuit (which passes the largest value among its inputs), is then converted to a constant "d" (here 0.9), and fed back to all the inputs of the F1 layer. But before the feedback signal can pass through the network again the network is reset by a large inhibition of the active node in the F2 layer.

This immediately causes the gate comparitor to output the signal of the top convergent subcircuit which manages to surge through the network before the reset circuit can react again. So the workings of the ART 2 network require precise signal timing which is why it is often described as a real time network. The addition of the feedback signal now forces the F1 nodes to output their maximum value of one which prevents any reset.

The key to making the ART 2 work is the reset rule. The simplest and most intuitive reset rule (and one which allows superset patterns to be presented first) would be to calculate the average of the non-zero inputs into the F2 layer. Since these lines have a maximum value of 1 a perfect match would have an average of 1. A vigilance parameter between 0 and 1 (0.9 used in the example) can then be defined to allow less than perfect matches to be averaged together by the convergent subcircuit (a vigilance parameter of 1 requires a perfect match by each convergent subcircuit). Yet this rule is an abrupt and discontinuous rule which makes an all-or-nothing cut-off in not considering 0 valued input lines. What about lines having a 0.1 value? So what ART 2 really does is use a directional average taken from vector mathematics in the form $U / ||U||$ which relative to a baseline coordinate equals the cosine of the angle between the vector U and its norm or length $||U||$. But with this formula the smaller line values actually have a greater impact so the vigilance parameter is divided by the result so that the small values will not "weighted" more than the larger values. In either case both rules would be difficult to accomplish in real neural systems.

Learning must take place when no convergent subcircuit matches some input pattern. When no match occurs the network cycles through several resets testing the most likely convergent subcircuits. This resetting will continue unless some other circuit eventually brings it to a halt allowing the weights to change on the final selected convergent subcircuit. The extra circuit in the ART 2 network is a slow positive feedback loop on the front end which increases in value after a period of time and in so doing shuts down the resetting. Repeated normalization's of the signals keep the values in bounds and a three level circuit assures that when the input pattern is removed the positive feedback signals stop as well.

The First Multivalued Logic Neural Network (1990)

Given the limitations of summation based neural networks and the need for logic like operations for higher level intelligence as shown by the artificial intelligence field some sort of analog based logic network seemed worth investigating (the next logical step :))The first multivalued logic neural network was a simple adaptive switching network meant to represent the reticular formation of the brain which was published (Olmsted, 1990). It was later incorporated as the second stage in a very successful hybrid network.

Bibliography for History section

Neurocomputing Foundations of Research (1988) Edited by James A. Anderson and Edward Rosenfeld. MIT Press, Cambridge, Massachusetts

Bain, A. (1873). *Mind and body. The theories of their relation.* London: Henry King.

Bain, A. (1904). *Autobiography.* London: Longmans, Green

Ferrier, D. (1876). *The Functions of the Brain.* London: Smith, Elder and Co.

McHenry, Lawrence, C. (1969). *Garrison's History of Neurology.* Springfield, IL:Charles C. Thomas

Wilkes, Alan L. and Wade, Nicholas, J. (1997). *Bain on Neural Networks.* Brain and Cognition 33:295-305

James, William (1890), *Principles of Psychology,* Henry Holt, New York

Sherrington, C.S. (1898) Experiments in Examination of the Peripheral Distribution of the Fibers of the Posterior Roots of some Spinal Nerves, *Proc. of the Royal Society of London, 190:45-186*

McCulloch, W.S. & Pitts, W.H. (1943). A Logical Calculus of the Ideas Immanent in Nervous Activity, *Bulletin of Mathematical Biophysics, 5:115-137*

Rashevsky, N (1938). *Mathematical Biophysics,* University of Chicago Press, Chicago, IL

Winder, R.O. (1964). Threshold Logic in Artificial Intelligence, *Artificial Intelligence, IEEE Publication S-142, pages 107-128.*

(Rashevsky, McCulloch, and Pitts also played a role in the development of systems science. See chapter 2 of the Macroscope by Joel de Rosnay)

Hebb, D.O. (1949). *The Organization of Behavior,* John Wiley & Sons, New York

Farley, B., and Clark, W.A. (1954). "Simulation of self-Organizing Systems by Digital Computer", *IRE Transactions on Information Theory 4:76-84*

Farley, B.G. (1960) Self-Organizing Models for Learned Perception; *in M.C. Yovits and S. Cameron (editors), Self-Organizing Systems, Pergamon Press, Oxford*

Rochester, N., Holland, J.H., Haibt L.H. and Duda, W.L. (1956). Tests on a Cell Assembly Theory of the Action of the Brain Using a Large Digital Computer, *IRE Transaction of Information Theory IT-2:80-93*

Rosenblatt, F. (1958). The Perceptron: A Probabilistic Model for Information Storage and Organization in the Brain, *Psychological Review, 65:386-408*

Rosenblatt, F (1960). Perceptron Simulation Experiments, *Proceedings of the IRE, 48:301-309*

Duda, R.O. & Hart, P.E. (1973). *Pattern Classification and Scene Analysis,* John Wiley & Sons, New York

Minsky, M. & Papert, S. (1969). *Perceptrons, An Introduction to Computational Geometry,* MIT Press, Cambridge, MA

Rosenblatt, F (1962). *Principles of Neurodynamics: Perceptrons and the Theory of Brain Mechanisms,* Spartan books, Washington D.C.

Anderson, James A (1972). A Simple Neural Network Generating an Interactive Memory, *Mathematical Biosciences 14:197-220*

Kohonen, Teuvo (1972). Correlation Matrix Memories, *IEEE Transaction on Computers, C-21:353-359*

Kohonen, Teuvo (1984). *Self-Organization and Associative Memory,* Springer-Verlag, Berlin

Fukushima, K (1975). Cognitron: A Self-organizing Multilayered Neural Network, *Biological Cybernetics, 20:121-136*

Fukushima, K (1980). Neocognitron: A Self-organizing Neural Network Model for a Mechanism of Pattern Recognition Unaffected by Shift in Position, *Biological Cybernetics 36, 193-202*

Reilly, Douglas L.; Cooper, Leon N.; Elbaum, Charles (1982). A Neural Model for Category Learning, *Biological Cybernetics 45, 35-41*

Parker, D.B. (1986). A Comparison of Algorithms for Neuron-Like Cells. In J.S. Denker (Ed.), Neural Networks for Computing (pp 327-332) *New York: American Institute of Physics.*

Rumelhart, David D., Hinton Geoffrey E., and Williams, Ronald J. (1986). Learning Representations by Back-Propagating Errors,*Nature 323:533-536*

Werbos, Paul (1974) Beyond regression: New Tools for Prediction and Analysis in the Behavioral Sciences. *Unpublished doctoral thesis, Harvard University, Cambridge, MA*

Carpenter, G.A. & Grossberg, Stephen (1987a). A Massively Parallel Architecture for a Self-Organizing Neural Pattern Recognition *Machine, Comput. Vision Graphics Image Processing 37:54*

Carpenter, G.A. & Grossberg, Stephen (1987b). ART 2: Self-Organization of Stable Category Recognition codes for Analog Input Patterns, *Applied Optics 26:4919-4930*

Grossberg, Stephen (1976). Adaptive Pattern Classification and Universal Recoding: II Feedback, Expectation, Olfaction, Illusions, *Biological Cybernetics 23:187-202*

□□□

2 ANNEXURE TWO

MORE ANN NETWORKS (FOR MORE ENTHUSIASTIC STUDENTS)

CHAPTER OUTLINE

A.2.1 Skeleton Neural Network
A.2.2 Wavelet Neural Network
A.2.3 Classification of Wavelet Neural Network
A.2.4 Design of Neural Network

There are some other networks and research thereof, which are not important to the Undergraduate students, yet, they are very important to create an enthusiasm in the students who wish to pursue ANN as their passion.

A.2.1. SKELETON NEURAL NETWORK

A problem with fully connected neural networks is their size. When it comes to the practical implementation of neural networks size becomes an important factor. In hardware implementation this becomes a problem of scaling and the number of components required. In the case of computer simulation the problem comes with the large computational cost. The larger and more complex the network the longer it takes to train and once trained it takes longer for the network to perform its recognition task. For this work the concern is with the computational cost.

In general, it is not known what size of network works best for a given task and this problem is unlikely to be resolved since each task demands different capabilities from the network. As with the decision about the number of hidden units, the size of network tends to be decided on a trial and error basis and the designers own experience. The approaches used to decide the size of the network may be:

1. One such approach is to start with a fully connected network and then gradually prune it. This involves removing those weights which are close to zero and contribute little to the solution.
2. Another approach is to base the network architecture on prior knowledge of the structure of the input data. The network topology can then be arranged to reflect the

data structure. All of these structures are similar in that the input nodes are grouped into zones which are connected to one node or a group of nodes in the hidden layer. Krause and Hackbarth used 'skeleton' type architecture to reduce the number of connections in a network. They showed that performance of a neural network does not necessarily improve as the number of connections between the nodes is increased. This architecture uses overlapped zones and was shown to support high recognition rates for isolated word recognition. This is the approach adopted for the work here since the localized structure of the input zones is somewhat analogous to the cochlear processing which occurs in the human ear. A preliminary investigation into the ability of a skeleton architecture neural network was carried out by A.D. Smith at the University of Newcastle Upon Tyne. Smith's work suggested that further investigation of this network was required to better determine the effect of changing the parameters of the network. Similarly Brian Davis in his homework series suggests that the network architecture can be varied to see better results. The work in this thesis is the continuation of that preliminary work.

It should be mentioned that several new methods of determining the optimal number of hidden units and the interconnectivity of units within a neural network have been developed since the work in this thesis was begun.

The use of rule-based knowledge was demonstrated to determine the structure of the network prior to training. With this approach the network already has some knowledge of the task for which it is being developed that can be refined and altered with training.

Neural networks have been developed where the network architecture is determined by the learning algorithm during the training phase. They use incremental learning algorithms which modify the architecture by removing or adding hidden units and/or the connections between them. The "Grow And Learn" (GAL) algorithm starts from a simple network structure and adds units and/or connections to decrease the error. As new units are added some of the previously added units may become redundant and can be eliminated. To get the classification without clearing a hidden layer, Polynomial Neural Network (PNN) were first created, later, researchers added a fuzzy front end to develop a FPNN and created hidden linear and non-linear layers and cascade of them to further improve them.

A.2.2. WAVELET NEURAL NETWORKS

Due to temporal behaviour of speech, not only it is important to have frequency information of the signal, but also the time dependent frequency information is needed. With only time information available in hand, we can discriminate whether those frequencies were available spread and mixed all the time or it was these frequencies available at different time separately. An effort was made by Dennis Gabor (1946) by taking small spans of random or non-stationary signal and considering it stationary for a small time, thereby generating, Short Time Fourier Transform (STFT).

The problem with STFT is given by Heisenberg Uncertainty Principle; in quantum physics, which says that location of a particle and its second moment can not be identified at the same time. With reference to signal, it says that, one can not know what spectral components exist at what instance of tine. However, the band of frequencies at a time interval can be found. A narrow window gives good time resolution but poor frequency resolution, whereas, wide window gives a poor time resolution but good frequency resolution. There are two main differences with STFT and Continuous Wavelet Transform (CWT):

1. The Fourier transform of the windowed signal are not taken, therefore single peak will be seen corresponding to a sinusoid.
2. The width of the window is changed as the transform is computed for every single spectral component, which is probably the most significant characteristic of the wavelet transform.

The window used can be simple gaussian function *i.e.*, $w(t)$ = exp (– a * t ^ 2) / 2, where a is the length of window and t is time. The problem, of course, is that of choosing a window, one and for all, and use that for the entire analysis.

The Discrete Wavelet Transform is the sampled version of the CWT, and the information it provides is highly redundant as for as the reconstruction of the signal is concerned. This redundancy, on the other hand, requires a significant amount of computation time and resources. This turns out, rather remarkably, that if we choose scales and positions are based on powers of two so called *dyadic scales and positions* then the analysis will be more accurate. This Discrete Wavelet Transform provides sufficient information both for analysis and synthesis of the original signal, with a sufficient reduction in the computation time.

For speech recognition, wavelets can be used for feature extraction or as activation function in network neurons. Some feature extraction techniques like DFT, FFT assume speech to be stationery; whereas, generally used LPC approach assumes an all pole linear model, which strictly speaking is not the case. Other approaches based on Cohens general class of time-frequency distribution such as Cone-Kernel and Choi-Williams methods have also found use in speech recognition but have the drawback of introducing unwanted cross-terms into the representation. The wavelet transform overcomes some of these limitations; it can provide an analysis of a given signal by projection onto a set of basis functions that are scale variant with frequency. Each wavelet is shifted scale version of an original mother wavelet. These families are usually orthogonal to each other, which is important as this yields computational efficiency and ease of numerical implementation. The wavelets have advantage of capturing local features.

These features are extracted by Best-Basis Algorithm. The idea is to do transform coding on a signal by choosing a wavelet basis which is best suited for the given problem, resulting in an adaptive time-scale analysis. There are two possibilities: First is the smooth local trigonometric transforms which essentially performs local Fourier analysis on the signal, and its frequency domain conjugate, the wavelet packet which similarly partitions the frequency axis smoothly. Since these transforms operate on recursively partitioned intervals on the respective axis, the basis whether wavelets packets or local trigonometric are said to form a library of orthonormal basis. If these bases are ordered by refinement, they form a tree which can be efficiently searched to result in only those coefficients which contain the most information.

Wavelet Neural networks have proved to be a popular predictive tool in nonlinear time sequences due to their approximation ability to learn rather complicated function and time sequence. The multi-layer perceptron is the most frequently used one. This network is complicated and converges too slowly. ANNs have a limited ability to characterize local features such as discontinuity in curvature, jumps in time and /or frequency, typically embody important process-critical information such as modes and defaults. Bottou noted that improved localized modelling can aid data reduction / compression and subsequent classification tasks that rely on an accurate representation of these local features. Wavelet Neural Networks were first proposed by Zhang and Benveniste [88] as an alternative to classical feedforward network for approximating nonlinear functions. Recently, Li and Chen [66] developed a Robust Wavelet Neural Network model based on theory of robust regression for dealing with outliers in the framework of function approximation. The LMS-based algorithms an identically independent and gaussian error distribution and provides an asymptotically optimal solution with minimum variance. Thus, by adaptively adjusting the number of training data involved during training, the efficiency loss in presence of gaussian noise is accommodated.

The Radial Basis Network (RBF), as an alternative to MLP has a simpler structure i.e. only one hidden layer, with some preprocessing on training data such as clustering can make RBF network learn much easier than MLP networks. However, pure RBF network sometimes cannot perform well as expected because the basis function is non orthogonal. Wavelet Neural Networks combines the advantages of MLP and Wavelets. In contrast to classical sigmoidal-based ANNs, WNNs provide efficient network construction technique, faster training times, and multiresolution analysis capabilities. However, they suffer from 'curse of dimensionality' and its application is usually limited to the case of small dimension. To the nonlinear speech signal, better predictive results can be attained by nonlinear methods based on neural network. Still, the unsteady property of speech signal requires an adaptive model, which then requires a network with fast convergence. Also, the long-term dependencies of this network require that the prediction order not to too low. Thus a dynamic wavelet network with a feedback from output of network, moved by the fact that feedback units introduce an inner memory of infinite length but of fading nature in addition to the external memory with the size determined by the prediction order.

A.2.3. CLASSIFICATION OF WAVELET NEURAL NETWORK

Wavelet analysis and neural networks have been combined in numerous ways. There are two distinguishable categories. In first one, the signal is decomposed on some wavelet and the wavelet coefficients are furnished to a neural network (feature extraction). In the second category, wavelet theory and neural network are combined into a single method.

Wavelets are used in networks in many ways

1. Wavelet Self Organising Maps (WSOM)

WSOM is a feed forward four layer network that estimates optimized wavelet bases for the Discrete Wavelet Transform on the basis of the distribution of input data. The function representation using WSOM requires lesser number of non-zero weights, and it supports methods for reducing the noise and recovering signal.

2. Wavelet Neural Network (WNN)

Is a neural network where wavelet basis functions are used to decompose the input signal under analysis. This introduces an additional layer in the neural network model. A linear WNN model, where the connection weights between the hidden and output layer of the original WNN was replaced by a local linear model was also tried. Dynamic Wavelet Neural Network: In this three layer network the hidden layer that consists of wavelet for nonlinear dynamic system identification is used. The output is fed back to the separate input nodes also and dynamic and PID back propagation learning algorithm is used. A recent study gave the idea of multi-input wavelet neural network in sequence of work procedure in large-scale industrial production process.

3. Wavelons

The combination of translation, rotation, dilation, and wavelet lying on the same line are called wavelon in sequel. Least square method is used to train this network.

4. Wavenets or Dyadic Wavelet Network

Orthogonal wave-nets were developed for approximation and classification based on multiresolution analysis. The basic feature of the wavenets is that the position. Dilation of wavelets are fixed and weights are optimized.

5. Fuzzy Wavenets

Recently, simple and efficient wavelet-based methods have been developed for online learning. These fuzzy wavenets methods combine wavelet theory to fuzzy logic and neural networks. They permit to determine and validate adaptively appropriate fuzzy rules in online problems. The model is refined as more data are furnished to the system. With only a few datapoints, the information on the underlying surface is small and a low resolution description of the system is appropriate, while with an increase in number of datapoints, a higher resolution may be justified. New rules are added to the description of the surface with increasing datapoints. The rules are validated through an automatic procedure based on fast wavelet decomposition and reconstruction. Learning is fully automated and does not require any external intervention, thereby, making these methods useful in practical applications. The major challenge to this is to convert the collection of datapoints into linguistically interpretable fuzzy rule. A solution to this problem is to use multiresolution techniques. In order to associate a linguistic interpretation to each membership function, the membership functions are selected among the family of scaling functions that have the property to be symmetric, everywhere positive and with a single maxima. Hotal also investigated with FWN for approximating arbitrary non-linear functions, each rule corresponding to sub-wavelet neural network consisting of single-scale wavelets.

The advantage of wavelet emerges from the fact that they, unlike Fourier Transform, do not loose the time information. The properties of wavelet transforms emerging from multi-scale decomposition of the signals allow the study of both stationary and non-stationary signals. On the other hand, the neural networks perform satisfactory analysis of nonlinear as well as linear dependencies due to different possible structures and activation functions, thereby creating multidimensional neural networks with shorter computation time. A wavelet network is such a technique, which absorbs the advantages of high resolution of wavelets and the advantages of learning and approximations of feedforward network.

A.2.4. DESIGN OF WAVELET NEURAL NETWORK

Like Fourier transform, wavelet transform is also a technique that consists in decomposing an arbitrary signal into elementary contributions, the so called wavelets that constructed from one single function È by means of dilations and translations:

$$\psi_{a,\,b}(t) = \frac{1}{\sqrt{a}}\,\psi\left(\frac{t-b}{a}\right) \cdots \text{ for } (t, a, b\ \varepsilon,\ \mathrm{R},\ a > 0) \qquad \ldots (1.1)$$

where $\psi(t)$ is wavelet prototype (which can be thought of as a band-pass function), the parameters a, b correspond to dilations and translations respectively. A very commonly used wavelet is Mexican Hat (laplacian of the gaussian), described by:

$$W(x) = (1-x^2)e^{-\frac{x^2}{2}} \tag{1.2}$$

If the embedded dimension is d, we will choose $\psi(x) = w(x_1)\, w(x_2)\, w(x_3) \ldots .\, w(x_d)$, where x_i for $i = 1, 2, 3, \ldots, d$ are the d components of the vector x.

The wavelet transform of a signal $\phi(t)$ is defined by

$$W_f(a, b) = < f, y_{a,\, b} > = \frac{1}{\sqrt{c_\varphi}} \int f(t)\, \psi_{a,b}(t)\, dt \quad ...(1.3)$$

where * denotes complex conjugate and $c_j = 2\pi \int \frac{|\hat{\psi}(\omega)|^2}{\omega} d\omega$ denotes the Fourier Transform:

$$\hat{\psi}(\omega) = (2\pi)^{\frac{1}{2}} \int e^{i\omega t}\, \psi(t)\, dt \quad ...(1.4)$$

Then the inverse transform gives the original signal as follows:

$$f(t) = \frac{1}{\sqrt{c_\varphi}} \iint \frac{1}{a^2} W_f(a, b)\, \psi^*{}_{a,b}(t)\, dadb \quad ...(1.5)$$

This is known as the wavelet decomposition of $\phi(t)$.

Additionally, it is convenient in practical application that the parameters a, b corresponding to wavelets are sampled on a so-called 'dyadic' grid in the space-scale plane (b, a). A common definition of such wavelets is

$$\psi_{j,\, k}(t) = 2^{\frac{j}{2}}\, \psi(2^{-j} t - k),\ j, k\, \varepsilon\, Z \quad ...(1.6)$$

The discrete wavelet transform turns out to be:

$$W_f(j, k) = \frac{1}{\sqrt{c_\varphi}} \int f(t)\, \psi_{j,k}(t)\, dt \quad ...(1.7)$$

Unlike the previous case, the inverse transform of this equation is not easy to perform. This is because this inverse transform is related to orthogonality of the wavelet basis $\{y_{j,\,k}(t), j, k \text{ e } Z\}$. Now if we take this as an orthogonal basis then the inverse transform can be given as:

$$f(t) = \frac{1}{\sqrt{c_\varphi}} \sum_{j,\, k \varepsilon Z} W_f(j, k)\, \psi_{j,k}(t),\ M, N\, \varepsilon\, Z \quad ...(1.8)$$

This called the discrete wavelet decomposition of $\phi(t)$. This is only the scalar wavelet. A high dimensional wavelet can be built by the product of scalar wavelets as follows. Let ψ refer to n-dimensional wavelet function, then

$$g_{M',\, N}(t) = \sum_{j=1}^{M} \sum_{k=1}^{N} c(j, k)\, \psi_{j,k}(t),\ M, N\, \varepsilon\, Z \quad ...(1.9)$$

forms the required discrete wavelet transform or wavelet decomposition. Here only the weights $c(j, k)$ will be identified, while the dilations and translations will follow the regular grid structure.

Thus, the wavelet network proposed is:

$$g(t) = \sum_{j=1}^{M} w_i \, \varphi(D_i \, (t - t_1)) + \overline{g} \qquad \text{...(1.10)}$$

Which is a combination of translation, rotation, dilation and wavelet lying on the same line will be called a wavelon in sequel. Here a rotation is adopted weights, dilations and translations will jointly be fitted from data. Obviously, this provides better flexibility in approximating arbitrary functions than the wavelet decomposition. Furthermore in order to compensate for the orientation selective nature of the dilations, a rotation is adopted for each transform. Thus the wavelet network shown in Fig.1.8 forms the following structure:

$$g(t) = \sum_{j=1}^{M} w_i \, \varphi(D_i R_i \, (t - t_i)) + \overline{g} \qquad \text{...(1.11)}$$

where R is the rotation matrix.

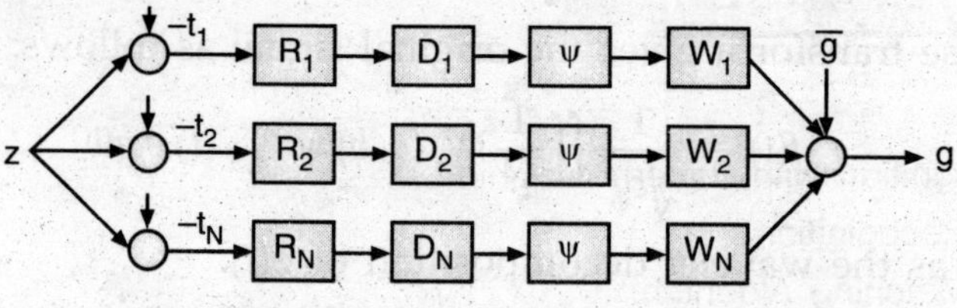

Fig. A.1. Structure of Wavelet Neural Network

Where the t_i are arbitrary translation vectors, the D_i are diagonal matrices built from arbitrary dilation vectors and the parameter $\overline{g}$ is introduced to help deal with nonzero mean functions on finite domains (since the wavelet È has non zero mean). A local linear wavelet neural network was tried. Here, the output of the k_{th} unit in the output layer is given by:

$$y_k = \sum_{i=l}^{N} (w_i + w_{i1} x_1 + w_{i2} \, x_2 + w_{iN} x_N) \, \psi_i(x) \qquad (1.12)$$

Particle Swarm Optimization (PSO) for optimizing a local linear network is used here [64].

□□□

3 ANNEXURE THREE

RESEARCH AREAS IN SOFT COMPUTING (FOR MORE AMBITIOUS STUDENTS)

CHAPTER OUTLINE

A.3.1. Speech Recognition with Pattern Matching Approach
A.3.2. Finger Print Recoginition
A.3.3. Finger Print Matching Algorithm
A.3.4. Electric Load Forcasting using Artifical Neural Network

A.3.1. SPEECH RECOGNITION WITH PATTERN MATCHING APPROACH

The objective of Speech Recognition is to determine the sequence of sound units from speech signal so that the linguistic message in the form of text can be decoded from the speech signal for recognition, understanding and synthesis.

In sub word recognition, The main thrust is to extract features of a given phrase or word or alphabet or phoneme. The solutions remain the same *i.e.,* extracting the best features, using the best distortion criterion and then to find the optimum neural network. In this work, LPC coefficients of framed and windowed training speech signal are analysed through Independent Component Analysis to extract features and create classes. These features are stored as templates. Then the similar processing is done with new or distorted pattern. These features are compared with the stored templates and Neural Network is used as pattern matching device. The time limiting behaviour of speech makes the neural network very complex. This fully connected feed forward neural network is reduced to skeleton neural network to find out the optimum trade off between performance and number of connections. To account for the highly temporal or local behaviour of speech, wavelet is used as activation function.

The pattern recognition approach for speech recognition is illustrated in Fig. A.1. Neural network may be used for training network with known patterns and create templates. Trained network is used as classifier for unknown patterns.

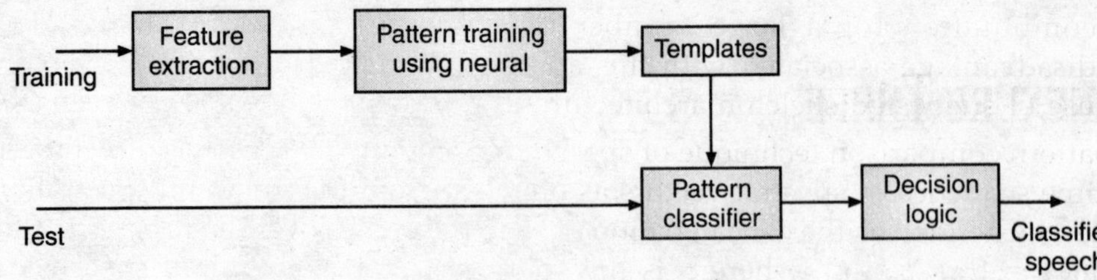

Fig. A.1. Block diagram of speech recognizer using Pattern recognition approach

The process is defined as follows:

(*a*) **Feature Extraction:** For speech signals, the feature extraction is the output of some type of spectral analysis technique, such as a filter bank analyzer, linear predictive coding (LPC), or Discrete Fourier Transform (DFT) or Sast Fourier Transform (FFT)[32]..

(*b*) **Pattern Training:** Training patterns corresponding to speech utterance of the same class are used to create a pattern representative to speech utterances of the same class. The resulting reference pattern can be a template derived from an averaging technique, or it can be a model that characterizes the statistics of the features of the reference pattern.

(*c*) **Pattern Classification:** The test pattern is compared with each class of reference pattern. A measure of similarity or dissimilarity between reference and test pattern is computed using a local distance measure, or 'spectral distance' between two measures. A global time alignment procedure as Dynamic Time Warping or Trace Segmentation is used to compensate for different rates of speaking of the two patterns [39, 59].

(*d*) **Decision Logic:** The similarity of test pattern with reference pattern is used to decide, which reference pattern or sequence of reference pattern is the best fit.

The salient features of pattern recognition approach are as follows:

(*a*) The performance of the system is sensitive to the amount of training data available for creating good reference pattern. Usually, more is the training, better is the performance of the system.

(*b*) The reference patterns are sensitive to the speaking environment and transmission characteristics of the medium used to create the speech. This is because the spectral characteristics are affected by the transmission and background noise [52].

(*c*) No speech-specific knowledge is used explicitly in the systems. Hence, the method is relatively insensitive to choice of vocabulary words, task, syntax, and task semantics.

(*d*) Since the system is insensitive to the sound class; the basic techniques and algorithms are widely applicable to a variety of sound class, *i.e.*, phrases, whole words, and sub-word units.

(*e*) It is relatively straightforward to incorporate syntactic and semantic constraints directly into the pattern-recognition structure, thereby improving recognition accuracy and reducing computation.

(*f*) The computational load for both pattern training and pattern classification is generally linearly proportional to the number of patterns being trained or recognized. Therefore, computation for a large number of classes becomes prohibitive. This is one disadvantage associated with this approach. This disadvantage is overcome by the use of suggested skeleton architecture in this study.

In pattern comparison technique of speech recognition, the speech is directly represented by the time sequence of spectral vectors as obtained from the front-end spectral analysis. A test pattern is defined as the concatenation of spectral frames over the duration of the speech, such that $T = \{t_1, t_2, t_3, \ldots t_I\}$, where t_i is the spectral vector of the input speech at any time instant i, and I is the total number of frames of speech. Similarly, reference pattern is defined as:

$R = \{R_1, R_2, R_3, \ldots R^V\}$, where, $R^j = \{r_j^1, r_j^2, \ldots, r_j^i\}$. The purpose of pattern comparison is to determine the distance or dissimilarity of T to each of R^j, where $j = 1$ to V, to identify the reference patter with minimal distance or dissimilarity.

The following problems are to be taken care to determine global dissimilarity:

(*a*) T and R^j are of unequal lengths, because of different dialects and different speaking rates by different speakers. This may also vary from time to time for the same speaker.

(*b*) T and R^j are not time aligned in any simple or well prescribed manner. This is because different sounds cannot be varied in duration to the same degree. Thus vowels are easily lengthened and shortened; whereas, consonants may not vary a lot.

(*c*) There is a requirement to compare pairs of spectral vectors, *i.e.* a local dissimilarity to evaluate the global dissimilarity. The spectral dissimilarity is equally applicable to feature-based and template-based approach.

The problem at hand can be divided into two parts:

1. The pattern classification: Different patterns are identified to map to a given class. On identification of an utterance that is not close to any class, a new class is created. The closeness to any class can be computed using either Euclidian distance or Centroid computation.
2. Pattern matching: The test pattern is not one of those out of which the reference classes are formed. The identification of test pattern is done by the distortion measure from reference classes.

The Classification and matching methods can be off-line or on-line. In off line methods, the classes are created first and the test patterns are tested out of these classes only. In on-line methods, if there is no desired closeness to the existing classes, a new class is created.

2. Blind Source Separation: During 1989-91 the problem of separation of two and more than two sources was addressed by various researches. In his doctoral thesis, Fety, addressed the problem of identifying the dynamic model of using second order cumulants. This was addressed as signal operation problem. This problem was later addressed by Tond in 1991. This has applications in famous 'Cocktail Party problem' and used in channel identification [17]. Other applications are antenna array processing, estimation of radiating sources form the unknown arrays, jammer rejection, noise reduction and two stage localization procedure, even if array are perturbed / noisy or ill- calibrated [4]. Later, Blind Source separation problem was widely addressed by Polynomial Neural Network and radial basis function network.

A.3.2. FINGERPRINT RECOGNITION

All the digits (fingers, thumbs and toes) have epidermal ridges, furrows and patterns. These patterns are widely believed to provide a friction surface to assist in gripping and handling objects and walking. The skin excretes oils and perspiration through sweat glands, flowing along the tops of the ridges. When a surface is touched the fingerprint is transferred. Smooth, clean surfaces record better quality fingerprints but fingerprints can also be found on irregular surfaces. There are three basic categories of fingerprint:

- **Visible** prints, such as those made in oil ,ink or blood
- **Latent** prints which are invisible under normal viewing conditions
- **Plastic** prints which are left in soft surfaces such as new paint.

Fig A.2. A fingerprint image

Henry Classification System

Sir Edward Henry developed a classification system which was officially adopted by British India in 1897. The fingerprint patterns are divided into five basic groups, with various subgroups:

Arch: a ridge that runs across the fingertip and curves up in the middle.

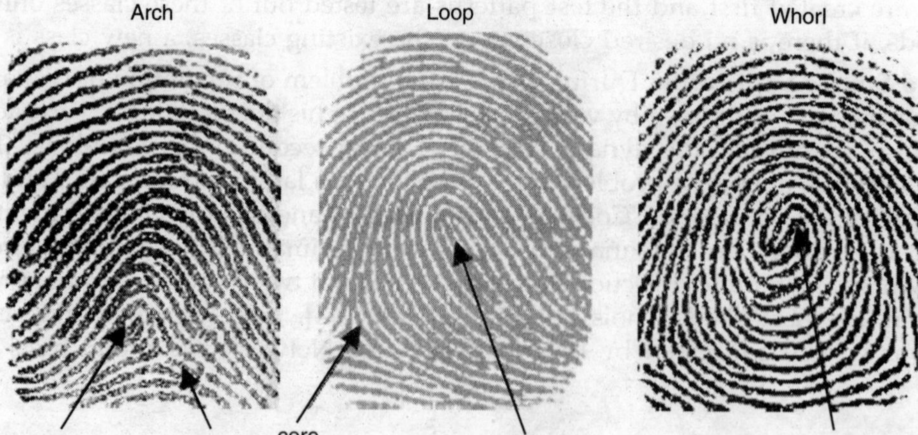

Fig. A.3. Fingerprint patterns: arch, loop and whorl

Whorl: an oval formation, often making a spiral pattern around a central point. Principal types are plain whorl and central pocket loop whorl.

Loops: These have a stronger curve than arches, and they exit and enter the print on the same side. Radial loops slant towards the thumb and ulanar loops away from the thumb.

Composites are a mix of other patterns;

Accidentals form an irregular pattern that's not classifiable as an arch, loop or Whorl.

Minutiae

Several other characteristics can be present within the fingerprint patterns. At local level, a total of 150 different local ridge characteristics, called minute details, have been identified. The minutiae or interruptions to the smooth flow of ridges and are the basis for most fingerprint identification. Minutiae are at their most rudimentary ridge endings, the points at which a ridge stops, bifurcations, at the point at which one ridge divides into two, and dots or small ridges.

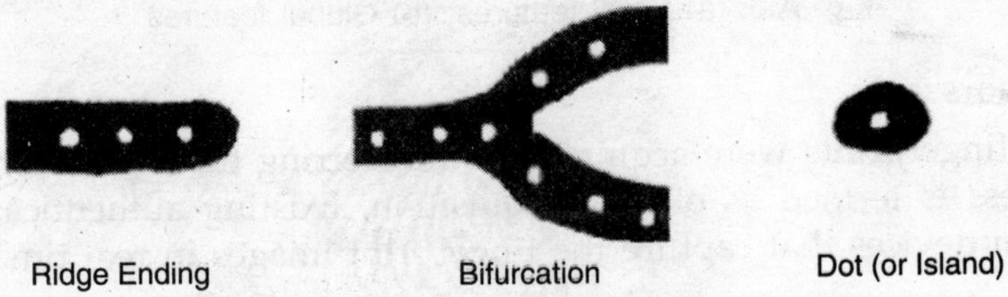

Fig. A.4. Minutiae: ridge ending, bifurcation and dot

Many types of minutiae are categorized and in addition to ridge endings, bifurcation and dots include: Islands, Ponds, Spur, Bridges, and Crossovers which are Level 2 detail. Each minutiae may be described by a number of attributes such as position (x, y), its orientation and its quality.

At the very fine level, intra-ridge details can be detected. These are sweat pores whose position and shape are considered highly distinctive. However, extracting pores is feasible only in high resolution fingerprint images (1000dpi) of good quality. These are Level 3 details.

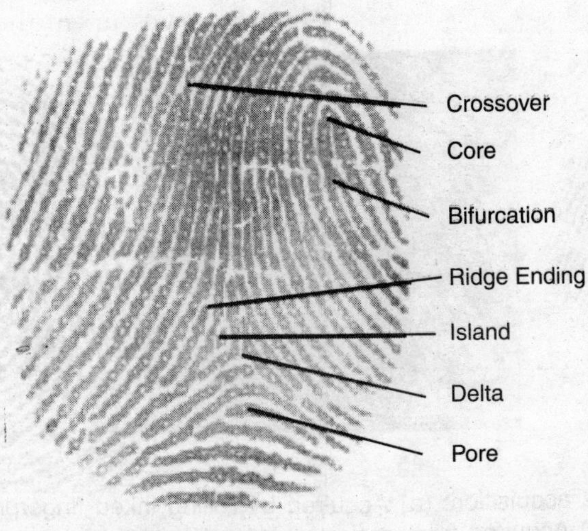

Fig. A.5. Various minutiae

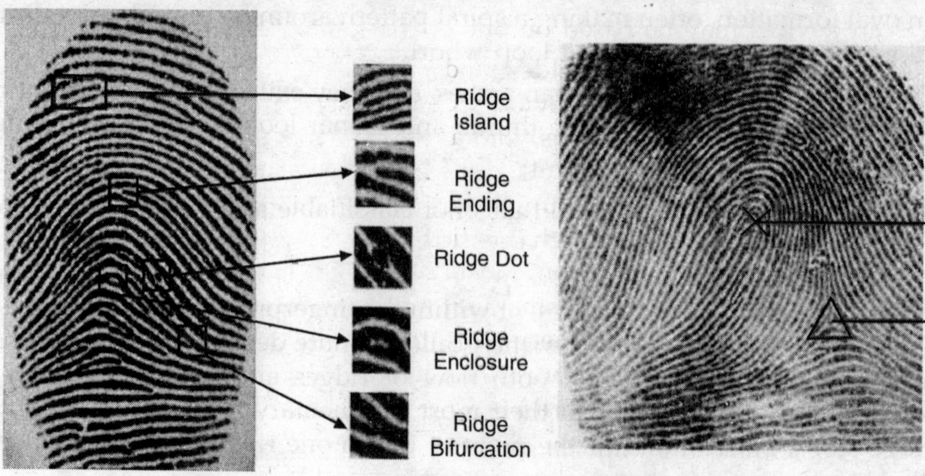

Fig. A.6. (a) Local features; (b) Global features

Fingerprint Sensors

Traditionally fingerprints were acquired by transferring the inked impression onto the paper. This process is termed as off-line acquisition. Existing authentication systems are based on live-scan devices that capture the fingerprint images in real time.

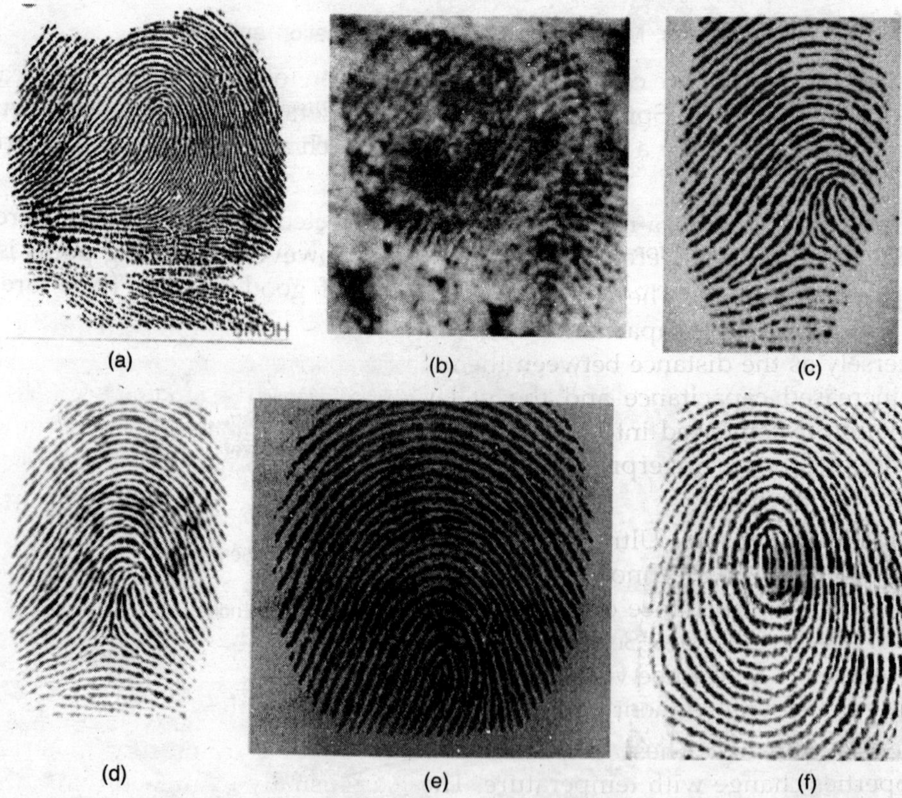

Fig. A.7. Offline and Online acquisition: (a) Acquired by rolling inked fingerprint acquired from a crime scene, (c, d, f) Acquired from optical, (e) Acquired from a capacitive sensor

The live scan devices may be based on one of the following sensing schemes:

1. **Optical Sensors:** These are the oldest and most widely used technology. In most of the devices, a charged couple device (CCD) converts the image of the fingerprint, with dark ridges and light valleys, into a digital signal. They are fairly inexpensive and can provide resolutions up to 500 dpi. Most sensors are based on FTIR (Frustrated total internal Reflection) technique to acquire the image. In this scheme, a source illuminates the fingerprint through one side of the prism as shown Fig. A.8. (*a*). Due to internal reflection phenomenon, most of the light is reflected back to other side where it is recorded by a CCD camera. However, in regions where the fingerprints surface comes in contact with the prism, the light is diffused in all directions and therefore does not reach the sensor resulting in dark regions. The quality of the image depends on whether the fingerprint is wet or dry.

   Another problem faced by optical sensor is the residual pattern left by the previous fingers. Furthermore it has been shown that fake fingers are able to fool most commercial sensors. Optical sensors are also among the bulkiest sensor due to the optics involved.

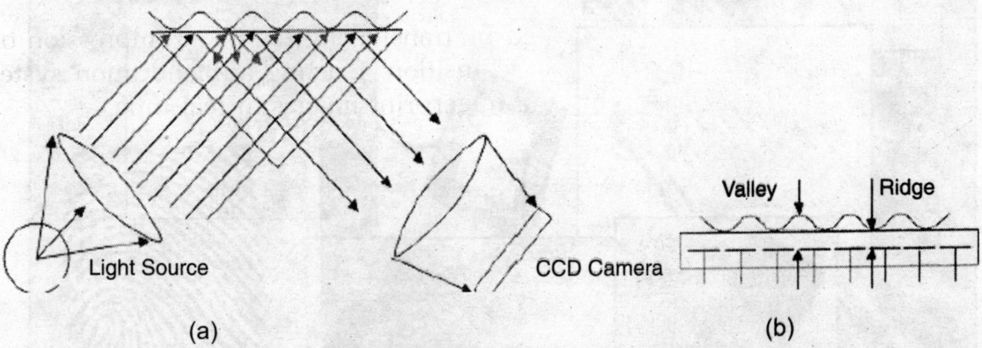

Fig. A.8. (a) General schematic for an FTIR based optical sensor (b) schematic of a capacitive sensor

2. **Capacitive Sensors:** The silicon sensor act as one plate of capacitor, and the finger as other plate. The capacitance between the sensing plate and the finger depends inversely as the distance between them. Since the ridges are closer, they correspond to increased capacitance and the valleys correspond to smaller capacitance. This variation is converted into an 8 bit gray scale digital image. Most of the electronic devices featuring fingerprint authentication use this form of solid state sensors due to its compactness.

3. **Ultrasound Sensors:** Ultra sound technology is perhaps the most accurate of the fingerprint sensing technologies. It uses ultrasound waves and measures the distance based on the impedance of the finger, the plate, and air. These sensors are capable of very high resolution. Sensors with 1000dpi or more are readily available .However these sensors tend to be very bulky and contain moving parts making them suitable only for law enforcement and access control applications.

4. **Thermal sensors:** These sensors are made up of pyro-electric materials whose properties change with temperature. These are usually manufactured in the form of strips .As the fingerprint is swipe across the sensor, there is differential conduction of heat between the ridges and the valleys that is measured by the sensor. Full size

thermal sensors are not practical since skin reaches thermal equilibrium very quickly once placed on the sensor leading to loss of signal. This would require us to constantly keep the sensor at a higher or lower temperatures making it very energy inefficient. The sweeping action prevents the finger from reaching thermal equilibrium leading to a good contrast images. However, since the sensor can acquire only small strips at a time, a sophisticated image registration and reconstruction scheme is required to construct the whole image from the strips.

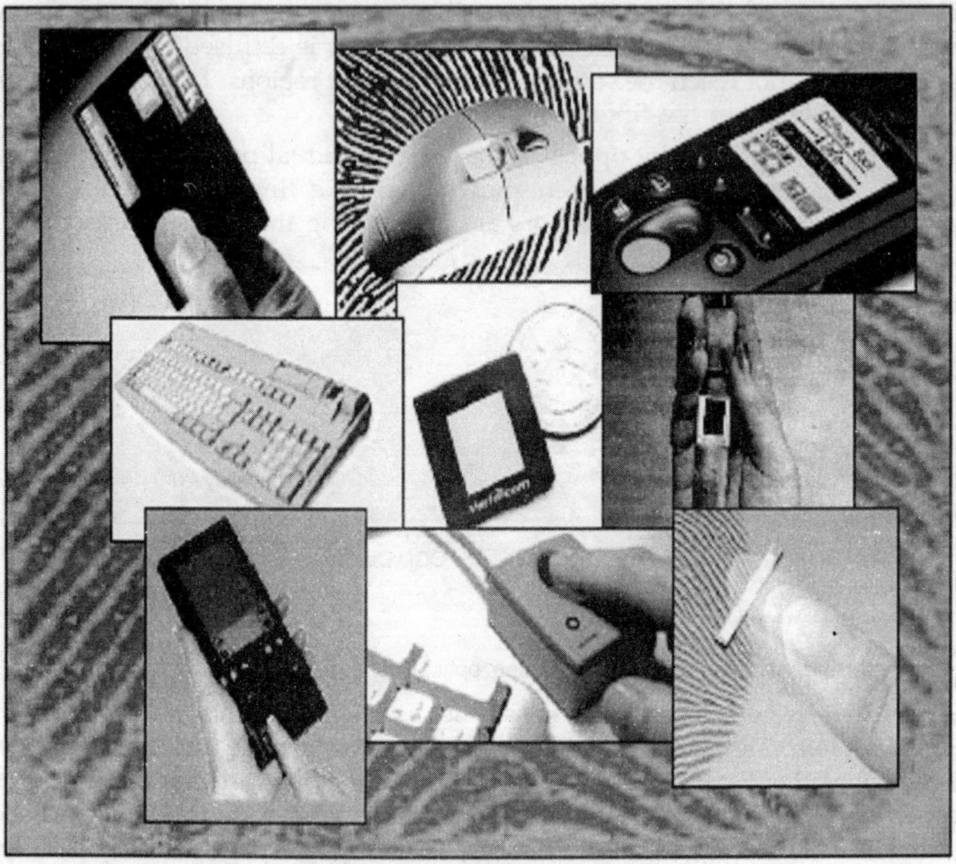

Fig. A.9. Various commercial sensors

A.3.3. FINGERPRINT MATCHING ALGORITHMS

Fingerprint feature extraction and matching methods may be broadly classified into three categories: minutiae-based, image-based and hybrid. Minutiae based methods use a feature vector extracted from fingerprints as sets of points in a multi-dimensional plane. The feature vector may comprise several characteristics of minutiae such as type, position, orientation etc. Most minutiae-based methods suffer from several shortcomings like extracting minutiae from a poor quality fingerprint image may result in low matching accuracy. These methods involve time consuming steps of fingerprint enhancement, directional filtering, segmentation and thinning and may erroneously introduce false minutiae and reject some real minutiae points. In addition these methods may not fully utilize the rich discriminatory information available in the fingerprint with high computational complexity.

There are a number of verification methods that do not use minutiae points for matching and are called image-based methods. They extract features for matching by applying a certain type of filters banks or using special transformations. The features other than print ridge pattern, such as local orientation and frequency, ridge shape, and texture information are also used. They usually require less preprocessing effort than minutiae based methods using global information from a fingerprint. Hybrid methods using features from both approaches have recently been researched. But these methods have mostly the same problem as the minutiae based methods.

Prior Related Work

Image based methods are frequently used for fingerprint recognition. A wavelet based fingerprint recognition method has been presented where recognition rate is improved with an increase in number of fingerprint images stored in database per user and hence larger memory size is needed for performance improvement. Also, verification is achieved using the features extracted by applying eight Gabor filters around the core point. Two alternations of Gabor filter method are present in various literature at first a sub sampling at the block level is performed on fingerprint image to improve the efficiency and then Gabor filters are applied. Instead of storing the response of each filter at each sub sampling point, only the index of filter with highest response is used for fingerprint matching.

Hybrid fingerprint matching schemes based on both minutiae and feature maps, utilize merits from both approaches. However, it is very difficult to accurately detect all minutiae, which may significantly affect the robustness of these approaches. Moreover, the computation required for feature maps extracted on each minutiae with Gabor filters becomes very high. There is discriminatory information found in orientation of fingerprint ridges that can be used for fingerprint verificatio steepest descent algorithm for fingerprint registration and verification based on orientation field is used by various researches. Decision level fusion of different verification methods is a challenge for performance improvement in fingerprint verification especially in low quality images. In some cases the matching systems are designed based on fusion of minutiae points with textural features extracted from Gabor filters. Similarly, decision level fusion of global orientation field with minutiae is used for fingerprint verification.

Transform based methods using digital wavelet transform features or digital cosine transform show a high matching accuracy for inputs identical to templates on its database. However, these methods have not considered the invariance to an affine transform to deal with different input conditions. A fingerprint verification system based on the correlation of fingerprint is proposed in papers. Rotation of fingerprint is not included in this approach which may increase the computation complexity.

Proposed Approach

An image-based algorithm using variance feature of orientation field for fingerprint matching is proposed .It reliably finds a reference point with a proposed preprocessing and aligns the input image with the template using orientation, which makes the algorithm rotation invariant. As the area of interest for finding the orientation field is the area around the core point, hence the segmentation process to separate the foreground from background is not required. Since the area of interest is around the core point, the algorithm becomes translation invariant. A set of features based on orientation is extracted to represent the fingerprint as information of the global structure. Fingerprint matching is based on the Euclidean distance to match the two corresponding feature vectors of the input fingerprint image and the template fingerprint image.

Algorithm Steps

The proposed algorithm contains main steps as shown preprocessing using STFT analysis, determination of reference point, cropping, block orientation field(BOF) estimation, determination of rotation, feature vector calculation, feature vector rotation ,similarity measure using Euclidean distance.

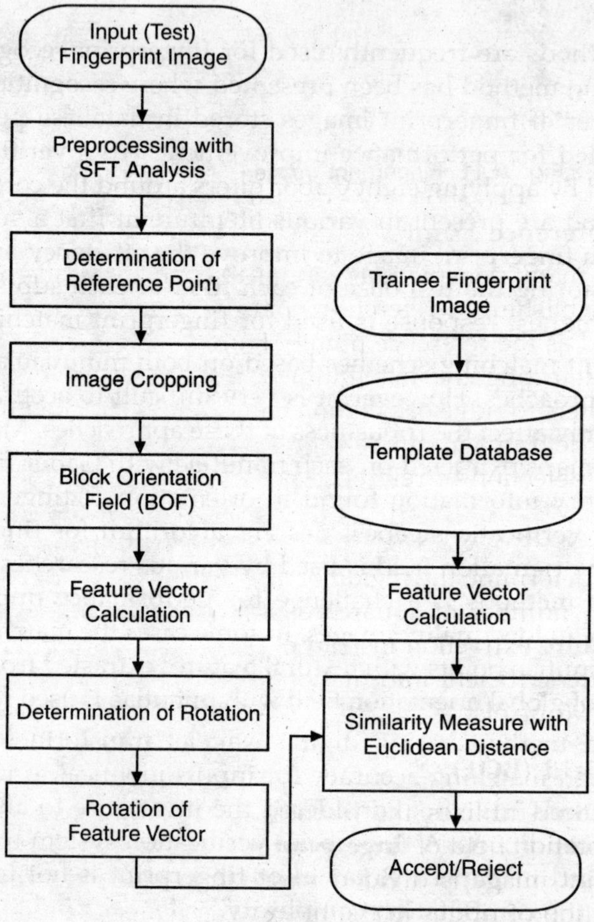

Fig. A.10.

Preprocessing with STFT Analysis

The fingerprint image can be electronically scanned with ranges of resolution. The quality of acquired images may vary in both the printed location and the clarity of the image itself. The uncertainty of first issue can be remedied by applying the guide frame when the scanning the finger. For the later case, image quality is highly dependant on finger condition. The enhancement process, therefore tries to level up the image condition to the state that it can be later processed with high degree of success. The STFT can be used to analyze the fingerprint image both in space and in frequency and hence enhance the fingerprint completely. The enhancement technique uses a Fourier domain-based block wise contextual filter approach for enhancing fingerprint images.

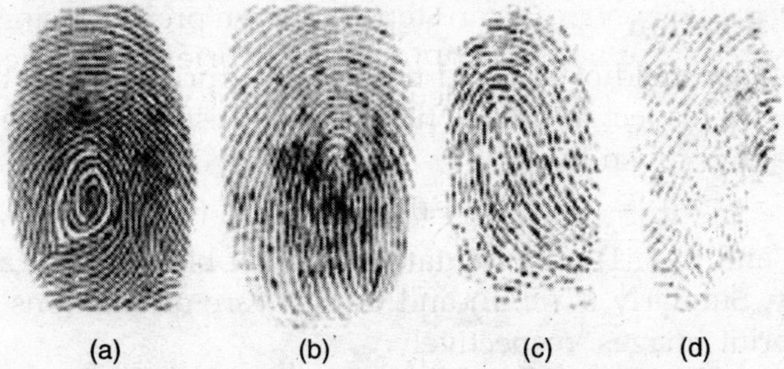

Fig. A.11. Fingerprint images of various qualities

Determination of Reference Point

Fingerprints have many conspicuous landmark structures and a combination of them could be used for establishing a reference point. A point of maximum curvature on the convex ridge is defined as the reference point. The reliable detection of a reference point can be accomplished by detecting the maximum curvature using complex filtering methods. They apply complex filters to ridge orientation field image generated from original fingerprint image. An orientation field image can be represented in multi resolution and it can be used to robust and reliable detection of reference point.

Image Cropping

The next step is to determine the region of interest (ROI) based on the reference point and tessellate the ROI into number of square cells. A predefined area (ROI) surrounding the reference point for feature extraction instead of entire fingerprint image is used. The size of area is chosen to ensure sufficient information for matching process. The fingerprint image is cropped into 96 by 96 pixels.

Block Orientation Field (BOF)

The proposed method utilizes the likeness of BOFs for similarity measuring of two fingerprints. The orientation field of fingerprint image represents the directionality of ridges The cropped fingerprint image is divided into number of non overlapping blocks and an orientation representation of ridges in the block is assigned to the block based on grayscale gradients in the block. The block orientation can be determined from pixel gradients by averaging. The orientation field of block (i, j) is given by

$$\theta(i, j) = 0.5 \tan^{-1}\left(\frac{V_x(i,j)}{V_y(i,j)}\right)$$

where $$V_x(i, j) = \sum_{u=i-w/2}^{i+w/2} \sum_{v=j-w/2}^{j+w/2} \left(G_x(u,v) G_y(u,v)\right)$$

and $$V_y(i,j) = \sum_{u=i-w/2}^{i+w/2} \sum_{v=j-w/2}^{j+w/2} \left(G_x^2(u,v) G_y^2(u,v)\right)$$

Where w is the size of block, G_x G_Y are the gradient magnitudes in x and y directions, respectively. Sobel operator has been used for computation of G_x and G_Y.

Determination of Rotation

From the orientation field of test and trainee fingerprint images the amount of rotation of test fingerprint with respect to trainee fingerprint is estimated. The amount of rotation in test fingerprint image is given by

$$\theta_r = (\theta'(1, 1) - \theta(1, 1)) + (\theta'(m, n) - \theta(m, n))/2$$

Where $\theta'(1, 1)$ and $\theta(1, 1)$ are orientations of first block of test and trainee fingerprint images respectively. Similarly $\theta'(m, n)$ and $\theta(m, n)$ are orientations of block (m, n) of test and trainee fingerprint images, respectively.

Feature Vector Calculation of Trainee Image

The variance feature vector of BOF of the cropped trainee is computed as follows:

$$\sigma^2{}_k = \sum_{i=1}^{n} \left(\theta(i,k) - \mu_k\right)^2 \quad \text{for } k = 1, \ldots, m,$$

Where μ_k is the mean of kth column of θ, m and n are the no. of columns and rows of θ respectively.

The feature vector is given by

$$\nu = [\sigma_1{}^2 \; \sigma^2{}_2 \ldots \; \sigma_m{}^2]$$

Feature Vector of Test image

The feature vector of test fingerprint image is calculated.

Rotation of Feature Vector

The feature vector obtained in previous step is rotated by ?r to obtain rotated feature vector of the test input image.

Similarity Measure Using Euclidean Distance

To measure the similarity between feature vectors of an input test fingerprint and of the trainee stored in template database, the Euclidean distance is employed. The less distance, the more similarity. A simpler distance measure substantially reduces the computational complexity of the system.

$$D = \|\nu' - \nu\|^2$$

Where ν' and ν are the feature vectors of test and trainee fingerprint images, respectively.

Experimental Results

The proposed algorithm is evaluated on fingerprint images taken from FVC2002 and Neurotechnologia database. By doing fingerprint matching, all possible genuine matching scores and impostor scores were obtained and then different threshold values were fixed to decide whether to accept or reject match. The decision of acceptance depends on the matching score. If the matching score is below certain threshold, then matching is labeled as accepted, otherwise labeled as rejected.

Performance of algorithm has been measured in terms of false accept rate (FAR) and false reject rate (FRR) for various thresholds. Different thresholds result in different values of FAR and FRR.

The FAR and FRR are computed as follows:

N is the number of subjects with five fingerprints each. Total fingerprint images are $T = 5 \times N$. A single template per subject has been considered for experimentation. Total trials carried out for finding true claims and imposter claims are $N \times (T - 1)$, out of which total true claims are $N \times 4$ and imposter claims are (total trials – true claims).

FRR % = (true claims rejected/total true claims) × 100

FAR% = (imposter claims accepted/total imposter claims) × 100

GAR% = 100 – FRR in percentage.

A receiver Operating Curve (ROC) is a plot of genuine acceptance rate against false acceptance rate for all possible values of thresholds. It measures overall performance of the system.

Neurotechnologia database consists of images of 10 subjects with eight images of each subject. Total 10 × 8 = 80 images. The size of the image is 480 × 504 and the resolution of database is 500dpi. Experimentation was done with six impressions of each finger. For a single threshold total trials carried out for computation of FAR and FRR were 10 × (60 – 1) = 590. Results reported for various thresholds are tabulated in Table A.1. and shown in Fig. A.12. The GAR reported 70% at 1.8% FAR and 96% at 20.9% FAR.

Table A.1. FAR% and GAR % at various thresholds (Neurotechnologia).

FAR%	1.85	2.04	2.59	3.33	3.7	6.11	6.48	9.26	19.3	20.9
GAR%	70	74	76	78	80	84	86	88	94	96

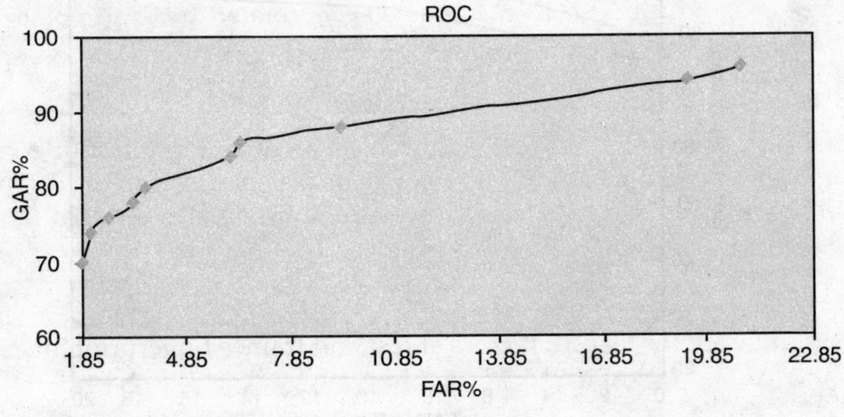

Fig. A.12. ROC for Neurotechnologia database(six impressions)

In next set of experimentation three out of eight impressions of each subject in database were rejected as the accurate core point could not be located or the rotation of the test fingerprint image with respect to the template was too high. With five impressions per subject the GAR reported 75% at 0% FAR and 95% at 21.1% FAR. ROC curve showing GAR and FAR for various values of threshold are shown here and the results are tabulated in Table A.2.

Table A.2. FAR% and GAR % at various thresholds (Neurotechnologia)

FAR%	0	0.22	1.56	9.33	16.7	19.3	19.6	21.1
GAR%	75.0	77.5	82.5	85.0	87.5	90.0	92.5	95.0

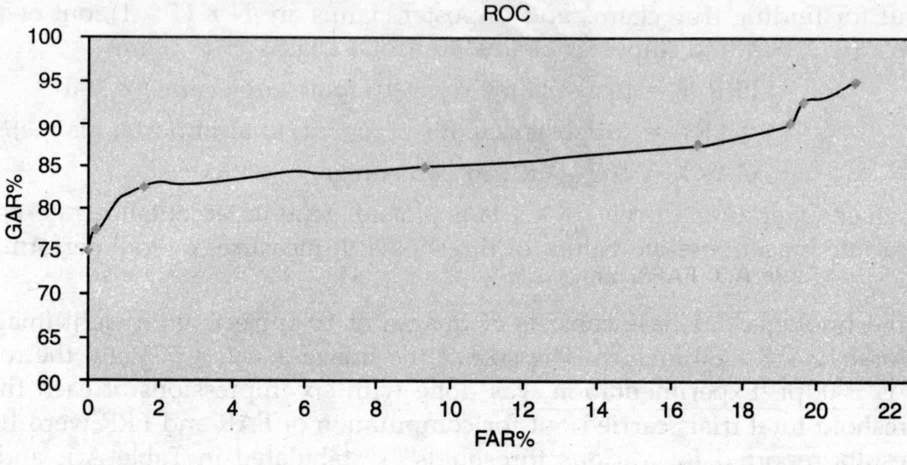

Fig. A.13. ROC for Neurotechnologia database (five impressions)

The performance of algorithm on FVC2002 and University of Bologna database with eight impressions per subject is shown in Fig. A.14.

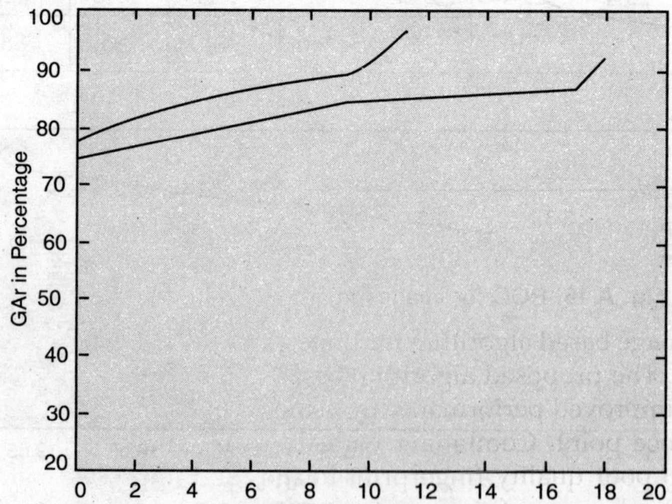

Fig. A.14. Performance curve on university of Bologna database (dashed line) and FVC2002 (solid line)

For experimentation eight subjects out of ten with five impressions of each subject of Db1_b of FVC2002 have been used. Three impressions have no core point due to the exaggerate displacement or the rotation between the test and trainee is too high. Also the core point detection is fairly accurate in some of the images. Results reported for various thresholds on FVC2002 database are given in Figure. The GAR reported 80% at 2.8% FAR and 97.5% for 8.66%FAR. Sample of cropped image based on reference point detection form FVC 2002 Db1_b is shown in Fig. A.15.

Fig. A.15. Cropped Image

Table A.3. FAR% and GAR % at various thresholds (FVC2002)

FAR%	2.8	3.77	4.22	4.44	7.33	8.66
GAR%	80	82.5	87.5	92.5	95	97.5

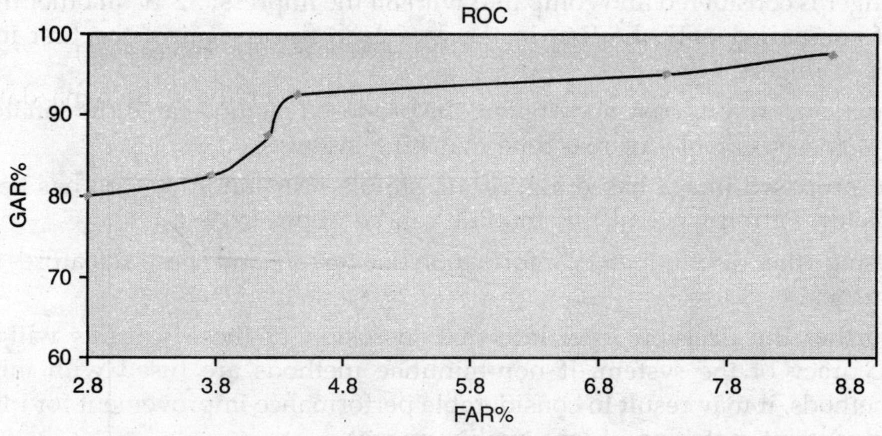

Fig. A.16. ROC for images from FVC2002 (five impressions)

An enhanced image based algorithm for fingerprint matching based on orientation feature has been proposed. The proposed algorithm basically uses orientation as a feature vector. It further pursues an improved performance by using the alignment and rotation after a reliable detection of reference point. Combining with the STFT analysis, this approach makes the algorithm robust to poor quality fingerprint images and improves matching accuracy. The orientation feature makes use of richer information without having to resort to using large feature vector as input. This work mainly addresses the use of combinational feature of orientation image information and singular point information to perform fingerprint verification. Fingerprint image quality has been improved using STFT analysis which helps in finding reference point reliably which result in accurate matching system .Matching with Euclidean distance results in low computational load.

The main features of the proposed algorithm are as follows:

(*i*) The proposed algorithm uses an enhancement method based on STFT analysis. A preprocessing with STFT can remove multi-spectral noises while enhancing structural information in images, and help to find a reference point accurately and reliably which improves the performance of fingerprint feature extraction and matching steps as both depend heavily upon the quality of the input fingerprint image.

(*ii*) As the performance of a fingerprint matching algorithm relies upon the efficiency of its feature extraction stage, the proposed algorithm rotates the feature vector generated by using orientation of fingerprint image for alignment to overcome the rotation between test and trainee fingerprint image instead of rotating the test fingerprint image.

(*iii*) Extracting orientation features from nonoverlapping tessellated cells significantly reduces the effect from noise and nonlinear distortions, and thus preserves the local information.

(*iv*) Fixed length vectors are used to represent orientation features and it reduces the computation load. Overall feature extraction and matching to a template are faster using fixed length vectors.

(*v*) The experimentation carried out for obtaining the performance of the proposed algorithm in this work is rigorous .For impostor comparisons each impression of a finger is considered and compared with all the impressions of all other fingers instead of comparing only the first instance of each finger against the first instance of all other fingers.

Experimental results have shown that, the proposed method can differentiate fingerprints accurately, and is suitable for real time matching systems.

In the proposed image based algorithm, simple orientation feature has been used as a feature vector. Performance of this module can be improved by:

- Using other discriminatory information like texture and spectral features of fingerprint image.
- Further the decision level information fusion of these features will improve the accuracy of the system. If non-minutiae methods are fused with minutiae based methods, it may result in considerable performance improvement for minutiae based fingerprint matching of low quality images.
- Using the efficient image alignment method.

All the above mentioned points will be considered in future work. Also, the shortcoming of using single biometric is that no one technology is suitable for all applications. Therefore, instead of using fingerprint alone, face and iris can be combined with fingerprint to gain the advantages of multimodal biometrics. Fusion of two (or more) biometric inputs against two (or more) different algorithms to arrive at a decision will be considered in future work.

The other area which can be investigated is designing hybrid approaches based on both minutiae and numerical features .Minutiae may be used to align images and a multi resolution analysis of fingerprint image performed for feature extraction and matching.

A.3.4. ELECTRIC LOAD FORECASTING USING AN ARTIFICIAL NEURAL NETWORK

Various techniques for power system load forecasting have been proposed in the last few decades. Load forecasting with lead-times, from a few minutes to several days, helps the system operator to efficiently schedule spinning reserve allocation. In addition, load forecasting can provide information which is able to be used for possible energy interchange with other utilities. In addition to these economical reasons, load forecasting is also useful for system security. If applied to the system security assessment problem, it can provide valuable information to detect many vulnerable situations in advance.

Traditional computationally economic approaches, such as regression and interpolation, may not give sufficiently accurate results. Conversely, complex algorithmic methods with heavy computational burden can converge slowly and may diverge in certain cases.

A number of algorithms have been suggested for the load forecasting problem. Previous approaches can be generally classified into two categories in accordance with techniques they employ. One approach treats the load pattern as a time series signal and predicts the future load by using various time series analysis techniques. The second approach recognizes that the load pattern is heavily dependent on weather variables, and finds a functional relationship between the weather variables and the system load. The future load is then predicted by inserting the predicted weather information into the predetermined functional relationship.

General problems with the time series approach include the inaccuracy of prediction and numerical instability. One of the reasons this method often gives inaccurate results is that it does not utilize weather information. There is a strong correlation between the behavior of power consumption and weather variables such as temperature, humidity, wind speed, a cloud cover. This is especially true in residential areas. The time series approach mostly utilizes computationally cumber-some matrix-oriented adaptive algorithms which, in certain cases, may be unstable.

Most regression approaches try to find functional relationships between weather variables and current load demands. The conventional regression approaches use linear or piecewise-linear representations for the forecasting functions. By a linear combination of these representations, the regression approach finds the functional relationships between selected weather variables and load demand. Conventional techniques assume, without justification, a linear relationshhip. The functional relationship between load and weather variables, however, is not stationary, but depends on spatio-temporal elements. Conventional regression approach does not have the versatility to address this temporal variation. It, rather, will produce an averaged result. Therefore, an adaptable technique is needed.

❑❑❑

4 ANNEXURE FOUR

ANN APPLICATIONS

CHAPTER OUTLINE

A.4.1 Trajectory generation
A.4.2 Camera-robot coordination

An important area of application of neural networks is in the field of ROBOTICS. Usually, these networks are designed to direct a manipulator, which is the most important form of the industrial robot, to grasp objects, based on sensor data. Another applications include the steering and path-planning of autonomous robot vehicles. In robotics, the major task involves making movements dependent on sensor data. There are four, related, problems to be distinguished:

Forward kinematics

Kinematics is the science of motion which treats motion without regard to the forces which cause it. Within this science one studies the position, velocity, acceleration, and all higher order derivatives of the position variables. A very basic problem in the study of mechanical manipulation is that of forward kinematics. This is the static geometrical problem of computing the position and orientation of the end-effector ('hand') of the manipulator. Specifi- cally, given a set of joint angles, the forward kinematic problem is to compute the position and orientation of the tool frame relative to the base frame.

An Exemplar Robot Manipulator

Inverse kinematics

This problem is posed as follows: given the position and orientation of the end-effecter of the manipulator, calculate all possible sets of joint angles which could be used to attain this given position and orientation. This is a fundamental problem in the practical use of manipulators. The inverse kinematic problem is not as simple as the forward one. Because the kinematic equations are nonlinear, their solution is not always easy or even possible in a closed form. Also, the questions of existence of a solution, and of multiple solutions, arise. Solving this problem is a least requirement for most robot control systems.

Dynamics

Dynamics is a field of study devoted to studying the forces required to cause motion. In order to accelerate a manipulator from rest, glide at a constant end-effector velocity, and finally decelerate to a stop, a complex set of torque functions must be applied by the joint actuators. In dynamics not only the geometrical properties (kinematics) are used, but also the physical properties of the robot are taken into account. Take for instance the weight (inertia) of the robotarm, which determines the force required to change the motion of the arm. The dynamics introduces two extra problems to the kinematic problems.

1. The robot arm has a 'memory'. Its responds to a control signal depends also on its history (*e.g.* previous positions, speed, acceleration).
2. If a robot grabs an object then the dynamics change but the kinematics don't. This is because the weight of the object has to be added to the weight of the arm (that's why robot arms are so heavy, making the relative weight change very small).

A.4.1. TRAJECTORY GENERATION

To move a manipulator from here to there in a smooth, controlled fashion each joint must be moved via a smooth function of time. Exactly how to compute these motion functions is the problem of trajectory generation. In the first section of this chapter we will discuss the problems associated with the positioning of the end-efector (in effect, representing the inverse kinematics in combination with sensory transformation).

End-efector positioning

The final goal in robot manipulator control is often the positioning of the hand or end-effector in order to be able to, e.g., pick up an object. With the accurate robot arm that are manufactured, this task is often relatively simple, involving the following steps:

1. determine the target coordinates relative to the base of the robot. Typically, when this position is not always the same, this is done with a number of fixed cameras or other sensors which observe the work scene, from the image frame determine the position of the object in that frame, and perform a pre-determined coordinate transformation;
2. with a precise model of the robot (supplied by the manufacturer), calculate the joint angles to reach the target (i.e., the inverse kinematics). This is a relatively simple problem;
3. move the arm (dynamics control) and close the gripper.

Involvement of neural networks. So if these parts are relatively simple to solve with a high accuracy, why involve neural networks? The reason is the applicability of robots. When 'traditional' methods are used to control a robot arm, accurate models of the sensors and manipulators (in some cases with unknown parameters which have to be estimated from the system's behaviour; yet still with accurate models as starting point) are required and the system must be calibrated. Also, systems which suffer from wear-and-tear (and which mechanical systems don't?) need frequent recalibration or parameter determination. Finally, the development of more complex (adaptive!) control methods allows the design and use of more exible (*i.e.*, less rigid) robot systems, both on the sensory and motory side.

A.4.2. CAMERA-ROBOT COORDINATION

The system we focus on in this section is a worker observed by a fixed cameras, and a robot arm. The visual system must identify the target as well as determine the visual position of the end-effector.

The target position Xtarget together with the visual position of the hand Xhand are input to the neural controller *N*(.). This controller then generates a joint position q for the robot:

$$\theta = N(X_{target}, X_{hand})$$

We can compare the neurally generated θ with the optimal θ0 generated by a fictitious perfect controller *R*(.):

$$\theta 0 = R(X_{target}, X_{hand})$$

The task of learning is to make the N generate an output 'close enough' to q0. There are two problems associated with teaching *N*(.):

1. generating learning samples which are in accordance with eq. θ0 = *R*(*X* target, *X* hand). This is not trivial, since in useful applications *R*(.) is an unknown function. Instead, a form of self-supervised or unsupervised learning is required. Some examples to solve this problem are given below;
2. constructing the mapping *N*(.)from the available learning samples. When the (usually randomly drawn) learning samples are available, a neural network uses these samples to represent the whole input space over which the robot is active. This is evidently a form of interpolation, but has the problem that the input space is of a high dimensionality, and the samples are randomly distributed.

Approach 1: Feed-forward networks: When using a feed-forward system for controlling the manipulator, a self-supervised learning system must be used. One such a system has been reported by Psaltis, Sideris and Yamamura (Psaltis, Sideris, & Yamamura, 1988). Here, the network, which is constrained to two-dimensional positioning of the robot arm, learns by experimentation. Three methods are proposed:

1. **Indirect learning:** In indirect learning, a Cartesian target point x in world coordinates is generated, e.g., by a two cameras looking at an object. This target point is fed into the network, which generates an angle vector θ. The manipulator moves to position q, and the cameras determine the new position x' of the end-effector in world coordinates. This *x*' again is input to the network, resulting in θ'. The network is then trained on the error ε1 = θ – θ'

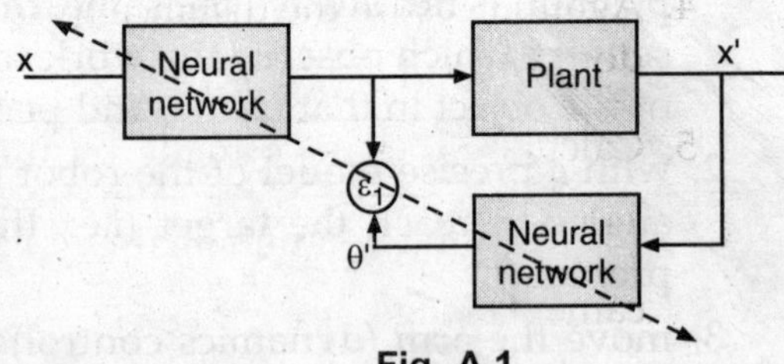

Fig. A.1.

Indirect learning system for robotics. In each cycle, the network is used in two different places: first in the forward step, then for feeding back the error.

However, minimisation of ε1 does not guarantee minimisation of the overall error

$\varepsilon = x - x'$. For example, the network often settles at a 'solution' that maps all x's to a single? (*i.e.*, the mapping).

2. **General learning:** The method is basically very much like supervised learning, but here the plant input q must be provided by the user. Thus the network can directly minimis $|\theta - \theta'|$. The success of this method depends on the interpolation capabilities of the network. Correct choice of q may pose a problem.
3. **Specialised learning:** Keep in mind that the goal of the training of the network is to minimise the error at the output of the plant: $\varepsilon = x - x'$. We can also train the neural network by 'backpropagating' this error trough the plant (compare this with the backpropagation of the error) . This method requires knowledge of the Jacobian matrix of the plant. A Jacobian matrix of a multidimensional function F is a matrix of partial derivatives of F, *i.e.*, the multidimensional form of the derivative. The learning rule applied here regards the plant as an additional and unmodifiable layer in the neural network. The total error $\varepsilon = x - x'$ is propagated back through the plant by calculating the δ.

Again a two-layer feed-forward network is trained with back-propagation. However, instead of calculating a desired output vector the input vector which should have invoked the current output vector is reconstructed, and back-propagation is applied to this new input vector and the existing output vector. The configuration used consists of a monocular manipulator which has to grasp objects. Due to the fact that the camera is situated in the hand of the robot, the task is to move the hand such that the object is in the centre of the image and has some predetermined size (in a later article, a biologically inspired system is proposed (Smagt, Kr • ose, and Groen, 1992) in which the visual flow-field is used to account for the monocularity of the system, such that the dimensions of the object need not to be known anymore to the system). One step towards the target consists of the following operations:

1. Measure the distance from the current position to the target position in camera domain, x.
2. Use this distance, together with the current state q of the robot, as input for the neural network. The network then generates a joint displacement vector $\Delta\theta$.
3. Send Dq to the manipulator.
4. Again measure the distance from the current position to the target position in camera domain, x'.
5. Calculate the move made by the manipulator in visual domain, $x - {}_{t}^{t+1}Rx'$ where ${}_{t}^{t+1}R$ is the rotation matrix of the second camera image with respect to the rst camera image;
6. Teach the learning pair $(x - {}_{t}^{t+1}Rx', \theta; \Delta\theta)$ to the neutral network.

This system has shown to learn correct behaviour in only tens of iterations, and to be very adaptive to changes in the sensor or manipulator.

By using a feed-forward network, the available learning samples are approximated by a single, smooth function consisting of a summation of sigmoid functions. As mentioned in section 4, a feed-forward network with one layer of sigmoid units is capable of representing practically any function. But how are the optimal weights determined in finite time to obtain this optimal representation? Experiments have shown that, although a reasonable representation can be obtained in a short period of time, an accurate representation of the function that governs the learning samples is often not feasible or extremely difficult. The reason for this is the global character of the approximation obtained with a feed-forward network with sigmoid units: every weight in the network has a global effect on the final approximation that is obtained. Building local representations is the obvious way out: every part of the network is responsible for a small subspace of the total input space. Thus accuracy is obtained locally (Keep It Small & Simple). This is typically obtained with a Kohonen neural network.

Approach 2: Topology conserving maps: The system described consists of a robot manipulator with three degrees of freedom (orientation of the end-effector is not included) which has to grab objects in 3D-space. The system is observed by two fixed cameras which output their (x; y) coordinates of the object and the end effector.

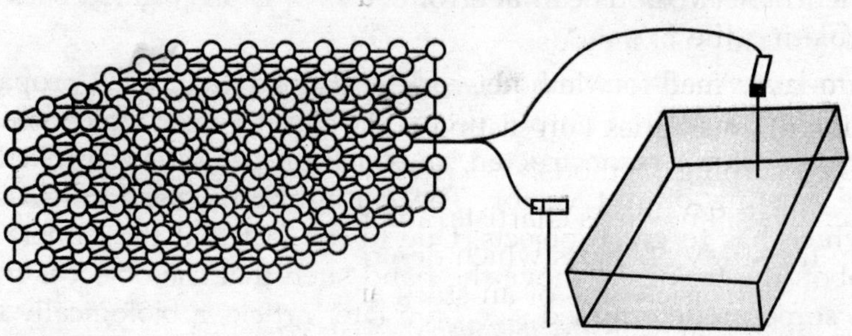

Fig. A.2. A Kohonen network merging the output of two cameras

Each run consists of two movements. In the gross move, the observed location of the object *x* (a four-component vector) is input to the network. As with the Kohonen neural network, the neuron *k* with highest activation value is selected as winner, because its weight vector *Wk* is nearest to *x*. The neurons, which are arranged in a 3-dimensional lattice, correspond in a 1 – 1 fashion with subregions of the 3D workspace of the robot, *i.e.*, the neuronal lattice is a discrete representation of the workspace. With each neuron a vector q and Jacobian matrix *A* are associated. During gross move ?k is fed to the robot which makes its move, resulting in retinal coordinates X g of the end-effector. To correct for the discretisation of the working space, an additional move is made which is dependent of the distance between the neuron and the object in space *wk* – *x*; this small displacement in Cartesian space is translated to an angle change using the Jacobian *Ak*:

$$\theta^{\text{final}} = \theta_k + A_k\,(x - w_k)$$

which is a first-order Taylor expansion of final. The final retinal coordinates of the end-effector after this fine move are in *xf*.

Learning proceeds as follows: when an improved estimate (θ, *A*)* has been found, the following adaptations are made for all neurons *j*:

Application 2: It is nowadays a common notion that vast amounts of capital are traded through the Stock Markets all around the world. National economies are strongly linked and heavily influenced of the performance of their Stock Markets. Moreover, recently the Markets have become a more accessible investment tool, not only for strategic investors but for common people as well. Consequently they are not only related to macroeconomic parameters, but they influence everyday life in a more direct way. Therefore they constitute a mechanism which has important and direct social impacts. The characteristic that all Stock Markets have in common is the uncertainty, which is related with their short and long-term future state. This feature is undesirable for the investor but it is also unavoidable whenever the Stock Market is selected as the investment tool. The best that one can do is to try to reduce this uncertainty. Stock Market Prediction (or Forecasting) is one of the instruments in this process.

The Stock Market prediction task divides researchers and academics into two groups those who believe that we can devise mechanisms to predict the market and those who believe that the market is efficient and whenever new information comes up the market absorbs it by correcting itself, thus there is no space for prediction (EMH). Furthermore they believe that the Stock Market follows a Random Walk, which implies that the best prediction you can have about tomorrow's value is today's value.

In literature a number of different methods have been applied in order to predict Stock Market returns. These methods can be grouped in four major categories:

(*i*) Technical Analysis Methods,
(*ii*) Fundamental Analysis Methods,
(*iii*) Traditional Time Series Forecasting and
(*iv*) Machine Learning Methods.

Technical analysis, known as chartists, attempt to predict the market by tracing patterns that come from the study of charts which describe historic data of the market. Fundamental analysts study the intrinsic value of an stock and they invest on it if they estimate that its current value is lower that its intrinsic value. In Traditional Time Series forecasting an attempt to create linear prediction models to trace patterns in historic data takes place. These linear models are divided in two categories: the univariate and the multivariate regression models, depending on whether they use one of more variables to approximate the Stock Market time series. Finally a number of methods have been developed under the common label Machine Learning these methods use a set of samples and try to trace patterns in it (linear or non-linear) in order to approximate the underlying function that generated the data.

Stock Markets and Prediction

Investment Theories:An investment theory suggests what parameters one should take into account before placing his (or her) capital on the market. Traditionally the investment community accepts two major theories: the Firm Foundation and the Castles in the Air. Reference to these theories allows us to understand how the market is shaped, or in other words how the investors think and react. It is this sequence of 'thought and reaction' by the investors that defines the capital allocation and thus the level of the market.

There is no doubt that the majority of the people related to stock markets is trying to achieve profit. Profit comes by investing in stocks that have a good future (short or long term future). Thus what they are trying to accomplish one way or the other is to predict 16 the future of the market. But what determines this future? The way that people invest their money is the answer; and people invest money based on the information they hold. Therefore we have the following schema:

$$\text{INFORMATION} \rightarrow \text{INVESTOR} \rightarrow \text{MARKET LEVEL}$$

The factors that are under discussion on this schema are: the content of the 'Information' component and the way that the 'Investor' reacts when having this info.

On the other hand, according to the Castles in the Air theory the investors are triggered by information that is related to other investors' behavior. So for this theory the only concern that the investor should have is to buy today with the price of 20 and sell tomorrow with the price of 30, no matter what the intrinsic value of the firm he (or she) invests on is.

Prediction Methods

The prediction of the market is without doubt an interesting task. In the literature there are a number of methods applied to accomplish this task. These methods use various approaches, ranging from highly informal ways (e.g. the study of a chart with the fluctuation of the market) to more formal ways (*e.g.* linear or non-linear regressions). We have categorized these techniques as follows:

- Technical Analysis Methods
- Fundamental Analysis Methods
- Traditional Time Series Prediction Methods
- Machine Learning Methods

The criterion to this categorization is the type of tools and the type of data that each method is using in order to predict the market. What is common to these techniques is that they are used to predict and thus benefit from the market's future behavior.

Technical Analysis

"Technical analysis is the method of predicting the appropriate time to buy or sell a stock used by those believing in the castles-in-the-air view of stock pricing". The idea behind technical analysis is that share prices move in trends dictated by the constantly changing attributes of investors in response to different forces. Using technical data such as price, volume, highest and lowest prices per trading period the technical analyst uses charts to predict future stock movements. Price charts are used to detect trends, these trends are assumed to be based on supply and demand issues which often have cyclical or noticeable patterns.

From the study of these charts trading rules are extracted and used in the market environment. The technical analysts are known and as 'chartists'. Most chartists believe that the market is only 10 percent logical and 90 percent psychological. The chartist's belief is that a careful study of what the other investors are doing will shed light on what the crowed is likely to do in the future.

This is a very popular approach used to predict the market, which has been heavily criticized. The major point of criticism is that the extraction of trading rules from the study of charts is highly subjective therefore different analysts might extract different trading rules by studying the same charts. Although it is possible to use this methodology to predict the market on daily basis we will not follow this approach on this study due to its subjective character.

Fundamental Analysis

"Fundamental analysis is the technique of applying the tenets of the firm foundation theory to the selection of individual stocks". The analysts that use this method of prediction use fundamental data in order to have a clear picture of the firm (industry or market) they will choose to invest on. They are aiming to compute the 'real' value of the asset that they will invest in and they determine this value by studying variables such as the growth, the dividend payout, the interest rates, the risk of investment, the sales level, the tax rates an so on. Their objective is to calculate the intrinsic value of an asset (*e.g.* of a stock). Since they do so they apply a simple trading rule. If the intrinsic 21 value of the asset is higher than the value it holds in the market, invest in it. If not, consider it a bad investment and avoid it. The fundamental analysts believe that the market is defined 90 percent by logical and 10 percent by physiological factors. This type of analysis is not possible to fit in the objectives of our study. The reason for this is that the data it uses in order to determine the intrinsic value of an asset does not change on daily basis. Therefore fundamental analysis is helpful for predicting the market only in a long-term basis.

Traditional Time Series Prediction

The Traditional Time Series Prediction analyzes historic data and attempts to approximate future values of a time series as a linear combination of these historic data. In econometrics there are two basic types of time series forecasting: univariate (simple regression) and multivariate (multivariate regression). These types of regression models are the most common tools used in econometrics to predict time series. The way they are applied in practice is that firstly a set of factors that influence (or more specific is assumed that influence) the series under prediction is formed. These factors are the explanatory variables xi of the prediction model.

Then a mapping between their values xit and the values of the time series yt (y is the to-be explained variable) is done, so that pairs {*xit* , *yt*} are formed. These pairs are used to define the importance of each explanatory variable in the formulation of the to-be explained variable. In other words the linear combination of xi that approximates in an optimum way y is defined. Univariate models are based on one explanatory variable ($I = 1$) while multivariate models use more than one variable ($I > 1$). Regression models have been used to predict stock market time series. A good example of the use of multivariate regression is the work of Pesaran and Timmermann (1994).

They attempted prediction of the excess returns time series of S&P 500 and the Dow Jones on monthly, quarterly and annually basis. The data they used was from Jan 1954 until Dec 1990. Initially they used the subset from Jan 1954 until Dec 1959 to adjust the coefficients of the explanatory variables of their models, and then applied the models to predict the returns for the next year, quarter and month respectively.

Machine Learning Methods

Several methods for inductive learning have been developed under the common label "Machine Learning". All these methods use a set of samples to generate an approximation of the underling function that generated the data. The aim is to draw conclusions from these samples in such way that when unseen data are presented to a model it is possible to infer the to-be explained variable from these data. The methods we discuss here are: The Nearest Neighbor and the Neural Networks Techniques. Both of these methods have been applied to market prediction; particularly for Neural Networks there is a rich literature related to the forecast of the market on daily basis.

Prediction Using Neural Networks

The experiment consisted of three phases (Figure A). In the first phase a genetic algorithm (GA) searched the space of NNs with different structures and resulted a generation with the fittest of all networks searched based on a metric which was either: TheilA or TheilB or TheilC or MAE. The GA search was repeated three times for each metric. Then the best three networks were selected from each repetition of the GA search and for each one of the metrics.

The output of the first phase was a set of thirtysix network structures. In the second phase for each one of the thirty-six resulting network structures we applied the following procedure. We trained (on Training1 set) and validated (on Validation1 set) the network. Then we used the indicated number of epochs from the validation procedure and based on it we retrained the network on the Training1 plus the Validation1 set. Finally we tested the performance of the network on unseen data (Validation2 set).

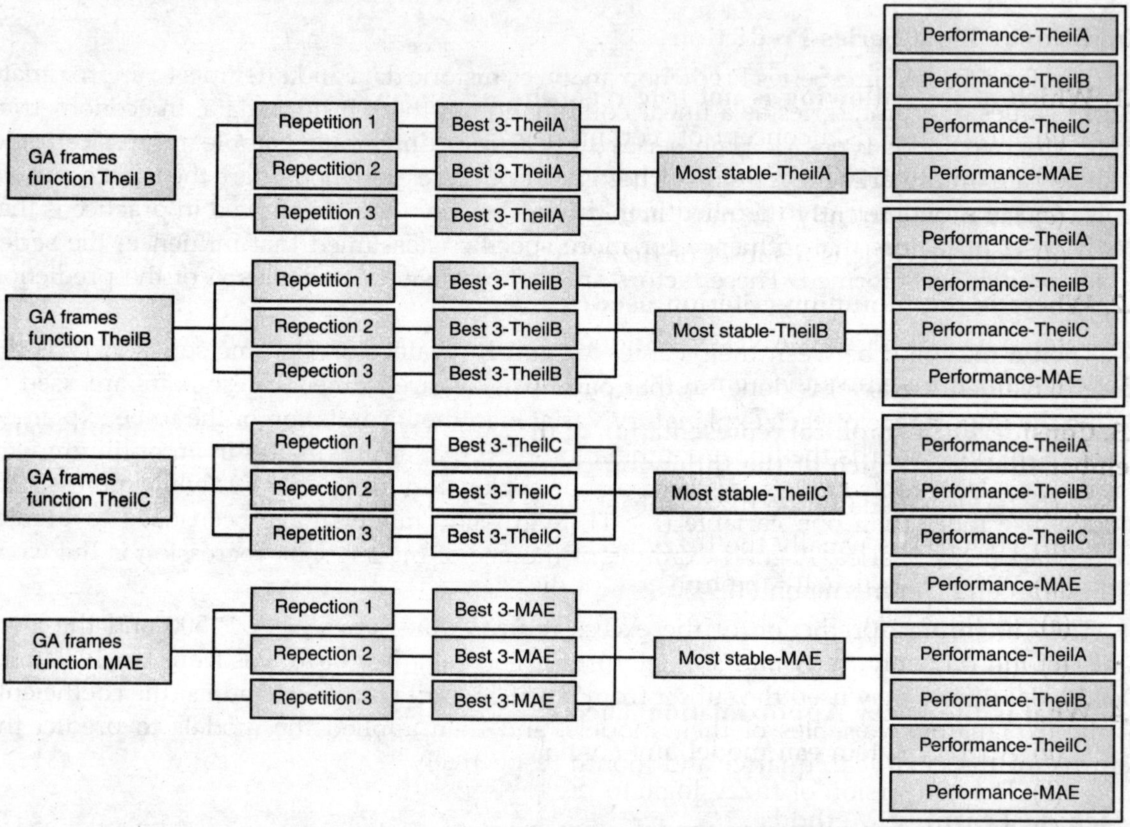

This procedure was repeated 50 times for each network structure for random initializations of its weights. From the nine networks for each performance statistic, we selected the most stable in terms of standard deviation of their performance. Thus the output of the second phase was a set of four network structures. During the third phase for each one of these four networks we applied the following procedure 50 times. We trained each network on the first half of the Training Set and we used the remaining half for validation. Then, using the indicated epochs by the validation procedure, we retrained the network on the complete Training Set. Finally we tested the network on the Test Set calculating all four metrics.

The performance for each network on each metric was measured again in terms of standard deviation and mean of its performance over 50 times that it was trained, validated and tested.

❑❑❑

5 ANNEXURE FIVE

MULTUPLE CHOICE QUESTIONS

Q.1. Which of the following is not true regarding the principles of fuzzy logic?

(*a*) Fuzzy logic is a concept of `certain degree'.

(*b*) Fuzzy logic follows the principle of Aristotle and Buddha.

(*c*) Japan is currently the most active users of fuzzy logic.

(*d*) Boolean logic is a subset of fuzzy logic.

Q.2. Where is the minimum criterion used?

(*a*) When there is an AND operation(*b*)When there is an OR operation

(*c*) In De Morgan's theorem (*d*) None of the above

Q.3. Considering a graphical representation of the `tallness' of people using its appropriate member function, which of the following combinations are true?

(*i*) TALL is usually the fuzzy subset.

(*ii*) HEIGHT is usually the fuzzy set.

(*iii*) PEOPLE is usually the universe of discourse.

(*a*). (*i*) (*ii*) & (*iii*) (*b*). (*i*) & (*ii*) only

(*c*). (*i*), (*iii*) only (*d*). (*ii*) & (*iii*).

Q.4. What is the Fuzzy Approximation Theorem(FAT)?

(*a*) A fuzzy sytem can model any continous sytem

(*b*) The conversion of fuzzy logic to probability.

(*c*) A continous sytem can model a fuzzy system.

(*d*) Fuzzy patches covering a series of fuzzy rules.

Q.5. What is the main difference between probability and fuzzy logic?

(*a*) Fuzzy logic is probability in disguise.

(*b*) Fuzzy logic is the likelihood of an event occuring and probablility is the extent of that event.

(*c*) Probability is ADDITIVE, meaning all its values must add up to one.

(*d*) Probability dissipates with decreasing information.

Q.6. In an adaptive fuzzy system

(*i*) The machine learns as more data are fed into it.

(*ii*) Nueral network is used to find the fuzzy rules.

(*iii*) The system creates rules without the intervention of human beings.

(*a*) (*i*) & (*ii*) & (*iii*) (*b*) (*i*) & (*ii*)

(*c*) (*i*), (*iii*) only (*d*) (*ii*) & (*iii*)

Q.7. What are the following sequence of steps taken in designing a fuzzy logic machine?

(*a*) Fuzzification->Rule evaluation->Defuzzification

(*b*) Rule evaluation->Fuzzification->Defuzzification

(*c*) Fuzzy Sets->Defuzzification->Rule evaluation

(*d*) Defuzzification->Rule evaluation->Fuzzification

Q.8. Fuzzy logic has rapidly become one of the most successful of today's technologies for developing sophisticated control systems. The reason for this is:

(*i*) Fuzzy logic resembles the human way of thinking.

(*ii*) Fuzzy logic enables the ability to generate precise solutions from certain or approximate information.

(*iii*) Fuzzy logic is easy to implement.

(*a*) *i* & *ii* & *iii* (*b*) *i* & *ii*

(*c*) *ii* & *iii* only (*d*) none of the above

Q.9. Given these fuzzy graphs for member functions A and B.

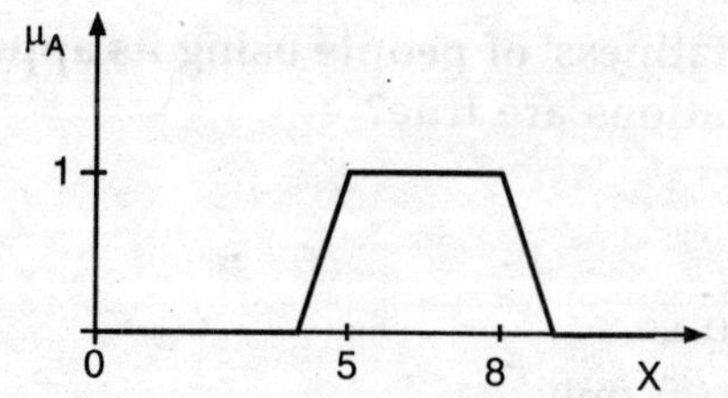

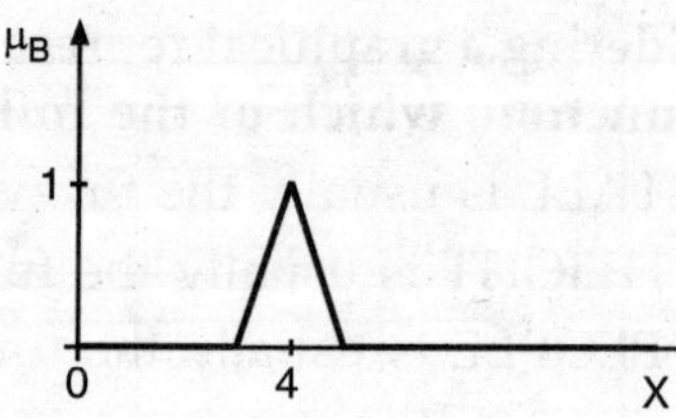

Which of the following graphs yields the result of the operation A OR B.

(*a*)

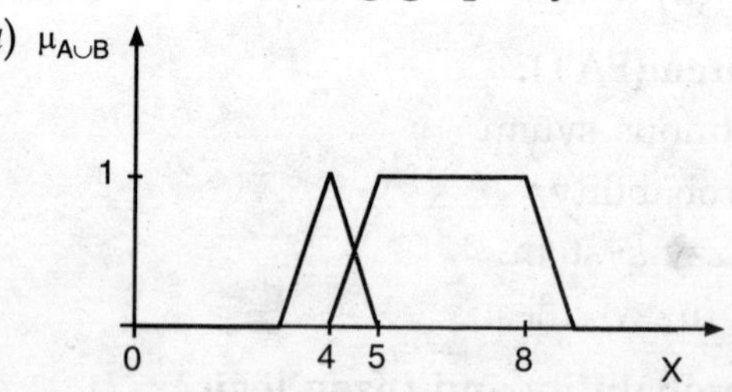

(*b*)

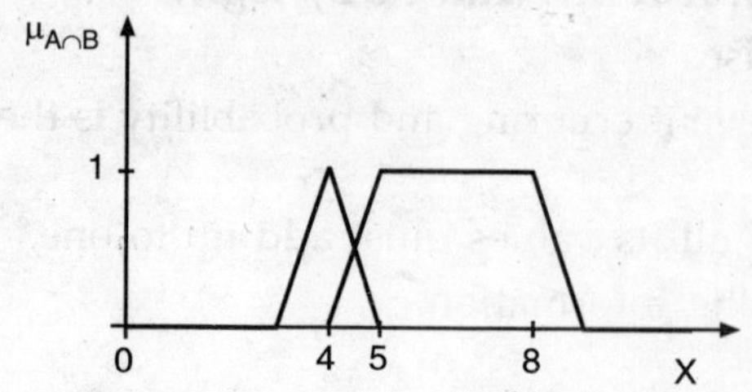

(c) 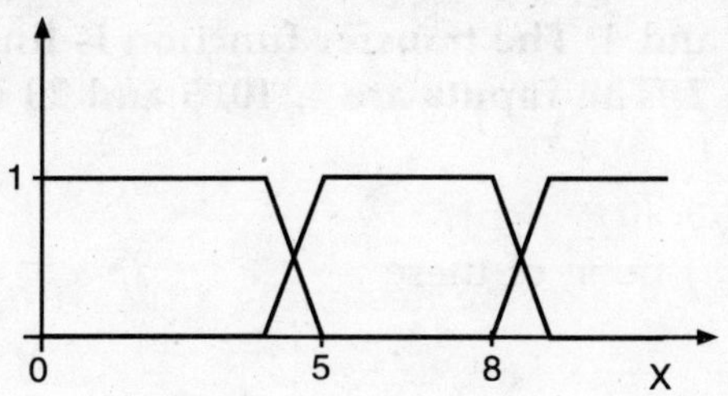

(d) None of the above.

Q.10. Who is the founder of fuzzy logic?

(a) Aristotle (b) Buddha
(c) Zader Lotfi (d) Bart Kosko

Q.11. The network of figure 1, is:

(a) a single layer feed-forward neural network
(b) an autoassociative neural network
(c) a multiple layer neural network

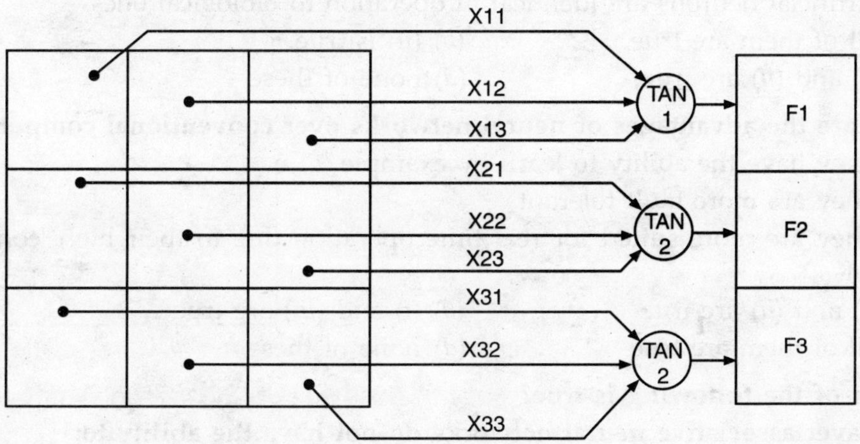

Q.12. A 3-input neuron is trained to output a zero when the input is 110 and a one when the input is 111. After generalisation, the output will be zero when and only when the input is:

(a) 000 or 110 or 011 or 101 (b) 010 or 100 or 110 or 101
(c) 000 or 010 or 110 or 100 (d) none of these

Q.13. A perceptron is:

(a) a single layer feed-forward neural network with preprocessing
(b) an autoassociative neural network

Q.14. An autoassociative network is:

(a) a neural network that contains no loops
(b) a neural network that contains feedback
(c) a neural network that has only one loop
(d) none of these

Q.15. A 4-input neuron has weights 1, 2, 3 and 4. The transfer function is linear with the constant of proportionality being equal to 2. The inputs are 4, 10, 5 and 20 respectively. The output will be:

(*a*) 238 (*b*) 76
(*c*) 119 (*d*) none of these

Q.16. Which of the following is true?

(*i*) On average, neural networks have higher computational rates than conventional computers.
(*ii*) Neural networks learn by example.
(*iii*) Neural networks mimic the way the human brain works.

(*a*) all of them are true (*b*) (*ii*) and (*iii*) are true
(*c*) (*i*), (*ii*) and (*iii*) are true (*d*) none of these

Q.17. Which of the following is true for neural networks?

(*i*) The training time depends on the size of the network.
(*ii*) Neural networks can be simulated on a conventional computer.
(*iii*) Artificial neurons are identical in operation to biological ones.

(*a*) all of them are true. (*b*) (*ii*) is true.
(*c*) (*i*) and (*ii*) are true. (*d*) none of these

Q.18. What are the advantages of neural networks over conventional computers?

(*i*) They have the ability to learn by example
(*ii*) They are more fault tolerant
(*iii*) They are more suited for real time operation due to their high 'computational' rates

(*a*) (*i*) and (*ii*) are true (*b*) (*i*) and (*iii*) are true
(*c*) all of them are true (*d*) none of these

Q.19. Which of the following is true?

Single layer associative neural networks do not have the ability to:

(*i*) perform pattern recognition
(*ii*) find the parity of a picture
(*iii*) determine whether two or more shapes in a picture are connented or not

(*a*) (*ii*) and (*iii*) are true (*b*) (*ii*) is true
(*c*) all of them are true (*d*) none of these

Q.20. The network shown in Fig. 1 is trained to recognize the characters H and T as shown below:

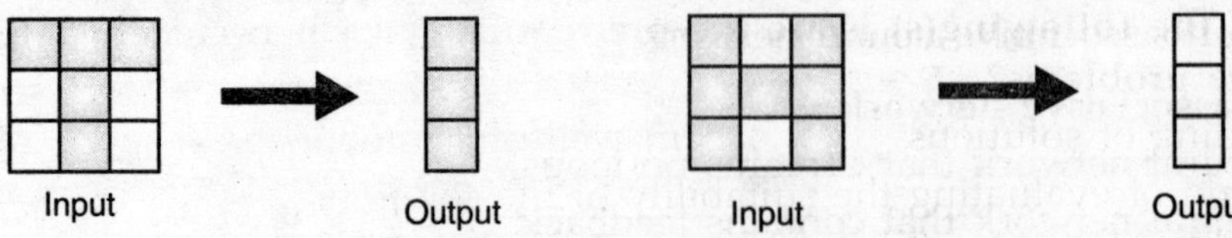

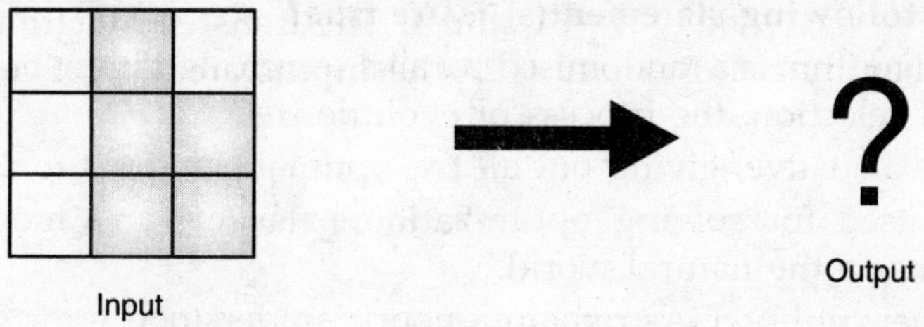

If the following pattern was given

What would be the output of the network?

(*a*) 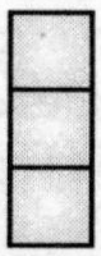

(*b*) 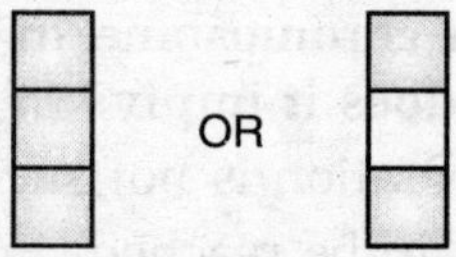

(*c*) 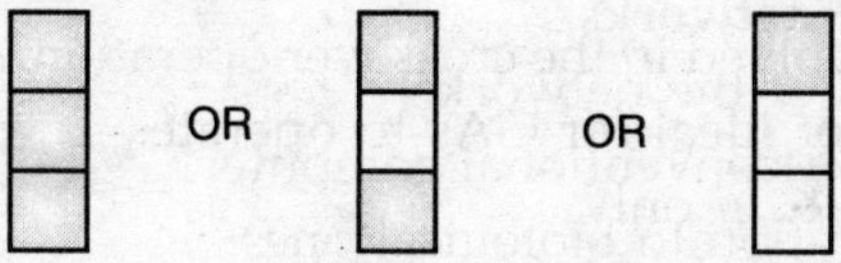

Q.21. Which of the following(s) is/are found in Genetic Algorithms?

(*i*) evolution (*ii*) selection

(*iii*) reproduction (*iv*) mutation

(*a*) *i* & *ii* only (*b*) *i, ii* & *iii* only

(*c*) *ii, iii* & *iv* only (*d*) all of the above

Q.22. Matching between terminologies of Genetic Algorithms and Genetics:

Genetic Algorithms	**Genetics (biology)**
(*a*) representation structures	(*i*) external disturbance, such as cosmic radiation
(*b*) crossover	(*ii*) chromosomes
(*c*) mutation	(*iii*) survivability
(*d*) selection	(*iv*) sexual reproduction

(*a*) .... (*b*) ....

(*c*) .... (*d*) ....

Q.23. Where are Genetic Algorithms applicable?

(*i*) real time application (*ii*) biology

(*iii*) Artificial Life (*iv*) economics

(*a*) *i, ii* & *iii* only (*b*) *ii, iii* & *iv* only

(*c*) *i, iii* & *iv* only (*d*) all of the above

Q.24. Which of the following(s) is/are the pre-requisite(s) when Genetic Algorithms are applied to solve problems?

(*i*) encoding of solutions (*ii*) well-understood search space

(*iii*) method of evaluating the suitability of the solutions

(*iv*) contain only one optimal solution

(*a*) *i* & *ii* only (*b*) *ii* & *iii* only

(*c*) *i* & *iii* only (*d*) *iii* & *iv* only

Q.25. Which of the following statement(s) is/are true?

(*i*) Genetic Algorithm is a randomised parallel search algorithm, based on the principles of natural selection, the process of evolution.

(*ii*) GAs are exhaustive, giving out all the optimal solutions to a given problem.

(*iii*) GAs are used for solving optimization problems and modeling evolutionary phenomena in the natural world.

(*iv*) Despite their utility, GAs remain a poorly understood topic.

(*a*) *i, ii* & *iii* only (*b*) *ii, iii* & *iv* only

(*c*) *i, iii* & *iv* only (*d*) all of the above

Q.26. If crossover between chromosome in search space does not produce significantly different offsprings, what does it imply? (if offspring consist of one half of each parent)

(*i*) The crossover operation is not succesful.

(*ii*) Solution is about to be reached.

(*iii*) Diversity is so poor that the parents involved in the crossover operation are similar.

(*iv*) The search space of the problem is not ideal for GAs to operate.

(*a*) *ii, iii* & *iv* only (*b*) *ii* & *iii* only

(*c*) *i, iii* & *iv* only (*d*) all of the above

Q.27. Which of the following comparison is true?

(*a*) In the event of restricted acess to information, GAs win out in that they require much fewer information to operate than other search.

(*b*) Under any circumstances, GAs always outperform other algorithms.

(*c*) The qualities of solutions offered by GAs for any problems are always better than those provided by other search.

(*d*) *G* As could be applied to any problem, whereas certain algorithms are applicable to limited domains.

Q.28. Which of the following statement(s) is/are true?

(*i*) Artificial Life is analytic, trying to break down complex phenomena into their basic components.

(*ii*) Alife is a kind of Artificial Intelligence (AI).

(*iii*) Alife pursues a two-fold goal: increasing our understanding of nature and enhancing our insight into artificial models, thereby providing us with the ability to improve their performance.

(*iv*) Alife extends our studies of biology, life-as-we-know-it, to the larger domain of possible life, life-as-it-could-be.

(*a*) *i* & *ii* only (*b*) *iii* & *iv* only

(*c*) *i, ii* & *iii* only (*d*) all of the above

Q.29. Where is Artificial Life applicable?

(*i*) film (movie, video) production (*ii*) biology

(*iii*) robotics (*iv*) air traffic control

(*a*) *i, ii* & *iii* only (*b*) *ii, iii* & *iv* only

(*c*) *i, iii* & *iv* only (*d*) all of the above

Q.30. Who can be benefited from Alife?

(*i*) children (*ii*) designers

(*iii*) artists (*iv*) patients

(*a*) *i, ii & iii* only (*b*) *ii, iii & iv* only

(*c*) *i, iii & iv* only (*d*) all of the above

ANSWERS

1. (*b*)	2. (*a*)	3. (*b*)	4. (*a*)	5. (*c*)	6. (*b*)	7. (*a*)
8. (*b*)	9. (*a*)	10. (*c*)	11. (*a*)	12. (*c*)	13. (*a*)	14. (*b*)
15. (*b*)	16. (*a*)	17. (*c*)	18. (*c*)	19. (*a*)	20. (*b*)	21. (*a*)
22. (*a*) (*ii*), (*b*) (*iv*), (*c*) (*i*), (*d*) (*iii*)		23. (*b*)		24. (*c*)	25. (*c*)	26. (*b*)
27. (*a*)	28. (*b*)	29. (*d*)	30. (*d*)			

HINTS AND SOLUTIONS

11. (*a*) The network of figure 1, is a single layer feed-forward neural network because there is only one neuron between any input and output. The network is not autoassociative i.e. doesn't have feedback because there are no loops in it.

12. (*c*) The truth table before generalisation is:

Inputs	Output
000	$
001	$
010	$
011	$
100	$
101	$
110	0
111	1

where $ represents don't know cases and the output is random.

After generalisation, the truth table becomes:

Inputs	Output
000	0
001	1
010	0
011	1
100	0
101	1
110	0
111	1

Therefore, the output will be zero when the input is 000 or 010 or 110 or 100.

13. (*a*) The perceptron is a single layer feed-forward neural network. It is not an autoassociative network because it has no feedback and is not a multiple layer neural network because the preprocessing stage is not made of neurons.

14. (*b*) An autoassociative network is equivalnet to a neural network that conatins feedback. The number of feedback paths(loops) does not have to be one.

15. (*b*) The output is found by multipling the weights with their respective inputs, summing the results and multipling with the trasfer function. Therefore:

$$\text{Output} = 2 * (1*4 + 2*10 + 3*5 + 4*20) = 238$$

16. (*a*) Neural networks have higher computational rates than conventional computers because a lot of the operation is done in parallel.

Note: that is not the case when the neural network is simulated on a computer. The idea behind neural nets is based on the way the human brain works. Neural nets cannot be programmed, they cam only learn by examples.

17. (*c*) The training time depends on the size of the network; the number of neuron is greater and therefore the the number of possible 'states' is increased. Neural networks can be simulated on a conventional computer but the main advantage of neural networks - parallel execution - is lost. Artificial neurons are not identical in operation to the biological ones. We don't know yet what the real neurons do in detail.

18. (*c*) Neural networks learn by example.

They are more fault tolerant because they are always able to respond and small changes in input do not normally cause a change in output.

Because of their parallel architecture, high computational rates are achived.here

19. (*a*) Pattern recognition is what single layer neural networks are best at but they don't have the ability to find the parity of a picture or to determine whether two shapes are connected or not.

20. (*b*)The top square of the output is black because the top pattern differs in two squares from a T and in 3 squares from an H.

The middle square is not defined because the middle row of the input defers the same amount from both T and H(differs in 1). Therefore, the output can be either black or white.

The bottom square is black because it differs from a T and in 2 from an H.

❑❑❑

MODEL QUESTION PAPERS AND PREVIOUS YEARS QUESTION PAPERS

MODEL QUESTION PAPERS

MODEL QUESTION PAPER-I

B.Tech.

SOFT COMPUTING

(Neural Networks, Fuzzy Logic and Genetic Algorithm)

Time: 3 Hours ***Total Marks: 100***

1. Attempt Any FOUR Parts:

(*a*) Explain Neural Network architecture and learning process.

(*b*) What's perceptron and explain perceptron convergence Algorithm.

(*c*) Explain Activation function and it's types.

(*d*) What is associative Memory? A hetro associative network is given as

S1 = (0 1 1 0) t_1 (1, 0)

S2 = (0, 1, 0, 0) t_2 (1, 0)

Find the weight matrix.

(*e*) Artificial Intelligence can be used in Neural Network or Not Justify your answer.

2. Attempt Any FOUR Parts:

(*a*) Give the comparison between Single layer Feed forward network, Multilayer feed forward network and Recurrent Network.

(*b*) Discus selection of Various parameters in BPN (Back Propagation Neural Network).

(*c*) What do you mean by Back propagation Learning Algorithm?

(*d*) Define Augmented BPN.

(*e*) How tuning parameters affect the BPN.

3. Attempt Any TWO Parts:

(*a*) What is single layer perceptron? Give its architecture.

(*b*) What is the difference between crisp logic and fuzzy logic? Also write down its Application.

(*c*) Define the following .

(*i*) Properties of the fuzzy set (*ii*) Various methods of Defuzzyfication.

4. Attempt Any TWO Parts:

(*a*) Define Membership function in detail. Also define its role and Applications.

(*b*) What are fuzzy quantifiers? Differentiate between absolute and relative quantifiers.

(*c*) Discus one of the application of fuzzy logic in detail.

5. Attempt Any TWO Parts:

(*a*) Draw and discus the flow chart of the genetic algorithm. Also write its importance.

(*b*) Define the following.

(*i*) Fitness function (*ii*) Genetic Operator

(*iii*) Crossover

(*c*) Write the working principle of Genetic Algorithm and application of Genetic algorithm.

❑❑❑

MODEL QUESTION PAPER-II

B.Tech.

SOFT COMPUTING

(Neural Networks, Fuzzy Logic and Genetic Algorithm)

Time: 3 Hours **Total Marks: 100**

1. Attempt Any FOUR Parts:

(*a*) What is neuron nerve structure and synapse.

(*b*) Define artificial neuron and its model.

(*c*) Define Auto associative and hetro associative memory.

(*d*) What is perceptron and explain perceptron convergence algorithm.

(*e*) Define single layer and multilayer feed forward networks.

2. Attempt Any FOUR Parts:

(*a*) What is back propagation algorithm. Discuss its application.

(*b*) Describe the perceptron model and X-OR problem in context of perceptron.

(*c*) Describe the method of stiepust discent.

(*d*) What are the limitations of perceptron model.

(*e*) Discuss the factors affecting the Back Propagation Algorithm.

3. Attempt Any FOUR Parts:

(*a*) Explain Partition and Couring of a set.

(*b*) What is the difference between fuzzy set and crisp set explain.

(*c*) Describe the properties of crisp set.

(*d*) Given two fuzzy sets

$$A = \left\{\frac{1}{2} + \frac{0.5}{2} + \frac{0.3}{4} + \frac{0.2}{5}\right\}$$

$$B = \left\{\frac{0.5}{2} + \frac{0.7}{2} + \frac{0.2}{4} + \frac{0.4}{5}\right\}$$

Find: (*i*) $\overline{A}, \overline{B}$

(*ii*) $A \cup B$

(*iii*) $A \cap B$

(*e*) Explain the fuzzy sets operation. With examples.

4. Attempt Any TWO Parts:

(*a*) What is Membership functions, explain with example.

(*b*) Explain fuzzy (Rule based) systems with example.

(*c*) What is the use of Inference of fuzzy logic. Discuss if then rules.

5. Attempt Any TWO Parts:

(*a*) Write basic concept of genetic algorithm

(*b*) Define Genetic operators

(*c*) Describe the process genetic algorithm through the flow chart.

□□□

MODEL QUESTION PAPER-III

B.Tech.

SOFT COMPUTING

(Neural Networks, Fuzzy Logic and Genetic Algorithm)

Time: 3 Hours ***Total Marks: 100***

1. Attempt Any FOUR Parts:

(*a*) Differentiate between Neural Network and Expert system.

(*b*) Give the comparison between artificial and biological neural network.

(*c*) Discuss the activation function like: Heaviside function, Identity function.

(*d*) Discuss supervised, reinforced and unsupervised learning process.

(*e*) A hetero associative network is given as

$S_1 = (1, 1, 0, 0)$ t_1 (1, 0)

$S_2 = (0, 1, 0, 0)$ t_2 (1, 0)

$S_3 = (0, 0, 1, 1)$ t_3 (0, 1)

$S_4 = (0, 0, 1, 1)$ t_4 (0, 1)

Find weight matrix.

(*f*) Differentiate between Hetero associative and Auto associative memory.

2. Attempt Any FOUR Parts:

(*a*) Discuss the effect of learning rule coefficient in BPN.

(*b*) Discuss perceptron model with its types.

(*c*) Derive the back propagation algorithm.

(*d*) Explain the training algorithm for Auto associative memory network.

(*e*) What are the different typical Non-linear activation operators.

(*f*) Briefly describe the method of steepest descent.

3. Attempt any TWO parts:

(*a*) Discuss the characteristic and graphs for membership function.

(*b*) Discuss crisp relations and operation of crisp relation.

(*c*) Elaborate the properties of fuzzy sets.

4. Attempt Any TWO Parts :

(*a*) Discuss fuzzy rule based system with fuzzy inferences.

(*b*) Discuss fuzzification and defuzzification.

(*c*) In brief explain the membership value assignments.

5. Attempt Any TWO Parts:

(*a*) Discuss the following: (*i*) Creation of offspring

(*ii*) Search space

(*b*) What are the different methods for encoding in Genetic algorithm.

(*c*) Discuss the operators of Genetic algorithm like Reproduction, Crossover and Mutuation.

❑❑❑

PREVIOUS YEARS QUESTION PAPERS

B.Tech.

(SEM VIII) EVEN SEMESTER THEORY EXAMINATION, 2006-07

NEURAL NETWORKS

Time: 3 Hours ***Total Marks: 100***

Note: *Attempt all questions. Each question carries 20 marks. Write suitable assumptions, if any.*

1. Attempt Any TWO Parts:

(*a*) Discuss analogy between biological and artificial neural network. 10

(*b*) What do you understand by the terms :

1. Learning
2. Generalization
3. Function approximation in context of ANN. 3+3+4

(*c*) With help of a suitable diagram discuss functioning of a simple artificial neuron. Explain how the functionality is affected if two such neuron are connected in series. 6+4

2. Attempt Any TWO Parts:

(*a*) Discuss the functioning of a perceptron. Consider the classification problem defined below

$\{p_1 = [-1, 1], t_1 = 1\}$ $\{p_2 = [0, 0], t_2 = 1\}$ $\{p_3 = [1, -1], t_3 = 1\}$

$\{p_4 = [1, 0], t_4 = 0\}$ $\{p_5 = [0, 1], t_5 = 0\}$

$p_i \rightarrow$ *i*-th input

$t_i \rightarrow$ *t*-th target

Design a single neuron perceptron to solve this problem. You need to describe architecture of perceptron, its decision boundary. 10

(*b*) Discuss back propagation algorithm for a multilayer network. 10

(*c*) Discuss the functioning of recurrent networks. What are strengths and limitations of such network. 6+4

3. Attempt Any TWO Parts:

(*a*) Consider LVQ network shown below :

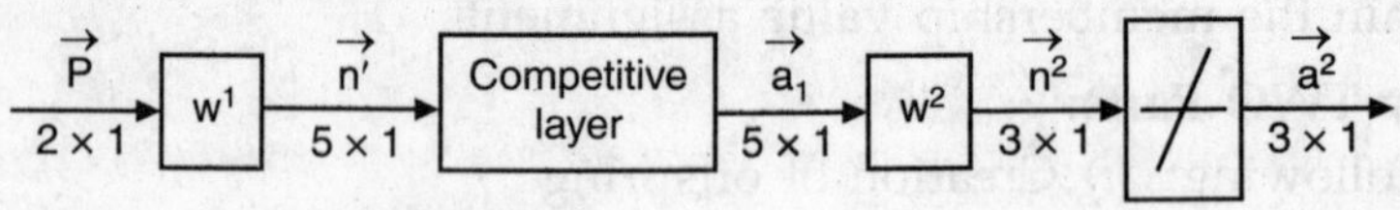

$n_i' = -\|_i w' - P\|$

$a' = \text{competitive net } (n')$

$$a_i' = \begin{cases} 1 & i = \text{neuron with largest net input} \\ 0 & \text{otherwise} \end{cases}$$

$$a^2 = w^2 a'$$

$$w^1 = \begin{bmatrix} 0 & 0 \\ 1 & -1 \\ 1 & 1 \\ -1 & 1 \\ -1 & -1 \end{bmatrix} \quad w^2 = \begin{bmatrix} 1 & 0 & 0 & 0 & 0 \\ 0 & 1 & 0 & 0 & 0 \\ 0 & 0 & 1 & 1 & 1 \end{bmatrix}$$

Determine regions of each class and draw a graph to illustrate this. 10

(*b*) Write short notes on :

(1) Vector quantization

(2) Unsupervised learning and its relative merits and limitations with respect to supervised learning 4+6

(*c*) Discuss Hebb rule in context of :

(1) Supervised learning

(2) Unsupervised learning. 4+6

4. Attempt Any TWO Parts:

(*a*) What do you understand by associative networks? In this context explain working of a Hopfield network. 10

(*b*) Describe brain-in-a-box model. Compare it with a recurrent network. 10

(*c*) Derive expression for Boltzmann learnng rule. 10

5. Attempt Any TWO Parts:

(*a*) Define gradient. Using steepest descent rule to the following function

$$f(x) = x_1^2 + 5x_1x_2 + 10x_2^2$$

determine first three points of trajectory starting from

$$x_0 = [0.5, 0.5]$$ 10

(*b*) Discuss the role of

(1) Selection

(2) Cross-over

(3) Mutation

in context of genetic algorithm.

(*c*) Write an algorithm to implement simulated annealing. 10

□□□

B.Tech.

(SEM IV) EVEN SEMESTER THEORY EXAMINATION, 2009-10

INTRODUCTION TO SOFT COMPUTING

(Neural Networks, Fuzzy Logic and Genetic Algorithm)

Time: 3 Hours **Total Marks: 100**

Note: *(1) Attempt all questions.*
(2) All questions carry equal marks.

1. Attempt Any FOUR Parts:

Q.1. (*a*) Define Neural Network Architecture.

Ans. An artificial neural network is defined as a data processing system consisting a large number of simple highly interconnected processing elements (artificial neuron) in an architecture inspired by the structure of the brain. Generally an artificial neural network structure can be represented by directed graph.

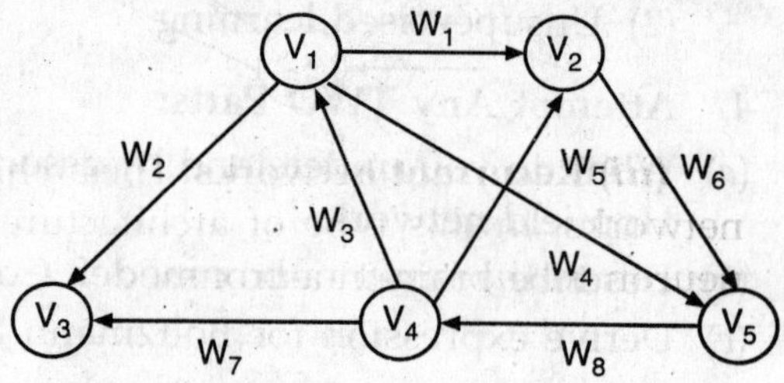

Graph G is a set of vertices V 4 edges E. In all graph the vertices may represents neurons either input or output. The edges of graph may represent the links. The edges are labelled by related weights attached to the synaptic links.

We have several classes of neural networks classified are to the learning mechanism. There are three different classes on neural network. They are–

1. Single layer field forward networks
2. Multi layer feed forward networks
3. Recurrent networks

Q.1. (*b*) Give a comparison between single layer feed forward networks, Multi-layer Feed Forward Network and Recurrent Network.

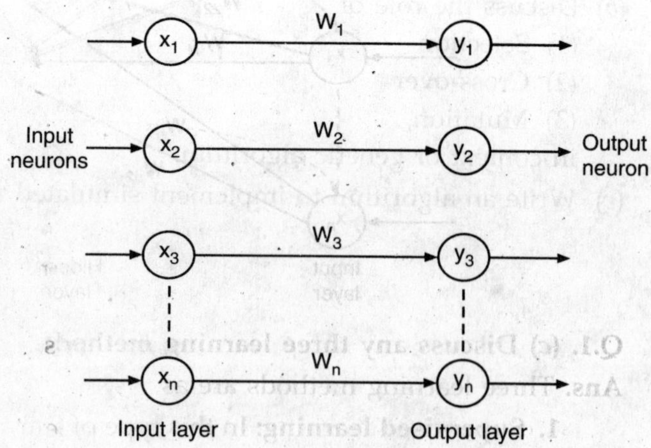

Ans.(*i*) Single layer feed forward networks: In this type of network we have single input layer 4 single output layer. The output layer receives the input from input layer and provide output as per the input neurons with their related weights. There is only one output layer that is why we called this type of network architecture as single layer feed forward networks.

***(ii)* Multilayers feed forward networks:** This layer as its name indicated have multiple layers. In this architecture we have a hidden output layer in between input and output layer. Hidden layer used to do some intermediate computation before directing the input to the output layer.

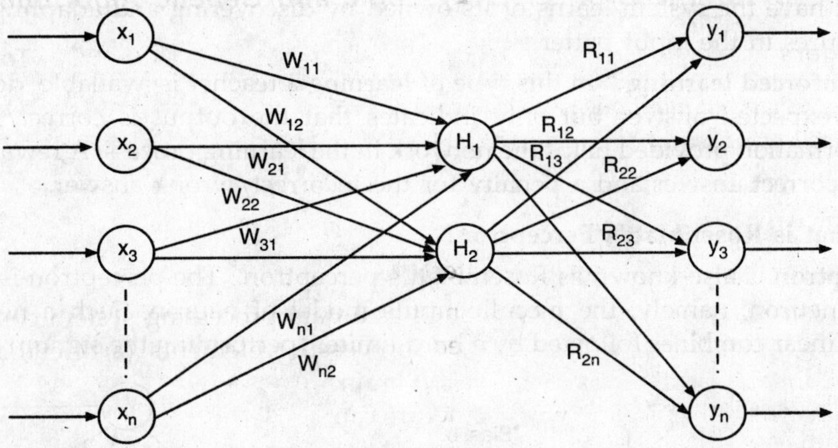

***(iii)* Recurrent networks:** The architecture of recurrent networks differ from feed forward networks. In this type of architecture there is atleast one feedback link from output layer neuron to the input neurons.

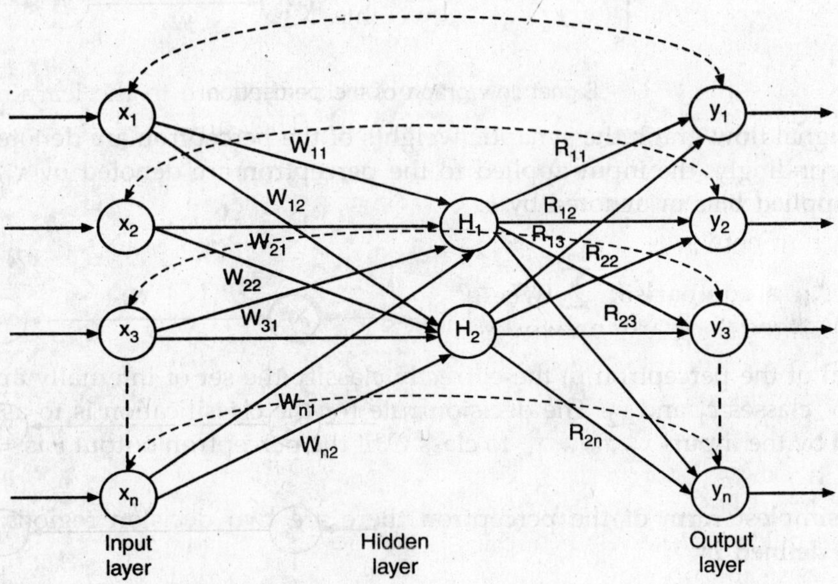

Q.1. *(c)* Discuss any three learning methods.

Ans. Three learning methods are as:

1. **Supervised learning:** In this type of learning, every input pattern is used to train the network is associated with an output pattern which is the target or the desired pattern. A teacher or supervisor is assumed to be present. During the learning process, when a comparison is made between the network computed output and the correct

expected output to determine the error. The error then can be used to change the network parameter, which result in an improvement in performance.

2. **Unsupervised learning:** In this type of learning the target output is not presented to the network. It is as if there is no teacher/supervisor to present the desired pattern and have the system learns of its owned by discovering and adapting to structural features in the input patterns.
3. **Reinforced learning** " In this type of learning a teacher is available, doesnot present the expected answer but only indicates that the output is correct/incorrect. The information provided helps the network in the learning process. A reward is provided for correct answer and a penalty for the incorrect/wrong answer.

Q.1. (*d*) What is Rosenblatt's Perceptron?

Ans. Perceptron is also known as Rosemblatt's perceptron.. The perceptron is build around non-linear neuron, namely, the mccullochputs model of neuron. Such a neuronal model consists of linear combines followed by a hard limiter (performing the signum fert) as shown in Fig.

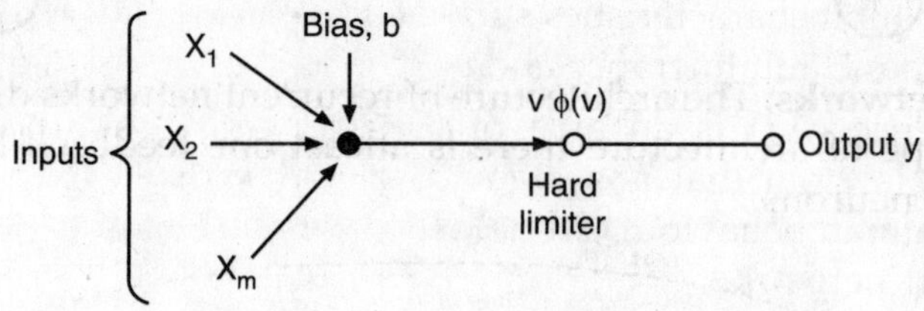

Signal flow graph of the perceptron

In the signal flow graph the synaptic weights of the perceptron are denoted by w_1, w_2 ... w_m. Correspondingly, the input applied to the perceptron are denoted by x_1, x_2 ... x_m. The internally applied bias in denoted by b.

$$V = \sum_{i=1}^{m} w_i x_i + b$$

The goal of the perceptron in the correctly classify the set of infernally applied x_1, x_2 ... x_m into two classes c_1 and c_2. The decision rule for the classification is to assign the point represented by the inputs x_1, x_2, ... x_m to class C_1 if the perceptron output y is + 1 and to class C_2 if it is – 1.

In the simplest form of the perceptron there are two decision regions separated by hyperplane defined by

$$\sum_{i=1}^{m} w_i x_i + b = 0$$

This is shown in Fig.

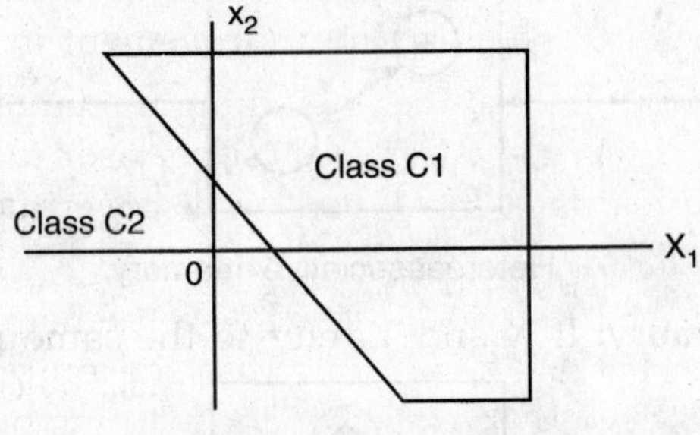

Illustration of the hyperplane decision boundary for a two dimensional, two class pattern classification.

Q.1. (*e*) Artificial Intelligence can be used in Neural Networks or not. Justify your answer.

Ans. Neural networks are simplified model of the biological nervous system and therefore have drawn their motivation from the kind of computing performance by a human brain. It consists of neuron also known as processing elements.

Neural network exhibit characteristics such as mapping capabilities or pattern association generalization, lobustness fault tolerance and parallel and high speed information processing.

Artificial intelligence (A1) is an area of computer science concerned with designing intelligence computer system that is systems that exhibits that characteristic we associate with intelligence in human behaviour. AI helps in learning inhalation, creativity and solving the problem of neural network.

Q.1. (*f*) Define Associative Memory and write down its applications.

Ans. An associated memory is a storehouse of associated patterns which are encoded in some form. When the storehouse is triggered or incited with a pattern the associated pair is recalled or output.

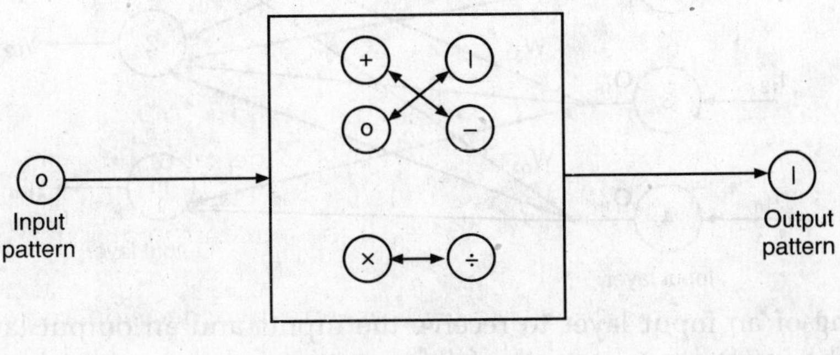

Associated pattern pairs

In this Figure (+, –), (0, 1), (*x*, ÷) are associated pattern pairs. The association represented using '↔' symbol are stored in memory. When the input pattern O, is triggered its associated pattern pair *i.e.*, is recalled (output).

The associated net is consists of two types:

1. Heteroassociative memory: If the associated pattern pair (*x*, *y*) are different 4 if the model recalls a *y* for given *Y* or vice versa.

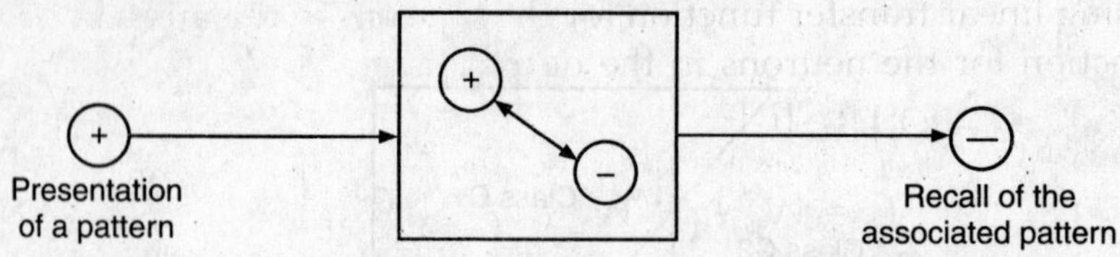

Heteroassociative memory

2. Autoassociative memory: If X and Y refer to the same pattern.

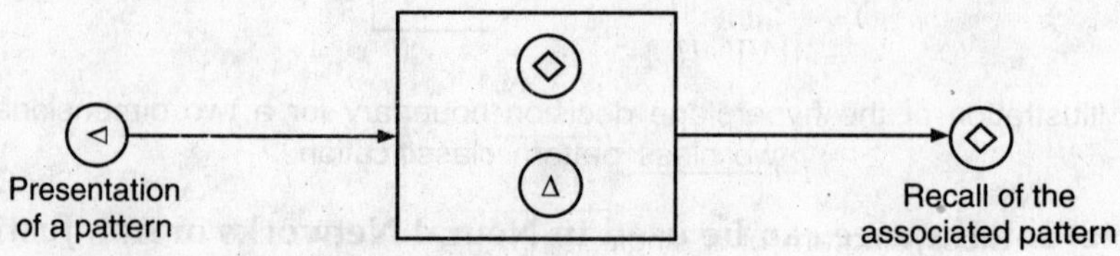

Autoassociative memory

Applications of Associative memory

1. Autoassociative memory are useful for image refinement that is given a distorted or a partial pattern than the output will be whole pattern stored i its perfect form.
2. Heteroassociative memories are useful for association of patterns.

2. Attempt any four parts:

Q.2. (*a*) Define Single Layer Artificial Neural Network.

Ans. A single layer feedforward neural network shown in figure 1.

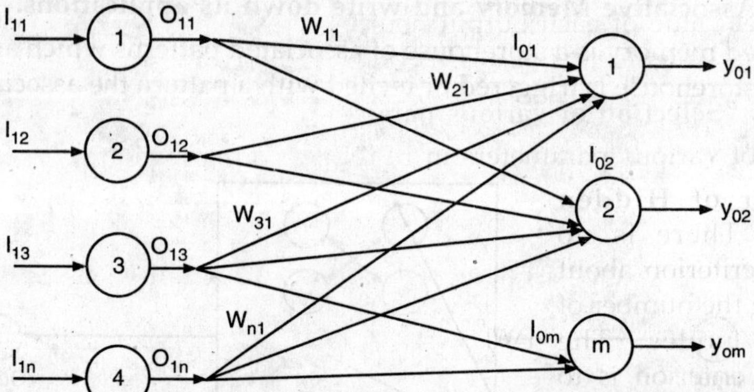

It consisting of an input layer to receive the inputs and an output layer to output the vectors sep. The input layer consists of 'n' neurons and the output layer consists of 'm' neurons. Indicate the weight of the synapse connecting with input neuron to the ith output neuron as w_{ij}. The inputs of the input layer and the corresponding output of the output layer are given as–

$$I_1 = \begin{Bmatrix} I_{I1} \\ I_{I2} \\ \vdots \\ I_{In} \end{Bmatrix}_{n\times 1} \quad O_o = \begin{Bmatrix} O_{o1} \\ O_{o2} \\ \vdots \\ O_{om} \end{Bmatrix}_{m\times 1}$$

We are using linear transfer function for the neurons in the input layer and the unipolar sigmoidal function for the neurons in the output layer.

$$\therefore \quad \underset{n\times 1}{\{O_I\}} = \underset{m\times 1}{\{I_I\}}$$

$$I_{oj} = W_{1j}\, I_{I1} + W_{2j}\, I_{I2} + \dots W_{nj}\, I_{IN} +$$

Hence, the input to the output layer can be given as

$$\underset{m\times 1}{\{I_O\}} = \underset{m\times n}{[W]^T \{O_I\}}$$

$$= \underset{n\times n}{[W]^T \{I_I\}}$$

Block diagram of a single layer feed forward neural network

Q.2. (b) What do you understand by the term 'training of Neural Network'?

Ans. The term 'training of neural network' is to adjust the weights to that the application of a set of input produces the desired outputs. To accomplish this the network is usually trained with a large number of input output pairs.

The training of an artificial neural network involves two passes.

1. In the forward pass, the input signals move forward from the network input to the output. In the backward pass, the calculated error signals propagate backward through the network, where they are used to adjust the weights.
2. In the reverse pass, the weights of the output neuron layer are adjusted first, since the target value of each output neuron is available to guide the adjustment of the associated weights.

Q.2. (c) Discuss "Selection of various parameter in BPN."

Ans. Selection of various parameters in BPN (Back Propagation Network).

1. **Number of Hidden Nodes:** There is no general criterion about deciding the number of hidden nodes. The guiding criterion is to select the minimum modes which would not impair the network performance so that the memory demand for storing the weights can be kept minimum. Mischandani and cao (1989) have proved that then number of separable regions in the input space, M is a function of the number of hidden nodes H in BPN and H = M – 1. While Huang and Huang (1991) argue that the in terms of learning efficiency, the optimal number of hidden neurons to be $H = K - 1$, where k is the number of elements in the learning set.

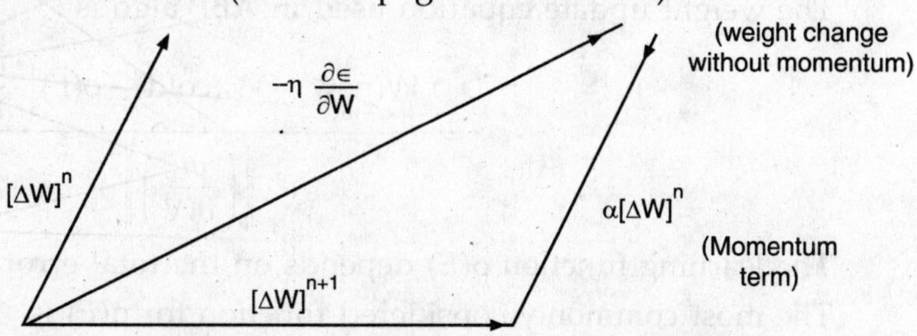

Influence of momentum term on weight change

2. **Momentum coefficient α:** The method of reducing the training time is the use of momentum factor α because it enhances the training process. The influence of the momentum on weight change in shown in figure 1. Where,

$$[\Delta W]^{n+1} = \eta \frac{\partial E}{\partial W} + \alpha [\Delta W]^n$$

3. **Sigmoidal gain λ:** When the weights become large and force the neuron to operate in a region where sigmoldal function is very flat, a better method of coping the network palalysis is to adjust the sigmoidal gain. By decreasing this scaling factor, we effectively spread out sigmoidal function on wide range so that the training proceeds faster.

4. **Learning coefficient η** = The learning coefficient cannot be negative because this would cause the change of weight vector to move away from ideal weight vector position. If the learning coefficient is zero, no learning takes place and hence the learning coefficient must be positive. If learning coefficient is equal to 2 than the network is unstable and if the η is greater than 1, the weight vector will overshoot from its ideal position and oscillate, the learning coefficient must be between zero 4 one.

Q.2. (*d*) What is adaptive Back Propagation?

Ans.Adaptive back propagation (Accelerated learning): The use of adaptive backpropagation (ABP) is developed by Alexender et al (1994) of a fault detection and identification system for a process composed of direst current motor, a centrifugal pump and the associated piping system. In the ABP learning algo, the network weight update rule is chosen such that the error function is forced to behave in a certain manner that accelerate convergence.

In the standard BPN, the weight update is given by

$$\Delta W = W(\text{new}) - W(\text{cold}) - \eta \frac{\partial E}{\partial W}$$

The weight update equation used in ABP algo is

$$= \frac{W(\text{new}) - W(\text{cold}) - \rho(E)\left(\dfrac{\partial E}{\partial W}\right)}{\left\| \left(\dfrac{\partial E}{\partial W}\right) \right\|^2}$$

The learning function $\rho(E)$ depends on the total error E.

The most commonly considered function for $\rho(E)$ is

$$\boxed{\rho(E) = \begin{bmatrix} \eta \\ \eta E \\ \eta \tan h\,(E) \\ \eta \tan h^{-1}\left(\dfrac{E}{E_0}\right) \end{bmatrix}}$$

where η & $\in_0$ are constant non negative no. rep. the learning rate 4 the error normalization factors resp. Since $\rho(\in) = \eta\in$ seens to provide most acceleration it is normally used.

Hence the weight update rule is given by

$$\Delta\omega = \frac{-\eta\epsilon\left(\frac{\partial\varepsilon}{\partial\omega}\right)}{\left\|\frac{\partial\epsilon}{\partial\omega}\right\|}$$

Q.2. (*e*) How Tuning Parameters affect the Basic Propagation Neural Network?

Ans. Effect of tuning parameters of the back propagation neural network.

A proper selection of tuning parameters such as

Momentum factor

Learning coefficient

Sigmoidal gain

Thrushold value

are required for efficient learning and designing of a stable network weight adjustment is made based on the momentum method. The momentum factor has a significant role in deciding the values of learning rate that will produce rapid learning. It determines the step size of change in weight or biases. If momentum factor is zero, the smoothing is minimum and the entire weight adjustment comes from the newly calculated change. If this is 1, new adjustment is ignore and the previous one is repeated.

The choice of learning coefficient is a tricky task in BP algo. The range of learning coeff that will produce rapid training depends on the number and types of input pattern. An emperical formula is select learning coefficient is given as –

$$\eta = \frac{1.5}{(N_1^2 + N_2^2 + \ldots N_m^2)}$$

where N_1 is the number of patterns of type and m is the number of different pattern types.

Sigmoidal gain: If sigmoidal function is selected, the input-output relationship of the neuron can be set as

$$O = \frac{1}{\left(1 + e^{-\sqrt{(1+\theta)}}\right)} \quad \ldots(1)$$

Where λ = scaling factor k is as sigmoidal classmate gain.

Threshold value: θ in the sigmoidal function is commonly called as threshold value of a neuron, or the bias or the noise factor. A neuron fires or generates an output if the weighted sum of the input exceeds the threshold value.

Q.2. (*f*) Define augmented BPN.

Ans. Augmented BPN: The principle of augmenting the network:

1. The augmented neurons are highly sensitive in the boundary domain, thereby facilitating the construction of accurate mapping in the models boundary domain.
2. The network denotes each input variable with multiple input neuron thus allowing a highly interactive functions on hidden neutron to be easily formed.

The architecture of the augmented neural N is shown in figure 1.

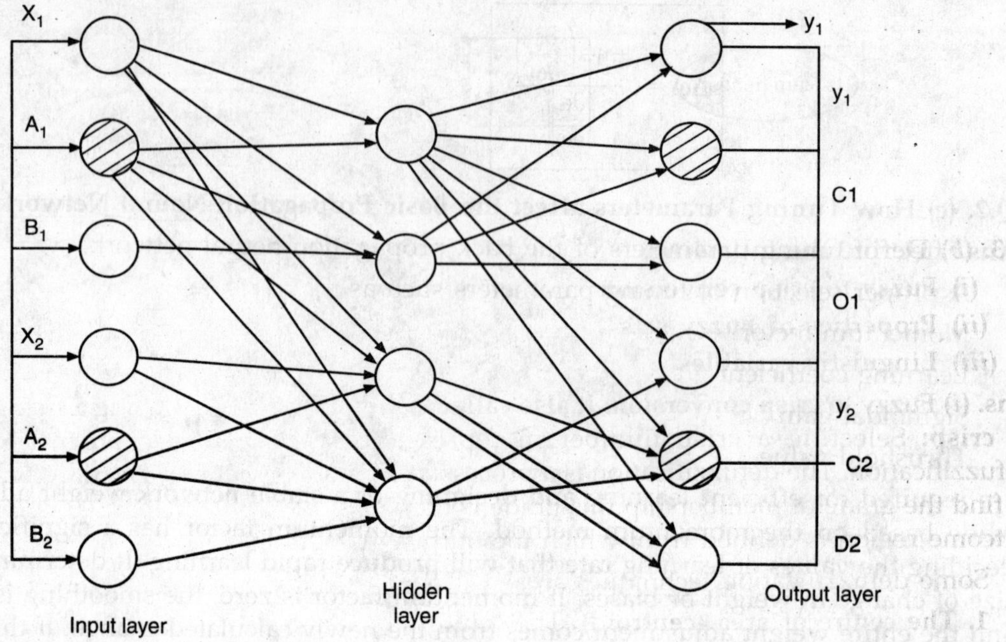

3. Attempt any two parts:

Q.3. (*a*) What is Time dependent Fuzzy Logic? Discuss in detail. How Crisp Logic is different from Fuzzy Logic define?

Ans. Time dependent fuzzy logic is called temporal fuzzy logic (TFL) and it is an extension of basic time independent fuzzy logic with the inclusion of time dependent membership function.

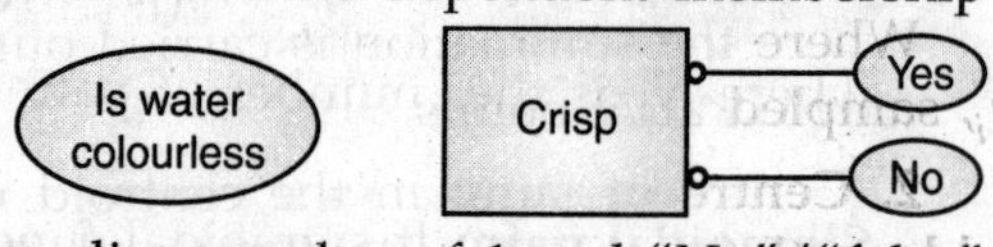

Crisp set: Consider a query Is water colourless? The answer to this is a definite yes/true or no/false, as warranted by the situation. If "yes"/ "true" is according a value of 1 and "No"/"false" is accorded a value of O, this statement results in a O/1 type of situation.

Such a logic which demands a binary (O/1) type of handling is turned "crisp" in the domain of fuzzy set theory.

A classical set is defined by crisp boundaries *i.e.,* there is no uncertainty in the prescription or location of the boundaries of the set.

Fuzzy set: Consider the statement- Is Ram honest? The answer to this query need not be a definite "yes" or "no". There may be variety of answer spanning a range such as "extremely honest" extremely dishonest, honest at times very honest. Could be generated. The situation is therefore so fluid that it can accept values between 0 and 1 such as situation is termed fuzzy.

A fuzzy set, then is a set of containing elements that have varying degrees of membership in the sets.

A fuzzy set is prescribed by vague or ambiguous properties, hence its boundaries are ambiguously specified.

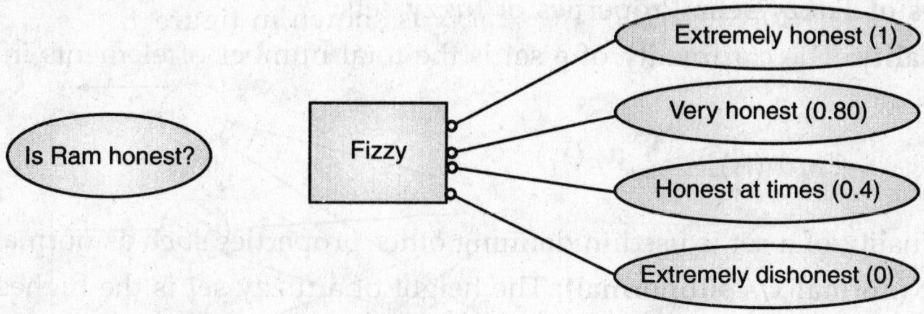

Q.3. (*b*) Define the following:

(*i*) Fuzzy to crisp conversion

(*ii*) Properties of Fuzzy sets

(*iii*) Linguistic variables

Ans. (*i*) Fuzzy to crisp conversion: It also called defuzzification converting a fuzzy quantity to crisp: Selecting a crisp number m representative of $\mu(u)$ is a process known as defuzzification. The defuzzification is a process in which membership functions are sampled to find the grade of membership this grade is then used in the fuzzy logic equations and an outcome region is defined from which the output is deduced.

Some defuzzification techniques are:

1. The centre of area (centroid): The centre of area defuzzification the crisp value u is taken to be the geometric centre of the output fuzzy value $\mu_{out}(u)$. It is defined as

$$u = \frac{\sum_{i=1}^{N} \mu_i \, \mu_{out}(u_i)}{\sum_{i=1}^{N} \mu_{out}(u_i)}$$

Where the summation is carried out over (discrete) values of the universe of discourse u_i, sampled at N points.

2. Centre of sum: In the centroid method, the overlapping area is counted once. To address the problems associated with COA and to take into account the overlapping areas of multiple rules more than once, a varient of COA caked centre sums (COS) is used. The defuzzification output is defined as:

$$u = \frac{\sum_{i=1}^{N} \mu_i \sum_{k=1}^{N} \mu_k(u_i)}{\sum_{i=1}^{N} \sum_{k=1}^{N} \mu_k(u_i)}$$

3. Mean of maxima (MOM) defuzzification: When we have more than one element in the universe of discourse with the maximum value, we take the mean value of the maximum or randomly select one of them. The crisp output can be obtained as

$$u = \sum_{m=1}^{M} \frac{u_m}{M}$$

where, u_m is the mth element in the universe of discourse.

***(ii)* Properties of Fuzzy sets:** Properties of fuzzy sets:

1. Cardinality: The cardinality of a set is the total number of elements in the set. It can be defined as

$$\text{Card (A)} = \sum_{x_i} \mu_A(x_i)$$

The cardinality of a set is used in defining other properties such as normalization factor.

2. Height (normal v/s subnormal): The height of a fuzzy set is the highest membership value of its memberships function.

$$\text{Height } (A) = \max_{x_i} \mu_A(x_i)$$

A fuzzy set with height -1 is called normal fuzzy set and whose height is less than 1 is called subnormal fuzzy set.

3. Conversity: A fuzzy set is converse if its membership function does not have a valley. A convex fuzzy set can be described as follows:

Let X be the universe of discourse of a variable x. Let A be a fuzzy set (subset) of X. The set A is convex if and only if

$$\mu_A(\sqrt{a} + (1 - \sqrt{\ })b) \geq \min \{\mu_A(a), \mu_A(b)\} \text{ for all } a, b \ \varepsilon \text{ X and } 0 \leq A \leq 1$$

***(iii)* Linguistic variables:** Fuzzy lingustic descriptions are formal representation of system made through fuzzy IF-THEN rules. Fuzzy linguistic variables are formulated using fuzzy sets and relations. The linguistic terms are sued to express concepts and knowledge in human communication whereas the membership function are used for processing numeric input data. These descriptions encode knowledge about a system in statements of the form.

If (a set of condition is satisfied) THEN (a set of consequences can be infered).

Fuzzy logic is based on rules of the form IF-THEN that convert an input fuzzy set into an output fuzzy set.

Q.3. *(c)* Discuss Fuzzy Artificial Neural Network in detail. Write down its application.

Ans.A fuzzy neural network or neuro-fuzzy system is a learning machine that finds the parameter of a fuzzy system (*i.e.*, fuzzy sets, fuzzy rules), by exploiting approximation techniques from neural networks.

Combining fuzzy systems with neural networks: Both neural network and fuzzy system have some things in common. They can be used for solving the problem (eg., pattern recognition, regression or density estimation). They have certain advantages and disadvantages which almost completely disappear by combining both concepts.

Neural network can only come into play if the problem is expressed by a sufficient amount of observed examples. These observation are used to train the black box. On the one hand no prior knowledge about the problem needs to be given.

On the other hand, it is not straight forward to extract comprehensible rules from the neural networks structure.

On the century a fuzzy system demands linguistic rules instead of learning examples as prior knowledge.

Comparison of neural control and fuzzy control

Neural networks	*Fuzzy systems*
No mathematical model necessary	No mathematical model necessary
Learning from scratch	A prior knowledge is essential
Several learning algorithm	Not capable to learn

Applications of fuzzy neural networks:

1. In fire alarm system of high rise building.
2. Optimal design of methane drainage pipeline system in coal mine.
3. For image tunor description.
4. Fuzzy neural network tech can be used in developing graphical information system.
5. Fault self diagnosis system of automatic transmission.

4. Attempt any two parts:

Q.4. (*a*) What is Fuzzy Quantifiers? Discuss in detail. Differentiate between Absolute and Relative Quantifier. Also define Fuzzification.

Ans. Fuzzy quantifiers are fuzzy numbers that take part in fuzzy propositions. They are of two kinds:

1. Absolute
2. Relative quantifier

The fuzzy quantifier of the first kind are defined on *R* and characterize longuistic terms such as about 10, much more than 100, at least 5 and so on.

The fuzzy quantifier of second type are defined on [0 – 1] and characterize linguistic terms such as almost all, about half, most and so on.

Absolute quantifier are defined on *R* relative quantifier are defined over [0 – 1].

Example: There are two basic forms of proposition that contain fuzzy quantifier of the first kind.

P: there are Q, S in I such that *V*(*i*) is *F*.

where V = variable for each individual i

F = fuzzy set defined on values of v.

Q = fuzzy number on R.

Fuzzy quantifiers of the first kind may also appear in fuzzy proposition is the form.

P = There are Q_1's in I such that $V_1(i)$ is F_1, and $V_2(i)$ is F_2

where V_1, V_2 are variables

fuzzy propositions with quantifier of the second kind is

P: Among i's in I such that $v_1(i)$ is F_1 there are Q_1's in I such that $V_2(i)$ is F_2.

Q.4. (*b*) Define membership function in detail. Also define its role and application.

Ans. A membership is a curve that defines how the values of a fuzzy variable in a certain region are mapped to a membership values between O and 1. A membership function can have different shapes. The simplest and most commonly used membership function is the triangular type which can be symmetric or asymmetric. Different types of membership functions are:

1. Triangular
2. Trapezoid
3. Gaussian
4. Two sided gaussian
5. Sigmoid light
6. Sigmoid left
7. Difference sigmoid
8. Product-sigmoid
9. Polynomid *z*.

The membership function essentially embodies all fuzziness for a particularly fuzzy property or operation because the importance of the shape of the membership function a great deal of attention has been focused on development of these functions.

Features of membership function are:

1. **Core:** The core of a membership function for some fuzzy set A is defined as that region of the universe that is characterized by complete and full membership in the set A.

$$\mu_A(x) = 1$$

2. **Support:** The support of a membership function is that region of universe that characterized by non-zero membership function in set *A i.e.,*

$$\mu_A(x) > 0$$

3. **Boundaries:** The boundary of a membership function is defined as that region of universe containing elements that have non-zero membership but not complete membership.

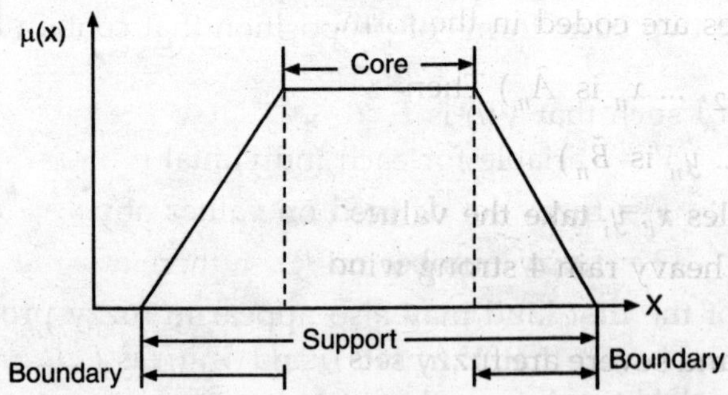

Applications:

1. Fuzzy logic enables a computer program to arrive at a decision based on multiple factor with different levels of importance.
2. It is used in fault diagnosis in transformers using deserved gas based on different standards.
3. In interpretability improvement of neuro fuzzy model.

Q.4. (c) Explain the following:

(i) Fuzzy interface

(ii) Formation of rule based matrix

(iii) Yamakawa's Air Conditioner controller

Ans.(i) Fuzzy interface: Fuzzy inference also called approximate reasoning refers to computational procedures used for evaluating or finding out linguistic description. There are two procedure for fuzzy inference:

1. Generalized modus ponens (GMP)
2. Generalized modus tollens (GMT)

If x is $\tilde{A}$ THEN y is $\tilde{B}$

GMP: $\dfrac{x \text{ is } \tilde{A}'}{y \text{ is } \tilde{B}'}$

$$\tilde{B}' = \tilde{A}' \, o \, \tilde{R}(x, y)$$

$$\mu \tilde{B}'(y) = \max(\min(\mu_{\tilde{A}'}(x), \mu_{\tilde{R}}(x, y)))$$

GMT: If x is $\tilde{A}$ THEN y is $\tilde{B}$

y is $\dfrac{y \text{ is } \tilde{B}'}{x \text{ is } \tilde{A}'}$

$$\tilde{A}' = \tilde{B}' o \, \tilde{R}(x, y)$$

$$\mu_{\tilde{A}'}(x) = \max(\min(\mu_{\tilde{B}'}(y), \mu_{\tilde{R}}(x, y)))$$

(ii) Formation of rule based matrix: Formation of rule based matrix. Fuzzy longuistic descriptions are formal representation of system made through fuzzy If-THEN rules. They encode knowledge about a system in a statement of the form.

If (a set of condition) are satisfied THEN (a set of consequents) can be inferred.

Fuzzy IF-THEN rules are coded in the form.

If (x_1 is $\tilde{A}_1$, x_2 is $\tilde{A}_2$, ... x_n is $\tilde{A}_n$) Then

(y_1 is $\tilde{B}_1$, y_2 is $\tilde{B}_2$ ... y_n is $\tilde{B}_n$)

where longuistic variables x_i, y_i take the values of fuzzy sets A_i and B_j resp.

Example: If there is heavy rain 4 strong winds then there must be severe flood warning .

Here heavy, strong and severe are fuzzy sets qualifying the variable rain, wind and flood warning resp.

A collection of rules referring to a particular system is known as a fuzzy rule base. If the conclusion C to be drawn from a rule base R is the conjunction of all the individual consequents C_i of each rule, then

$$C = C_1 \wedge C_2 \wedge \ldots C_n$$

where $\mu_c(y) = \min(\mu_{c_1}(y), \mu_{c_2}(y) \ldots \mu_{c_n}(y) \, \forall \, y \, \varepsilon y)$

where y is the universe of discourse, on the other hand, if the conclusion C to be drawn from a rule base R is the disjunction of the individual consequents of each rule, then

$$C = C_1 \cup C_2 \cup C_3 \ldots \cup C_n$$

where $\mu_C(y) = \max(\mu_{c_1}(y), \mu_{c_2}(y) \ldots \mu_{c_n}(y), \forall\, y\, \varepsilon\, y)$

***(iii)* Yamakawa's Air Conditioner controller:** Yamakawa's Air conditioner controller is one of the major application of fuzzy systems. The systems as illustrated in Figure.

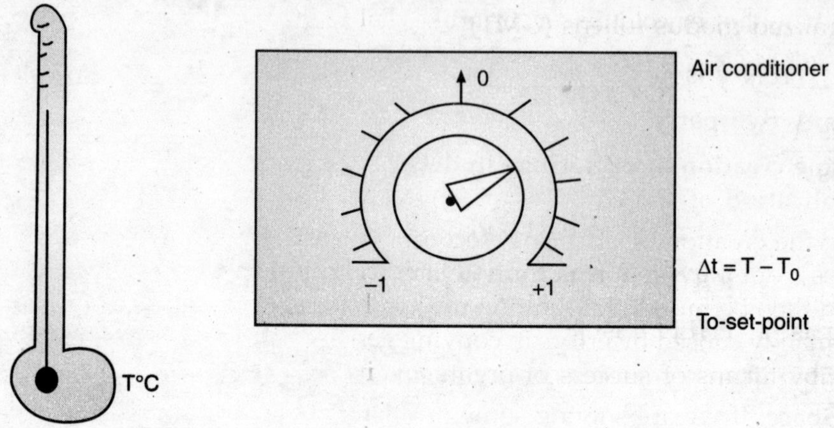

Air conditioner control system

It comprises a deal to control the flow of warm/hot or cool/cold air and thermometer to measure the room temperature (T°C). When the deal is turned positive, warm/hot air is supplied from the air conditioner and if it is turned negative, cool/cold air is supplied. If set to zero, no air is supplied.

Fuzzy rule base for air-conditioner controller

S.No.	Fuzzy rule (descriptive)	Fuzzy rule (Notational)
1.	If the room temperature is approx equal to the set point T°C, ΔT is approx zero ($Z\ \varepsilon$) and the temp is rapidly changing higher, *i.e.*, $\frac{d\Delta T}{dt}$ is positively large (PL) then below cold air rapidly *i.e.*, turn the dial negative large (NL)	If ΔT is $Z\ \varepsilon$ and $\frac{d\Delta T}{dt}$ is PL then dial should be NL
2.	If the room temp is high and there is no change in knp *i.e.*, ΔT is positive large (ρL) and $\frac{d\Delta T}{dt}$ is app zero ($Z\ \varepsilon$) then blow cold air at an intermediate level *i.e.*, turn the dial negative medium (*Nn*)	If ΔT is ρL and $\frac{d\Delta T}{dt}$ is $Z\ \varepsilon$ then deal should be *Nn*

3.	If the room temp is a little bit higher than the set point and the temp is gradually decreasing *i.e.;* ΔT is positively small (ρS) and $\frac{d\Delta T}{dt}$ is negatively small (*NS*) then there is no need to blow hot or cold air *i.e.,* turn like dial to app. zero ($Z\ \varepsilon$).	If ΔT is *PS* and $\frac{d\Delta T}{dt}$ is *NS* then dial should be $Z\ \varepsilon$

5. Attempt any two parts:

Q.5. (*a*) Define creation of offsprings in detail. Also write down the working principle of GA and application of GA.

Ans. During the creation of offspring, recombination occurs (due to cross over) and in that process genes from parents forms a whole new chemosomes in some way. The new created offspring can then be mutated. Mutation means that the elements of DNA is modified. These changes are mainly caused by errors in copying genes from parents. The fitness of an organism is measured by means of success of organisms in life.

Search Space: If we are solving some problems, we work towards some solution which is the best among others. The space for all possible feasible solutions is called search space. The search space can be known by the time of solving a problem and we generate other points as the process of finding the solutions.

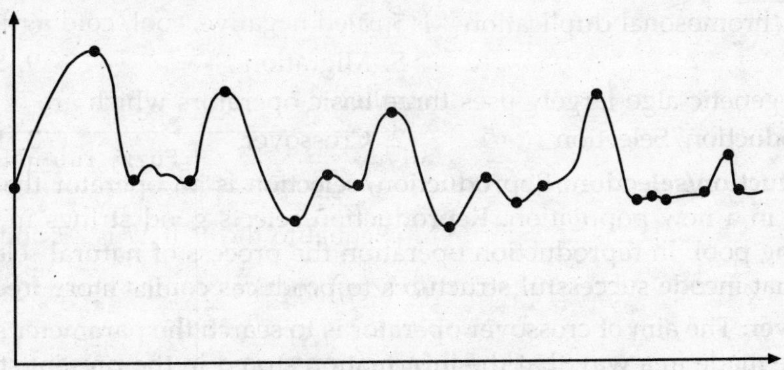

Examples of search space

Working principle: To illustrate the working principle of GA, we first consider unconstrained optimization problem. Let us consider the following maximization problem. Maximize $F(x)$

$$X_i^{(L)} \le X_i \le X_i^{(U)} \quad \text{for} \quad i = 1, 2 - N$$

If we want to maximize $F(x)$ for $F(x) > 0$, then the objective function as maximize $\cdot \frac{1}{1+F(x)}$

If $F(x) < 0$ instead of minimize $F(x)$, maximize $\{-F(x)\}$, hence both maximization and minimization problem can be handled by GA.

Application of GA:

1. GA have widely used in radar and communication system to synthesize multilayer ladar absorbing coating which maximizes the absorption of electromagnetic waves.
2. GA also optimises weights in neural network training.
3. Adapt infinite impulse response (IIR) filters.

Standard application of GA are –

The travelling salesman problem: TSP is one of the standard combination optimization problems. The application of GA to TSP is a very interesting permutation type problem where different type of crossover and mutation have been tested.

Q.5. (*b*) Define the following:

(*i*) **Fitness function** (*ii*) **Genetic operator**

(*iii*) **Cross over**

Ans.(*i*) Fitness function: The genetic algorithm is a randomized search and optimization technique guided by the principle of natural genetic systems.

"The objective or fitness" function is a judge of the GA'S attempts. GAS do not know how to derive a problems solution, but they do know, from the objective function, how close they are to a better solution.

(*ii*) Genetic operator: Genetic algorithm derives power from the genetic operators. Some low level operators are:

1. Inversion	2. Dominance	3. Deletion
4. Interachromosonal duplication	5. ??	6. Segregation
7. ??	8. Migration	9. Sharing

A simple genetic algo largely uses three basic operators which are

1. Reproduction/Selection	2. Crossover	3. Mutation

1. Reproduction/selection: Reproduction/selection is an operator that makes copies of better strings in a new population. Reproduction selects good strings in a population and forms a mating pool. In reproduction operation the process of natural selection cause those individuals that incode successful structures to produces copies more frequently.

2. Crossover: The aim of crossover operator is to search the parameter space. In addition, search is to be made in a way that the information stored in the present string is maximally preserved, because these parent strings are instances of good strings selected during reproduction crossover is a recombination operator, which proceeds in 3 steps. First, the reproduction operator selects at random a pair of two individual strings for mating, than a crosssite is selected at random along the string length and the position values are swapped between two strings following the cross site.

For instance, let the two selected strings in a mating pair be A = 11111 and B = 00000. If the random selection of a cross-site is two, than the new string following cross over would be A^* = 11000 and B^* = 00111.

3. Mutation: Mutation is the process of random modification of the value of a string with small probability. It is not a primary operator but it ensures that the probability of searching any region in the problem space is never zero and prevents complete loss of genetic material through reproduction and crossover.

(*iii*) Cross over: There exists many types of cross over operators which are discussed in the following sections:

(*a*)Single-site cross over: In a single site cross over a cross-site is selected randomly along the length of the mated strings and bits ment to the cross sites are unchanged as shown in Figure.

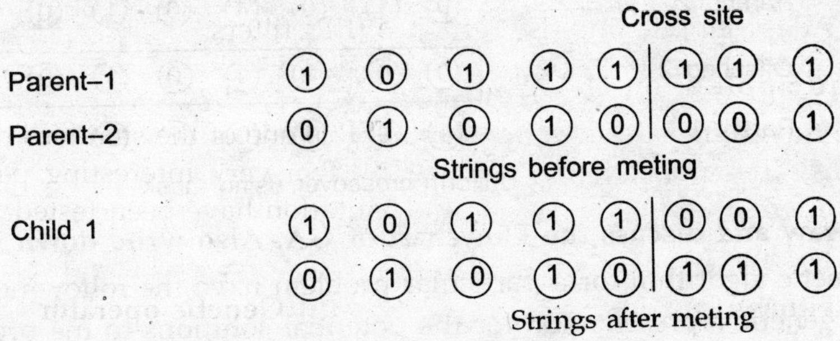

Single site cross-over

In short, we may conclude that "

- Randomly one position in the chromosomes is chosen.
- Child 1 is head of chromosome of parent 1 with tail of chromosome of parent 2.
- Child 2 is head of 2 with tail of 1.

The following process shows the generation of child from one point cross over.

Two-point cross over:

Randomly two positions in the chromosomes are chosen.

The following process shows the generation of child from two point cross over.

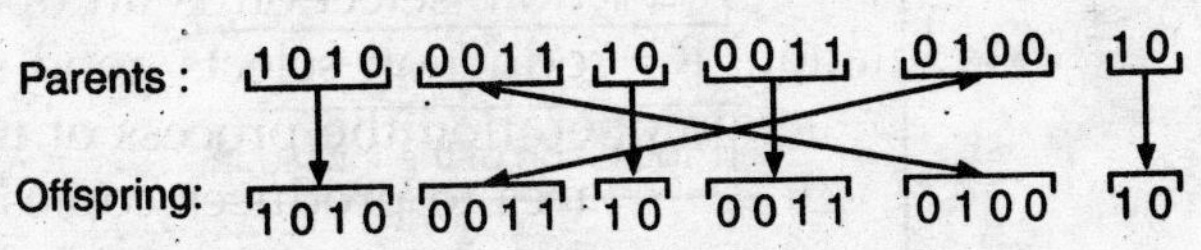

Uniform cross over: In this each bit from either parent is selected with probability of 0.5 and then interchanged as shown below:

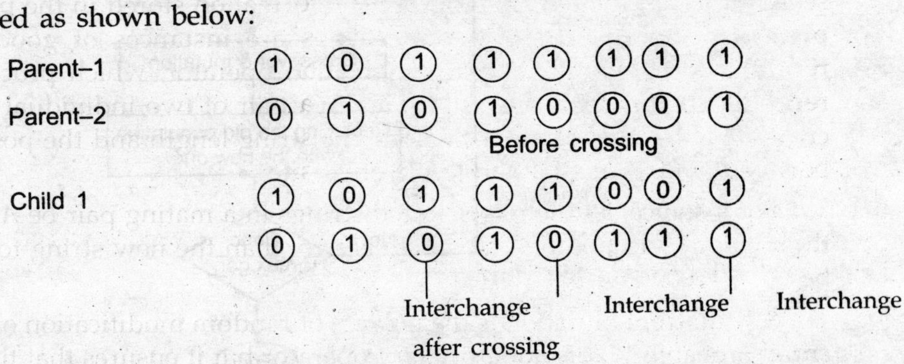

Uniform crossover without using mask

Cross over mask	1	0	0	1	0	1	1	1	0	0	1
Parent 1	1	0	1	0	0	0	1	1	1	0	1
Offspring 1	1	1	0	0	0	0	0	1	1	1	1
Parent 2	0	1	0	1	0	1	0	0	1	1	0
Offspring 2	0	0	1	1	0	1	0	0	1	0	0
Parent 1	1	0	1	0	0	0	1	1	1	0	1

Uniform crossover using mask

Q.5. (*c*) Draw and discuss the Flowchart of GA. Also write down its importance.

Ans. A genetic algorithm for a particular problem have the following five components:

1. A genetic representation for the potential solutions to the problem.
2. A way to create an initial population of potential solution.
3. An evaluation function that plays the role of the environment, rating solutions in terms of their fitness.
4. Genetic operators that after the composition of the offspring.
5. Values for the various parameters that the genetic algorithm uses (population size, probabilities of applying genetic operators).

The flow chart of GA is shown below

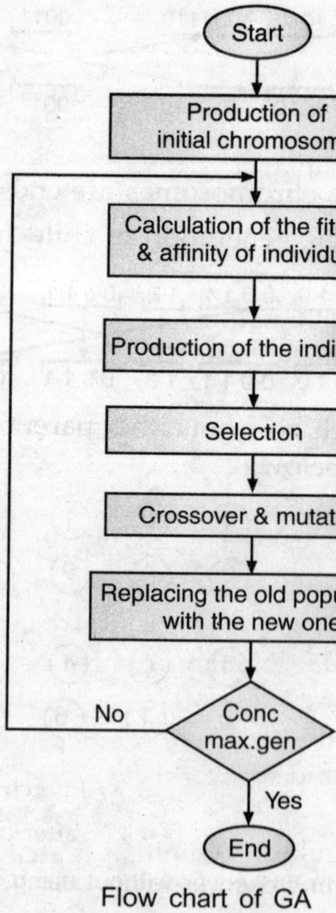

Flow chart of GA

□□□

References

[1] Yagyanarayan, *Artificial Neural Network*, TMH, 2003

[2] Lawrence Rabiner & Biing-Hwang Juang, *Fundamental of Speech Recognition*, Pearson Education, 2003.

[3] Bose & Liang, *Artificial Neural Networks*, TMH, 2001

[4] Simon Haykin, *Neural Networks*, Pearson Education, 2001

[5] Jacek M.Zurada, *Introduction to Artificial Neural Network*, Jaico Publishing House, 2003

[6] Rao & Rao, C++ *Neural Network & Fuzzy Logic*, BPB Publishing, 2004

[7] Potanitis L.,Fakotakis N., Kokkinakis G., "Independent Component Analysis,", *http:/wel.ee.upatras.gr*

[8] Montri Karnjanadecha and Stephen A. Zoharian, "Signal modeling for high-performance robust isolated word recognition," *IEEE Trans. On speech and audio processing*,vol.9, no.6, pp 647-654.

[9] Dinh Tuan Pham, "Blind separation of mixture of independent sources through a quasi-maximum likelihood approach," *IEEE Transaction on Signal Processing*,vol.45, no. 7, pp. 1712-1725.

[10] Woo W.L. and Khor L.C., "Blind restoration of nonlinearly mixed signals using multilayer polynomial neural network," *IEE Proc.-Vis. Image Signal Process.*,vol.151, no. 1, pp.51-61.

[11] Amari Shun-Ichi and Andrzej Cichocki, "Adaptive blind signal processing – Neural Network Approaches," *Proc. Of IEEE*, vol.86,no. 10, pp.2026-2048.

[12] ICA algorithm and program, *http:/koders.com*

[13] Adel A.M.Saleh, "Frequency-Independent and Frequency-Dependent nonlinear models of TWT amplifiers," *IEEE Trans. Of communications*, vol.com-29, no.11,1715-1720.

[14] Woo W.L. and Sali S., "General multilayer perceptron demixer scheme for nonlinear blind souce separation," *IEE Proc-Vis. Image Signal Process.*, vol. 149, No.5, pp.253-262.

[15] Mark D.Plumbley, " A Nonnegative PCA algorithm for Independent Component Analysis," *IEEE Trans. On Neural Networks*,15,(1), pp.66-76.

[16] Blaschke and Laurenz Wiskott, "CuBICA: Independent component analysis by simultaneous third and fourth order cumulant diagonalization," *IEEE Trans. on Signal Processing*, vol.52, no.3, pp.1250-1256.

[17] Shun-ichi Amari and Jean-Francois Cardoso, "Blind source separation-semiparametric statistical approach", *IEEE Trans. on Signal Processing*, vol.45, no.11, pp.2692-2700.

[18] Nikos Vlassis, "Efficient source adaptivity in Independent component analysis," *IEEE Trans. On Neural Networks*, vol.12, no.3, pp.559-565.

[19] Anisse Taleb and Chistian Jutten, "Source separation in post nonlinear mixtures," *IEEE Trans. on Signal Processing*, vol.47,no.10, pp.2807-2820.

[20] Ying Tan, Jun Wang, Jacek M.Zurada, "Nonlinear blind source separation using radial basis function network," *IEEE Trans. On Neural Networks*, vol.12,no.1, pp.124-133.

[21] Qi Li, Jinsong Zheng, Augustine Tsai and Qiru Zhou, "Robust endpoint detection and energy normalization for real-time speech and speaker recognition," *IEEE Trans. On speech and audio processing*, vol.10, no.3, pp146-156.

[22] Satya Dharanipragada and Salim Roukos, "A multistage algorithm for spotting new words in speech," *IEEE Trans. On speech and audio processing*,vol.10,no.8,pp.542-550.

[23] Nico Weymaere and Jean-Pierre Martens, "On the Initialization and Optimization of Multilayer Perceptrons", *IEEE Trans. On Neural Networks*, vol.5, no.5,pp 738-751.

[24] Jear-Pierre Martens and Nico Weymaere, "An equalized error backpropagation algorithm for the on-line training of multilayer perceptrons," *IEEE Trans. On Neural Networks*, vol.13,no.3, pp.532-541.

[25] Rita Singh, Bhiksha Raj and Richard M.Stern, "Automatic generation of subword units for speech recognition," *IEEE Trans. On speech and audio processing,* vol.10,no.2, pp 89- 99.

[26] Vijayan K.Asari, "Training of a feedforward multiple-valued neural network by error backpropagation with a multilevel threshold function," *IEEE Trans. On Neural Networks*, vol.12.no.5,pp 1519-1521.

[27] Ze'ev Roth and Yoram Baram, "Multidimensional density shaping by sigmoids," *IEEE Trans. on Neural Networks*, vol.7, no.5, pp 1291-1298.

[28] Panayiota Poirazi, Costas Neocleous, Costantinos S. Pattichis, and Christos N.Schizas, "Classification Capacity of a modular neural network implementing neurally inspired architecture and training rules," *IEEE Trans. On Neural Networks*, vol.15, no.3, pp.597-612.

[29] Masoud Karimi-Ghartemani and Alireza K.Ziarani, "A nonlinear time-frequency analysis method," *IEEE Trans. on Signal Processing,* vol.52, no.6, pp.1585-1595.

[30] Ozgur Yilmazans and Scott Rickard, "Blind separation of speech mixtures via time-frequency masking," *IEEE Trans. on Signal Processing,* vol.52, no.7,pp.1830-1847.

[31] Riccardo Boscolo, Hong Pan and Vwani P.Roychowdhury, "Independent component analysis based on nonparametric density estimation," *IEEE Trans. On speech and audio processing* vol.15, no.1, pp.55-65.

[32] Q. Zhu and A. Alwan, "Nonlinear feature extraction," *Computer Speech and Language*, vol.17, pp.381-402.

[33] Joseph W. Picone, "Signal Modeling Techniques in speech recognition," *Proceedings of the IEEE*, vol.81, no.9, pp 1215-1247.

[34] O. Farooq and S.Datta, "Speech recognition with emphasis on wavelet based feature extraction," *IETE Journal of Research,*vol.8, no.1, pp.3-13.

[35] Jean Hennebert, Martin Hasler and Herve Dedieu, " Neural Networks in speech recognition," *http:/circhp.eplf.ch/staff/henneber/jean/papiers/microcomp/document.html*

[36] Chistian Eitziger and Hartwig Plach, "A new approach to perceptron training," *IEEE Trans. On Neural Networks*, vol.14,no.1,pp.216-221.

[37] Edgar Osuna, Robert Freund, Federico Girosi, "Training Support Vector Machines: An Application to face recognition," *Neurocomputing*, 51 (2003) pp 130-136

[38] Wlodzimierz Kasprzak, Adam F.Okazaki, " Applying Independent Component Analysis for Speech Feature Detection," *grant no.4T11A 003 25, Warsaw University of Technology, Poland.*

[39] John B.Hampshire II, B.K.V.Vijaya Kumar, "Why Error Measures are Suboptimal for Taining neural network pattern Classifiers," *Neurocomputing*, pp.IV-220-227.

[40] A.J.Hussain, P.Liatsis, "A new Recurrent Polynomial neural network for predictive image coding," *Image Processing and its Applications, Conference Publication no. 465@IEE 1999*, pp. 82-86.

[41] Sung-Kwun Oh, Tae-Chao Ahn, Witold Pedrycz, "A Study on Self Organising Polynomial Neural Network," *Neurocomputing*,78, 2001, pp.1690-1695.

[42] Byoung-Jun Park, Witold Pedirycz, Sung-Kwun Oh, "Fuzzy Polynomial Neural Networks: Hybrid Architechtures of Fuzzy Modeling," *IEEE Transactions on Fuzzy Systems*, Vol.10, No.5, Oct.2002, pp 607-620.

[43] Yi-Qiong, Bi-Cheng Li, Bo Wang, "Face Detection and recognition using neural network and hidden markov models," *IEEE Int. Conf. Neural Network and Signal Processing*, Dec.2003,pp. 228-231

[44] Robert E. Parker and Murali Tummala, "Identification of Volterra Systems with polynomial neural network," *Neurocomputing*, pp.561-564

[45] Byoung-Jun Park, Sung-Kwun Oh, Witold Pedrycz, "The Hybrid Multi-layer Inference Architechture ad Algorithm for FPNN based on FNN and PNN," *Neurocomputing*, pp.1361-1366.

[46] Adam Kowalczyk, Hernman L.Ferra, "Developing Higher-Order Networks with Emperically Selected Units," *IEEE Transactions on neural networks*, Vol.5, No.5, Sept.1994, pp. 698-711.

[47] Dong-Won Kim, Jang-Hyun Park, Gwi-Tae Park, "Combination of Fuzzy Rule Based Model and Self-organising Approximator Technique: A new Approach to Nonlinear Modeling," *IEEE International Conference on Fuzzy Systems*, pp.1363-1376.

[48] W.L.Woo and L.C.Khor "Blind Restoration of nonlinearly mixed signals using multilayer polynomial neural network," *IEE Proceedings of Visual. Image Processing*, vol.151, no.1,Feb.2004, pp.51-60

[49] B.H.Juang and L.R.Rabiner, "Spectral Representations for speech recognition by neural network-ATutorial," *IEEE Transactions on Acoustics, Speech, and Signal Processing*, vol.ASSO-28,No.4,Aug. 1992. pp.214-222

[50] N.Nocerino, F.K.Soong, L.R.Rabiner, D.H.Klatt, "Comparative Study of Several Distortion Measures of Speech," *IEEE Transactions on Acoustics, Speech, and Signal Processing*, vol.ASSO-28,No.4,Aug. 1985, pp. 25-28.

[51] Robert M.Gray, Andres Buzo, Augustine H.Gray, Yasuo Matsuyama, "Distortion Measures for speech processing," *IEEE Transactions on Acoustics, Speech, and Signal Processing*, vol.ASSO-28,No.4,Aug.1980, pp.367-376

[52] Parham Zolfaghari, Shinji Watanabe, Atsushi Nakamura, Shigeru Katagiri, "Bayesian Modelling of the speech spectrum using mixture of guassians," *IEEE, ICASSP* 2004, pp.I553-I556.

[53] Fredrik Norden and Thomas Eriksson, "Time Evolution of LPC Spectrum Coding," *IEEE Transactions on Speech and Audio Processing*, Vol.12, No.3, May 2004, pp.290-301.

[54] Fredrik Norden and Thomas Eriksson, "A Speech Spectrum Distortion Measure with Interframe Memory," *IEEE, ICASSP*, 2001, pp.717-720.

[55] M.Ferhaoui, S.Van Gerven, "LSP Quantization in Wideband speech coders," *IEEE Transactions on speech and Audio Processing*, 1999, pp. 25-27.

[56] Roar Hagen, Erdal Paksoy, Allen Gresho, "Voicing-Specific LPC Quantization for Variable-Rate Speech Coding," *IEEE Transactions on Speech and Audio Processing*, vol.7,no.5,pp 485-494.

[57] Jozsef Vass, Yunxin Zhao, Xinhua Zhuang, "Adaptive Forward-Backward Quantizer for low bit rate high-quality speech coding," *IEEE Transactions on Speech and Audio Processing*, vol.5,no.6, Nov. 1997.

[58] D.C.Bateman, D.K.Bye, M.J.Hunt, "Spectral Contrast Normalization and other techniques for speech recognition in noise," *ICASSP 1992*, I241-I244.

[59] John E.Shore, David K.Burton, "Discrete Utternace Speech Recognition with time alignment," *IEEE Transactions on Information Theory*, vol.IT-29, no.4, July 1983.

[60] S.Horikawa, T.Furuhashi, and Uchigawa, "On Fuzzy modeling using Fuzzy Neural Networks with Back Propagation algorithm" IEEE Transactions on Neural Networks,vol.3, pp.801-806, Oct.1992.

[61] Abhinav Mathur, "Wavelets Self Organising Maps and Wavelet neural network: A study," Deptt. of Electrical and Computer Engineering Mississippi state Univ.

[62] Yonghong Tan, Xuanju Dang, Feng Liang, Chun-Yi Su,"Dynamic wavelet neural network for nonlinear dynamic system identification," *Proceedings of the 2001 IEEE International conference on control applications*, pp 214-219.

[63] Lan Gao, Ling Lu, Zhijun Li, "Prediction of chaotic time series based on wavelet neural network," *MTS 0-933957-28-9*, pp.2046-2050

[64] Yuehui Chen, Jiwen Dong, Bo Yang, Yong Zhang, "A local linear wavelet neural network," *Proceedings of 5th world congress on Intelligent Control and Automation*, 2005,pp.1954-2049

[65] An-Hong Wang, Zhi-Yi Sun, Zhi-Hong Li,Xiong Zhang, "A dynamic wavelet neural network for prediction of speech signal," *Proceedings of 3rd International conference on machine learning and cybernetics,* shanghai,26-29 Aug.2004,pp 3524-3528.

[66] Sheng-Tun Li, Shu-CHing Chen, "Function approximation using robust wavelet neural network," *Proceedings of 14th International Conference on tools with artificial intelligence.*

[67] Eric A.Rying, Griff L.Bilbro, Jye-Chyi Lu, "Focused Local Learning with wavelet neural networks," *IEEE Transactions on neural networks*, vol.13, no.2, march 2002, pp.304-319.

[68] Wang Ting, Yasuo Sugai, "A wavelet neural network for the approximation of nonlinear multivariable function," *IEEE, ICASSP-1999*, pp. III378-III383.

[69] R.Villant, C.Monrocq, and Y.Le Cun., "Original Approach for Localisation of Objects in Images," IEEE Proceedings of Visual Image Signal Processing, vol.141, no.4, August 1994, pp.389-397.

[70] Robi Polikar, "A Tutorial on The Wavelet Transform," at http://users.rowan.edu/~polikar/WAVELETS/WTtutorial.html

[71] C. Valens, "A Really Friendly Guide to Wavelets", at perso.wanadoo.fr/poly**valens**/clemens/**wavelets**/**wavelets**.html

[72] Zidan Bashir, El-Hawary, "Short term load forecasting by using wavelet neural network,", *0-7803-5957-7, 2002*, pp.163-166

[73] G. Hussonks, "Wavelets vs Fourier Transforms," on net

[74] C. J.Long, S.Datta, "Wavelet based feature extraction for phoneme recognition," citeseer.csail.mit.edu/506806.html

[75] Hyvarinen A., "Fast and robust fixed-point algorithms for independent component analysis," *IEEE Transaction on Neural Networks*, vol.10,no.3, pp 626-634.

[76] Daniel Clancy, Umit Ozguner, "Wavelet neural network: a design perspective," *IEEE International Symposium on Intelligent Control*, 1994,pp.376-381

[77] Marc Thuillard, "A Review of Wavelet Networks, wavelets, fuzzy wavelets and their applications," *ESIT 2000*, Germany, pp. 5-16.

[78] Huanqin Li, Baiwu WAN, "Multi-input-layer wavelet neural network and its application," *Proceedings of 5th International Conference on Computational Intelligence and Myultimedia Applications*, 2003

[79] Daniel W.C.Ho, Ping-An Zhang, Jinhua Xu, "Fuzzy wavelet network for function learning," *IEEE Transaction on Fuzzy Systems*, vol.9,no.1, feb.2001, pp.200-211.

[80] Renevey P., Drygajlo A., "Statistical estimation of unreliable features for robust speech recognition," Proceedings of the International Conference on Acoustics, Speech ans Signal Processing, vol.3., pp. 1731-1734.

[81] Mohamed Kamal Omar and Mark Hasegawa-Johnson, "Non-linear independent component analysis for speech recognition," *NSF award no. 0132900, Univ. of Illinois, Urbara.*

[82] Hyvarinen A. and Erkki Oja, "Independent Component Analysis: Algorithms and Applications", *http://www.cis.hut.fi/projects/ica/*

[83] Pierre Comon, "Independent Component Analysis, A new concept?," *Signal Processing*, 36(1994), pp.287-314.

[84] Michel, MFCC codes, *http:/www.icsl.ucla.edu/~spapl/qifeng/speechdemo.html*

[85] Davis S. and P.Mermelstein," Comparison of parametric representations for monosyllabic word recognition in continuously spoken sentences," *IEEE Trans. Acoust., Speech, Signal Processing*, vol. ASSP-28, pp.357-366,1980.

[86] Kreinovich V., Sirisaengtaksin O., "Wavelet Neural Networks are Asymptotically Optimal Appriximators for functions of one variable," *IEEE World Congress on Computational intelligence*, 1994, vol.1, pp.299-304.

[87] L.Bottu and V.Vapnik, "Local learning algorithms," *Neural Computing*, vol.4, pp.888-900, 1992.

[88] Q.Zhang and A.Benveniste, "Wavelet Networks," *IEEE Transactions on Neural Networks*, vol.3,pp.889-898,Nov.1992.

[89] Coifman, R.R. and Wickerhauser M.L. "Entropy based algorithms for best-basis selection, *IEEE Transactions on Information Theory*, vol.32, pp.712-718, March 1992.

[90] Jari Kangras, Kari Torkkola and mikko Kokkonen; "Using SOMS as feature extractors for speech recognition," *IEEE ASSP mag.*,II-341, pp.341-344.

[91] Populis, Probability Theory, Pearson Edition, 1999.

[92] J.P.Saini and M.C.Srivastava, "Recovery of noisy ECG signal using artificial neural network," *Proc. of International Conf. on Biomedical Engineering*, C.I.T., Coimbatore, India, vol.1, pp vii9-vi13, Dec.1996.,

[93] J.P.Saini and M.C.Srivastava, "Channel equalization and CCI suppression in 160QAM receiver system using a new multistage complex FSBLP model," *Proc. of NCC'98, IISC Bangalore, Communication 1998, TMH India*, pp 24-28, Jan 98.

[94] J.P.Saini and M.C.Srivastava, "Multi-stage complex fractionally spaced bilinear perceptron and its application to MQAM system for channel equalization in FSRF and CCI environment," *Proc. Of IEEE IJCNN'98*, Anchorage, Alaska, May 1998.

[95] J.P.Saini and M.C.Srivastava. "A modular polynomial perceptron approach to channel equalization and ACI suppression, " *Proc. IEEE International Conf. On Personal Wireless Communication, Jaipur, India*, vol.1, pp. 256-259, Feb.1999.

[96] Mazin G.Rahim, "A neural tree network for phoneme classification with experiments on TIMIT database," *IEEE ASSP mag.*,II, pp.345-348.

[97] Herve Bourlard, Nelson Morgan, Chuck Wooters ans Steve Renals, "CDNN: A context dependent neural network for continuous speech recognition,' *IEEE ASSP mag.*,II, pp.349-352.

[98] Helge B.D.Sorensen and Uwe Hartmann, "Self structuring control neural model for speech recognition," *IEEE ASSP mag.*,II, pp.353-356

[99] Thomas M.English and Lois C.Boggess, "Back-propagation training of a neural network for word spotting," *IEEE ASSP mag.*,II, pp.357-360

[100] Jie Zhang, A.J..Morris, "A sequential learning approach for single hidden layer neural network," *Neural networks*, vol.11, no.1,pp.60-80

[101] John-Paul Hosom, Ron Cole, Mark Fanty, "Speech recognition using neural network at the centre for spoken language understanding,", http://cslu.ogi.edu/tutordemos

[102] Richard P.Lippmann, "An introduction to computing with neural nets," *IEEE ASSP magazine*, Apr.1997, pp. 4-22

[103] Mary Jo Creaney-Stockton, "Isolated word recognition using reduced connectivity neural network with non-linear time alignment methods," http://www.moonstar.com

[104] L.R.Rabiner, B.H.Juang, C.H.Lee, "An Overview of Automatic Speech recognition," Automatic speech and speaker recognition, *a photocopied volume in IIT Kanpur Library,* Barcode A-128872, A/c no.0006.454 Aug2K, pp. 1-30

[105] Jean-Claude Junqua, "Robust speech recognition for embedded systems and PC applications", *a photocopied volume in IIT Kanpur Library,* barcode A 132286, 006.45J968r

[106] John M.Vincent, "Face Finding in Images," Application of Neural Networks, Kluwer Academic Pub., 1995

[107] Jan Robin Rohlicek, 'Word Spotting," Modern methods of speech processing, edited by Ravi P.Ramachandran, Richard Mammone, *a photocopied volume in IIT Kanpur Library,* Acc No. 0006.454M72r, pp. 123-158

[108] Stephen V.Kosonocky, "Speech recognition using neural networks," Modern methods of speech processing, edited by Ravi P.Ramachandran, Richard Mammone, *a photocopied volume in IIT Kanpur Library,* Acc No. 0006.454M72r, pp. 159-183

[109] Burr D.J., "Comparison of Gaussian and neural network classifiers on vowel recognition using discrete cosine transform," *IEEE ASSP mag.,*II, pp.365-368.

[110] Curtis K.M. and Burniston J., "A hybrid neural system for driving a parallel synthesizer so as to synthesis high quality accented speech," *IEEE ASSP*I, pp.369-371.

[111] Cardoso J.F., "Source separation using higher order moments," *ICASSP*, pp.2109-2112, 1989.

[112] Michael A. Frazini, "A new connectionist architecture for word spotting," *IEEE ASSP mag.,*II, pp.361-364.

[113] Zhu Q., Alwan A., "The effect of additive noise on speech amplitude spectra: a quantitative approach," *IEEE Signal Processing Letters*, no.9, vol.9, pp. 275-277.

[114] Alan Murray, "Neural architectures and algorithms," Applications of Neural networks, 1-33, Kluwer Academic Publishers,1995

[115] Daniel Klerfors, "Artificial neural networks," project, Saint Louis Univ.

[116] John-Paul Hosom, Ron Cole, Mark Fanty, Johan Schalkwyk, Yonghong Yan, Wei Wei, "Training neural network for speech recognition," www.ocf.berkeley.edu/~vincebot/ robots/moble/ SonicBOT

[117] Tony Robinson, "Conversion routine from linear to mue-law" www.ee.ic.ac.uk/hp/ staff/ dmb/ voicebox/voicebox.html

[118] E. Trentin, M.Gori, "A survey of hybrid ANN/HMM models for automatic speech recognition,", *Neurocomputing*, 37,2001 91-126

[119] K.R.Farrell, Richard J.Mammone, Allen L.Gorin, "Incremental algebraic learning for adaptive language acquisition," *Artificial Neural Networks for Speech and Vision*, edited by R.J.Mammone, Chapman & Hall, 1993,pp. 389-400

[120] E.Trentin, Y,Nengio, C. Furlanello, R.D.Mori, " Neural network for speech recognition," *Signal Processing and its applications-Spoken dialogues with computer*, edited by R.D.Mori, ISBN 0-12-2-9055-1

[121] Gary M.Kuhn, Raymond L.Watrous, "Comparison of feed-forward and recurrent sensitivities in speech recognition," *Artificial Neural Networks for Speech and Vision, edited by R.J.Mammone,* Chapman & Hall, 1993, pp 269-277.

[122] Rajesh Parekh, J.Yang, V.Honavar, "Constructive neural network learning algorithms for pattern classification," *IEEE Transaction on Neural Networks*, vol.11,no.2, march 2000

[123] B.B.Chaudhuri,U.Bhattacharya, "Efficient training and improved performance of multilayer perceptron in pattern classification," *Neurocomputing*, 34(2000) pp.11-27.

[124] M.G.Rahim, "A self-learning neural tree network for phone recognition", *Artificial Neural Networks for Speech and Vision*, edited by R.J.Mammone, Chapman & Hall, 1993,pp. 227-239

[125] Amir F. Atiya, Alexander G.Parlos "New results on recurrent network training unifying the algorithms and accelerating convergence," *IEEE Transactions on neural networks*, vol.11,no.3, May 2000, pp 697-705

[126] Chuan Wang J.C.Principe, "Training neural networks with additive noise in the desired signal," *IEEE Transactions on neural networks*, vol.10, no.6. Nov.1999, pp1511-1517.

[127] Sushmita Mitra Y.Hayashi, "Neuro-Fuzzy rule generation: Survey in soft computing framework," *IEEE Transactions on neural networks*, vol.11, no. 3, May 2000, pp 748-768

[128] J.M.Behitez, J.L.Castro, I.Requena, "Are Artificial neural network black boxes?," *IEEE Transactions on neural networks*, vol.8., no.5 Sept. 1997

[129] S.K.Behera, S.Dandapat, "Performance evaluation of speech features for speaker identification," *IE(I) journal-ET*, vol.83, July 2002, pp 1-6.

[130] G.P.Bansal, A.K.Tewari, P.K.Chande, "Genetic Algorithm based fuzzy expert system," *IETE technical review, vol.19, no.3*, May-June 2002, pp 111-118.

[131] E.Trentin, M.Gori, "Robust combination of neural network and hidden markov models for speech recognition," *IEEE Transaction on Neural Networks*, vol.14, no.6, Nov.2003

[132] Witold Pedrycz, George Vukovich, "Granular neural network," *Neurocomputing*, 36,(2001) 205-224

[133] Shaijun Wang, Yunxin Zhao, " Online Baysian Tree-structured transformation of HMMs with optimal selection for speaker adaption," *IEEE Transaction on speech and audio processing*, vol.9, no.6, sept.2001, pp 663-675

[134] Yali Amit, Alejandro Murua, "Speech recognition using randomized relational decision trees," *IEEE Transaction on speech and audio processing*, vol.9, no.4, May 2001, pp 333-356.

[135] Jim Y.F.Yam, Tommy W.S.Chow, "A weight initialization method for improving training speed in feedforward neural network," *Neurocomputing*, 30 (2000) 219-232.

[136] Enrico Bocchieri, Brian Kan-Wing Mak, "Subspace distribution clustering hidden markov model," *IEEE Transaction on speech and audio processing*, vol.9, no.3, March 2001, 264-273

[137] Omri Weisman, Ziv Pollack. "The Perceptron," –A tutorial on www.cs.bgu.ac.il/~**omri**/Perceptron

[138] S.Rajasekaran, Deepa Suresh, G.A.Vijayalakshmi Pai, "Sequential learning artificial fuzzy neural networks with single hidden layer,", *Neurocomputing*, 42 (2002) 287-310.

[139] Madan M.Gupta, D.H.Rao, "Neuro-Control Systems: A tutorial" on www.dbai.tuwien.ac.at/marchives/fuzzy-mail92/0021.htm.

[140] J.P. Saini, " Neural predictive architecture and algorithm for channel equalization in digital reception system," *A Ph.D. thesis at Dr. Ram Manohar Lohia Avadh University, Faizabad*, June 2000.

SOME INTERNATIONAL PAPERS BY THE AUTHOR:

1. Sumit Gupta, Dr. Neeta Awasthy, Dr. Deepankar Sharma, "Reliability Forecast for Telecommunication System", International Journal of Computational Intelligence and Information Security, Vil. 1 No. 9, November 2010, pp 110-118

2. Sumit Gupta, Dr. Neeta Awasthy, Dr. Deepankar Sharma, "Cost Estimation for ATM with Preemptive Resume Repair", International Journal of Computational Intelligence and Information Security, Vil. 1 No. 9, November 2010, pp 119-127
3. Bhasker Sharma, Dr. Neeta Awasthy, "State Feedback Controller Design via T-S Fuzzy Model", International Journal of Engineering Research and Industrial Applications, Volume 2, pp. 299-312, No. V, September 2009
4. Ruchi Varshney, Tarun Varshney, Dr. Neeta Awasthy, 'A neural network based approach for pattern recognition', International Conference on Computer Analysis of Images and Patterns (CAIP 2009) 2-4 September 2009, Münster (North Rhine-Westphalia), Germany,
5. Ruchi Varshney, Tarun Varshney, Dr. Neeta Awasthy, 'Noise Corrupted Pattern Recognition using Hopfield Neural Network', International conference of artificial intelligence and pattern recognition (AIPR 09), Orland, FL, USA, pp. 175-181
6. Gauri Katiyar, Dr. Neeta Awasthy, 'Character Recognition and its Interpretation using Neural Network', International conference of artificial intelligence and pattern recognition (AIPR 09), Orland, FL, USA, pp 112-116
7. Jaspreet Kour, Neeta Awasthy ' Nonminutiae Based Fingerprint Matching',International Conference on IACSIT Spring Conference 2009,17-20 April, 2009, Singapore
8. Neeta Awasthy, 'Fuzzy based Knowledge Management report on URBAN BASIC SERVICES FOR THE POOR (UBSP) PROGRAMME', International Conference on Data Management, ICDM2009: Feb 10-11, 2009, IMT Ghaziabad, India, pg. 307-314
9. Jaspreet Kour, Neeta Awasthy, Gauri Katiyar, 'Orientation Feature for Fingerprint Recognition', All India Conference on Recent Innovations in Computer Science & Engineering', 6-8 February 09 Chhatrapati Shivaji Institute of Technology, Durg, Pg.102-109
10. Neeta Awasthy, Dr. J.P.Saini & Dr. D.S.Chauhan, 'Wavelet Neural Network for Temporal Behavior of Speech', National Conference on Modern Trends in Electronics and Communication Systems MTECS-2008, March 8-9,2008, Aligarh Muslim University .
11. Neeta Awasthy, Jayanti Ranjan, 'Role of Knowledge Management to revamp the Professional institutes', presented in International Conference on Data Management, IMT Ghaziabad, India, 25-26 February 2008. Chaired a session also
12. Neeta Awasthy, Dr. J.P.Saini & Dr. D.S.Chauhan, 'Spectral Analysis of Speech: A New Technique', *IJSP International Journal. Of Signal Processing,* vol.2 number 1 2005, ISSN: 1304-4478, pp 19-28.
13. Neeta Awasthy, Dr. J.P.Saini & Dr. D.S.Chauhan, 'Better Spectral Analysis of Speech By Improved Preprocessing', *IEEE SPS DART 2005, 19-20 April, 2005, Antwerp, Belgium.*
14. Neeta Awasthy, Dr. J.P.Saini & Dr. D.S.Chauhan, 'Mfcc APplied Independent Component Analysis (MAPICA): A new approach to parametric representation of speech', *Enformatika world Congress held during 25-27 feb.,2005 at Istanbul, Turkey,.Enformatika proceedings (ISBN 975-98458-3-0),* Vol.1., pp 148-151
15. Neeta Awasthy, Dr. H.M.Srivastava, 'Identification of Best Neural Architecture for character recognition ' *Proceedings of National Conference on Electrical, Electronic and Control Technology, ELECT-2004,Crescent College, Chennai.* pp 193-198 in Feb. 2004.

□□□

INDEX

A...

Activation Functions, 19
Adaline, 17
Algorithm for Design of Stable Fuzzy Logic Controllers, 144
An Artificial Neuron, 9
Analysis and Design for T-S Fuzzy Model, 140
Artificial
- Neural Networks, 5
- Neuron Model, 11

Auto Associative Memory, 37

B...

Back Propagation
- Algorithm, 60
- Learning Methods, 59

Backpropagation Learning, 51
Basic Concepts of Fuzzy Logic, 73
Basic Fuzzy Set Operation, 85
Binary
- Encoding, 172
- Step Function, 20

Biological Neural Network, 4

C...

Characteristics of Crisp Sets, 75
Classification of Fuzzy Controllers, 137
Common Applications of Genetic Algorithm, 179
Conventional Control System, 134
Crisp and Fuzzy Relation, 92
Crisp Relation, 92
Crisp Sets, 75

D...

Design of Fuzzy Logic Controller, 139
Design of Stable Fuzzy Logic Controllers, 143

E...

Encoding, Initialization and Selection, 172

F...

Factors Affecting Back Propagation Learning, 62
Fuzzification and Defuzzification, 129
Fuzzy
- Control, 136
- Controller, 134
- IF- Then Rules, 120
- Implications and Fuzzy Algorithm, 122
- logic Control System, 142
- Logic Controller: Basic Concept, 137
- Modeling, 141
- PD Controller, 139
- PI Controller, 138
- Relations, 95
- Sets: Theory and Operations, 83
- to Crisp Conversion, 98

G...

Genetic
- Algorithm Flowchart, 171
- Operators, 176
- Representation, 172

H...

Hardlim Function, 20
Hetro Associative Memory, 37

I...

Important Heuristics for Training MLP, 56
Inference in Fuzzy Logic, 116

L...

Learning Methods, 26
Linear Functions, 19

M...

McCulloch and Pitts Model, 11
Membership Function, 112
Membership Function, 84
Model for Multilayer Perceptron, 51
Multi Layer Perceptron (MLP), 55
Multi layers Feed Forward Neural Network, 23
Mutation Rate P_m, 178
Mutation, 177

N...

Neural Network
- Architecture, 22
- Introduction, 1

O...

Operations of Crisp Sets, 77
Operations on Fuzzy Relations, 96

P...

Partition and Covering, 81
Perceptron and Convergence Rule, 33
Perceptron, 48
Permutation Encoding, 174
Problems with MLP, 56
Procedure of Genetic Algorithm, 170
Properties of
- Crisp Sets, 79
- Fuzzy Sets, 90

R...

Radial Basis Network, 23
Recurrent Neural Network (RNN), 25
Results (and Discussion), 148
Rosen Blatt's Perceptron Model, 14

S...

Saturated Linear Activation Function, 21
Sigmoidal Functions, 21
Single Layer Artificial Neural Network, 50
Single Layer Network, 22
Solution to the Neuron Model, 49
Some Application Domains, 64
Supervised Learning Algorithms, 30

T...

Time Delay Neural Networks (TDNN), 23
Tree Encoding, 175

V...

Value Encoding, 173
Various Learning Techniques, 30

W...

Working Principle, 162

□□□

Win Prizes

ATTENTION READERS

Win Prizes

We request for your frank assessment regarding various aspects of the book as given below :

SOFT COMPUTING

By: Dr. NEETA AWASTHY

Please Fill up the Sheet in Neat and Clean Writing

(i) What topics of your syllabus, important for your examination point of view, are not covered in this book ?

....

....

....

....

(ii) What are the chapters wherein the treatment of the subject matter is not systematic or organised or updated ?

....

....

....

....

(iii) Have you come across misprints/mistakes/factual inaccuracies in the book? Please specify the chapters and the page numbers.

....

....

....

....

(iv) What other books on the same subject have you found/heard better than the present book? Please specify in terms of price and quality

....

....

....

....

(v) Further suggestions and comments for the improvement of this book.

....

....

....

....

(PS: If need be please attach a separate sheet to write your views)

OTHER DETAILS

(i) Who recommended you this book ? (Please tick (✓) in the box near the option relevant to you.

☐ Teacher ☐ Friends ☐ Bookseller

....

....

....

....

(ii) Name of the recommending teacher, his designation and address :

....

....

....

....

(iii) Name and address of the bookseller you purchased the book from :

....

....

....

....

(iv) Name and address of your institution (Please mention the University or Board as the case may be)

....

....

....

(v) Your name and complete postal address :

....

....

....

(vi) Course you are enrolled in :

....

....

....

....

The best assessment will be awarded each month: The award will be in the form of our publications as decided by the Editorial Board.

Please mail the filled up coupon at your earliest to :

S.K.KATARIA & SONS®
4760-61/23, Ansari Road,
Darya Ganj, New Delhi-110002
☎: 23243489, 23269384
Email : katariabooks@yahoo.com
katson_sanjeev@yahoo.co.in